WITHDRAWN
UTSA Libraries

Digital Cognitive Technologies

Digital Cognitive Technologies

Epistemology and the Knowledge Economy

Edited by
Bernard Reber
Claire Brossaud

First published 2010 in Great Britain and the United States by ISTE Ltd and John Wiley & Sons, Inc.
Adapted and updated from *Humanités numériques volumes 1 et 2* published 2007 in France by Hermes Science/Lavoisier © LAVOISIER 2007

Apart from any fair dealing for the purposes of research or private study, or criticism or review, as permitted under the Copyright, Designs and Patents Act 1988, this publication may only be reproduced, stored or transmitted, in any form or by any means, with the prior permission in writing of the publishers, or in the case of reprographic reproduction in accordance with the terms and licenses issued by the CLA. Enquiries concerning reproduction outside these terms should be sent to the publishers at the undermentioned address:

ISTE Ltd
27-37 St George's Road
London SW19 4EU
UK

John Wiley & Sons, Inc.
111 River Street
Hoboken, NJ 07030
USA

www.iste.co.uk

www.wiley.com

© ISTE Ltd 2010

The rights of Bernard Reber and Claire Brossaud to be identified as the authors of this work have been asserted by them in accordance with the Copyright, Designs and Patents Act 1988.

Library of Congress Cataloging-in-Publication Data

Humanités numériques. English
  Digital cognitive technologies : epistemology and the knowledge economy / edited by Bernard Reber, Claire Brossaud.
    p. cm.
  Translated from French.
  Includes bibliographical references and index.
  ISBN 978-1-84821-073-8
  1. Social sciences--Information services. 2. Social sciences--Data processing. 3. Communication in the social sciences. I. Reber, Bernard. II. Brossaud, Claire. III. Title.
  H61.3.H8613 2009
  303.48'33--dc22
                                                                                2009020231

British Library Cataloguing-in-Publication Data
A CIP record for this book is available from the British Library
ISBN 978-1-84821-073-8

Printed and bound in Great Britain by CPI Antony Rowe, Chippenham and Eastbourne.

Table of Contents

Foreword xv
Dominique BOULLIER

Introduction xxi
Claire BROSSAUD and Bernard REBER

PART I. CAN ICT TELL HISTORY? 1

Chapter 1. Elements for a Digital Historiography 3
Andrea IACOVELLA

    1.1. Introduction. 3
        1.1.1. Epistemological mutations of historiography 4
        1.1.2. History and documentation 4
    1.2. Historiography facing digital document 5
        1.2.1. The digital document. 5
        1.2.2. Consequences related to a traditional document. 6
        1.2.3. Consequences on historiography. 7
    1.3. ICTS contributions to historiography methods 8
        1.3.1. Nomenclature and historical semantics 8
        1.3.2. The initiatives of formalisms 8
        1.3.3. A semiotics of the documentary object 11
    1.4. Conclusion 15
    1.5. Bibliography 16

Chapter 2. "In Search of Real Time" or Man Facing the Desire and Duty of Speed 23
Luc BONNEVILLE and Sylvie GROSJEAN

    2.1. Introduction. 23

2.2. Rate, speed and ICT: emergence of a new social temporality 24
2.3. Speed: stigma of a new socio-economic and socio-cultural reality . . . 26
   2.3.1. Emergence of a "speed economy" 26
   2.3.2. Ambivalence of "hypermodern Man" 29
2.4. Conclusion 31
2.5. Bibliography 32

Chapter 3. Narrativity Against Temporality: a Computational Model for Story Processing 37
Eddie SOULIER

3.1. Background: problems of temporality representation in social sciences 37
3.2. A theoretical framework for processing temporality 38
   3.2.1. Narrative theories of action 38
   3.2.2. Narrative explanation of social processes 40
   3.2.3. Social narrative ontology 42
3.3. A computational model for story processing 46
   3.3.1. Hyperstoria 46
   3.3.2. Ontostoria 47
   3.3.3. *Sum It Up* 48
   3.3.4. *MemorExpert* 49
3.4. Conclusion 50
3.5. Bibliography 51

PART II. HOW CAN WE LOCATE OURSELVES WITHIN ICT? 57

Chapter 4. Are Virtual Maps used for Orientation? 59
Alain MILON

4.1. Introduction 59
4.2. Orientation context 59
4.3. The flat sphere 60
4.4. Orientating 62
4.5. Nature of the map 63
4.6. The virtual map 64
4.7. Map territory 65
4.8. Program of the map 66
4.9. Map instruction 67
4.10. Bibliography 69

Chapter 5. Geography of the Information Society 71
Henry BAKIS and Philippe VIDAL

5.1. Introduction 71

5.2. Technological determinism of the facts 72
   5.2.1. Avoidance of travel and space contraction: two ideas promoted
   by policy makers and industrials. 73
   5.2.2. The research world: a frank rebuttal 73
5.3. From the "end of geography" to the "territorialization of ICT" 74
   5.3.1. The appropriation of ICT by urban participants 75
   5.3.2. ICT, tools of mobility and proximity 76
   5.3.3. ICT, instruments of competition of territories 77
5.4. The trivialization of ICT in territories in industrialized countries 78
   5.4.1. The geographical space of the 21st Century 79
   5.4.2. An integrated approach between space and ICT. 80
   5.4.3. A more complex geographical and social space 80
5.5. Conclusion 81
5.6. Bibliography 82

Chapter 6. Mapping Public Web Space with the Issuecrawler 89
Richard ROGERS

6.1. Introduction. 89
6.2. The death of cyberspace 89
6.3. Tethering websites in hyperspace through inlinks 90
6.4. The depluralization of the Web 92
6.5. The Web as (issue) network space 95
6.6. Conclusion 98
6.7. Bibliography 98

PART III. ICT: A WORLD OF NETWORKS? 101

Chapter 7. Metrology of Internet Networks 103
Nicolas LARRIEU and Philippe OWEZARSKI

7.1. Introduction. 103
7.2. Problems associated with Internet measurement 104
   7.2.1. Geographical and administrative dimension 105
   7.2.2. Distribution problems 106
   7.2.3. Measurements, estimations and analysis 107
7.3. Measurement techniques 108
   7.3.1. Active measurements 108
   7.3.2. Passive measurements 109
7.4. Characteristics of Internet traffic 111
7.5. Conclusion 115
7.6. Bibliography 117

Chapter 8. Online Social Networks: A Research Object for Computer Science and Social Sciences 119
Dominique CARDON and Christophe PRIEUR

    8.1. Introduction. 119
    8.2. A massively relational Internet 120
    8.3. Four properties of the social link on the Internet 121
    8.4. The network as a mathematical object 124
        8.4.1. Graphs 124
        8.4.2. Complex networks and "small worlds" 125
        8.4.3. Switching scale: working on large networks 127
    8.5. Structure of networks and relational patterns 128
        8.5.1. Clusters. 128
        8.5.2. Between communities: the "bridges" 130
        8.5.3. Relational patterns 131
    8.6. Conclusion 132
    8.7. Bibliography 133

Chapter 9. Analysis of Heterogenous Networks: the *ReseauLu* Project. . . 137
Alberto CAMBROSIO, Pascal COTTEREAU, Stefan POPOWYCZ,
Andrei MOGOUTOV and Tania VICHNEVSKAIA

    9.1. Introduction. 137
    9.2. The *ReseauLu* project 139
        9.2.1. Scientometric analysis 141
        9.2.2. Example of the scientometric analysis of a domain of biomedical research concerning migraine 144
    9.3. Conclusion 150
    9.4. Bibliography 150

PART IV. COMPUTERIZED PROCESSING OF SPEECHES AND HYPERDOCUMENTS: WHAT ARE THE METHODOLOGICAL CONSEQUENCES? 153

Chapter 10. Hypertext, an Intellectual Technology in the Era of Complexity 155
Jean CLÉMENT

    10.1. The hypertextual paradigm 155
    10.2. Cognitive activity and evolution of textual support 156
    10.3. The invention of hypertext 157
    10.4. Hypertext and databases 158
    10.5. Automatic hypertextualization 160
    10.6. The paradigm of complexity 161
    10.7. Writing of the complex 163
    10.8. Hypertextual discursivity 164

10.9. Conclusion 165
10.10. Bibliography. 166

Chapter 11. A Brief History of Software Resources for Qualitative Analysis 169
Christophe LEJEUNE

11.1. Introduction. 169
11.2. Which tool for which analysis? 170
   11.2.1. The felt-tip 171
   11.2.2. Text processing 171
   11.2.3. The operating system 172
   11.2.4. The cotext 174
   11.2.5. The co-occurrence. 177
   11.2.6. Data analysis. 178
   11.2.7. Segments of text. 179
   11.2.8. Dictionaries 182
11.3. Conclusion: taking advantage of software 182
11.4. Bibliography 183

Chapter 12. Sea Peoples, Island Folk: Hypertext and Societies without Writing 187
Pierre MARANDA

12.1. Introduction. 187
12.2. A concrete and non-linear approach 188
12.3. Type of hypermedia modeling and implementation 190
   12.3.1. A model derived from neurosciences 190
   12.3.2. Implementation: "attractors" and "attraction basins" 192
12.4. The construction of attractors and their basins 195
   12.4.1. An example of an attractor and of its basin: ancestors 195
12.5. Conclusion 197
12.6. Bibliography 198

PART V. HOW DO ICT SUPPORT PLURALISM OF INTERPRETATIONS? 203

Chapter 13. Semantic Web and Ontologies 205
Philippe LAUBLET

13.1. Introduction. 205
13.2. Semantic Web as an extension of current Web 207
13.3. Use of ontologies. 209
13.4. Metadata and annotations. 211
13.5. Diversity of ontologies debated 214

13.6. Conclusion	216
13.7. Bibliography	217

Chapter 14. Interrelations between Types of Analysis and Types of Interpretation 219
Karl M. VAN METER

14.1. Introduction	219
14.2. ICT and choice within the general outline of HSS research	220
14.3. Questions relating to initial data and presentation of results	222
14.4. Choice of analysis methods and general outline	224
14.5. Methodological choices, schools of thought and interlanguage	225
14.6. Conclusion	227
14.7. Bibliography	228

Chapter 15. Pluralism and Plurality of Interpretations 231
François DAOUST and Jules DUCHASTEL

15.1. Introduction	231
15.2. Diversity of interpretations	231
15.3. Interpretations and experimental set-ups	233
15.4. Exploratory analysis and iterative construction of grids of categories	234
15.5. Categorization process from a grid	237
15.6. Validation by triangulation of methods	240
15.7. Conclusion	242
15.8. Bibliography	242

PART VI. DISTANCE COOPERATION 245

Chapter 16. A Communicational and Documentary Theory of ICT 247
Manuel ZACKLAD

16.1. Introduction	247
16.2. Transactional approach of action	248
16.3. Transactional flows: rhizomatic and machinery configuration	249
16.4. ICT and documents: transition operators between situations of activity within a transactional flow	251
16.5. Four classes of documents within the information system	252
16.6. ICT status in the coordination and regulation of transactional flows	255
16.6.1. Coordination through access to situations of activity and their ingredients	256
16.6.2. Coordination by prior standardization of the primary transaction	256

16.6.3. Coordination through the use of regulating transactions and
resources in situation. 259
16.6.4. Internal structuring of DFA according to the nature of the
transaction 259
16.7. Conclusion: document for action and distributed transactions 261
16.8. Bibliography 262

**Chapter 17. Knowledge Distributed by ICT: How do Communication
Networks Modify Epistemic Networks?** 265
Bernard CONEIN

17.1. Introduction. 265
17.2. ICT and distributed cognition 266
17.3. Mailing lists as a distributed system 269
17.4. Evolution of communication networks. 272
17.5. Epistemic networks and discussion networks 275
17.6. Conclusion 278
17.7. Bibliography 279

**Chapter 18. Towards New Links between HSS and Computer Science:
the *CoolDev* Project** 283
Grégory BOURGUIN and Arnaud LEWANDOWSKI

18.1. Introduction. 283
18.1.1. A new support to cooperative activities of software
development. 284
18.2. Towards new links between HSS and computer science:
application to AT 287
18.2.1. Activity theory. 288
18.2.2. In search of generic tools for human activities 289
18.3. Generic links between AT and computer science: application in
CoolDev 290
18.3.1. Link between generic properties of AT and computer science
techniques. 290
18.4. Discussion: towards new developments 294
18.5. Bibliography 296

PART VII. TOWARDS RENEWED POLITICAL LIFE AND CITIZENSHIP. 299

Chapter 19. Electronic Voting and Computer Security. 301
Stéphan BRUNESSAUX

19.1. Introduction. 301
19.2. Motivations of electronic voting. 302
19.2.1. Reducing costs and time mobilized for elections 302

19.2.2. Improving the participation rate 302
19.2.3. Meeting requirements of mobility 303
19.3. The different modes of electronic voting 304
19.4. Prerequisites for the establishment of an Internet voting system 305
19.5. Operation of an Internet voting system 306
  19.5.1. Authentication of the voter 306
  19.5.2. The choice of candidates 308
  19.5.3. The vote 308
  19.5.4. Acknowledgement 308
  19.5.5. Counting of votes 309
19.6. Prevention of threats 309
  19.6.1. Threats 309
  19.6.2. Prevention 310
19.7. The different technical approaches 311
  19.7.1. The mix-net 311
  19.7.2. The blind signature 311
  19.7.3. The homomorphic encryption 312
19.8. Two examples of realizations in France 312
  19.8.1. The Chambers of Commerce and Industry vote 313
  19.8.2. French nationals voting from abroad 313
19.9. Bibliography 314

Chapter 20. Politicization of Socio-technical Spaces of Collective Cognition: the Practice of Public Wikis 317
Serge PROULX and Anne GOLDENBERG

20.1. Introduction 317
20.2. Forms of ICT politicization 319
  20.2.1. Approaching politicization of cognitive technologies 319
  20.2.2. Reflexivity and politicization of cooperation practices 321
20.3. Writing on wikis: a practice of deliberative cognition 322
  20.3.1. The plans of action of public wikis 322
  20.3.2. Setting discussion of a collective cognition process 324
20.4. Conclusion: the stakes of a cognitive democracy 326
20.5. Bibliography 327

Chapter 21. Liaising using a Multi-agent System 331
Maxime MORGE

21.1. Introduction 331
21.2. Motivations 332
21.3. Game theory 332
21.4. The principles 334
  21.4.1. Representation of the problem 335

21.4.2. Expression of preferences 336
21.4.3. Summary of judgments. 337
21.5. Multi-wizard system. 338
21.5.1. Joint elaboration of an argumentative scheme 338
21.5.2. Detection of conflicts and consensus. 340
21.6. Conclusion 340
21.7. Bibliography 341

PART VIII. IS "SOCIO-INFORMATICS" POSSIBLE? 343

Chapter 22. The Interdisciplinary Dialog of Social Informatics 345
William TURNER

22.1. Introduction 345
22.2. Identifying procedures for configuring collective action 346
22.3. Analyzing socio-technical change. 349
22.4. Improving the design of computer applications. 354
22.5. Bibliography 355

**Chapter 23. Limitations of Computerization of Sciences of Man
and Society** 357
Thierry FOUCART

23.1. Introduction 357
23.2. The scientific approach in sciences of man and society 358
23.3. Complexity and intricacy of mathematical and interpretative
approaches 359
23.4. Difficulties in application of methods 362
23.5. Quantification and loss of information 363
23.6. Some dangers of wrongly controlled socio-informatics 365
23.7. Socio-informatics and social technology. 368
23.8. Bibliography 369

**Chapter 24. The Internet in the Process of Data Collection and
Dissemination** 373
Gaël GUEGUEN and Saïd YAMI

24.1. Introduction 373
24.2. Construction of the questionnaire 375
24.3. Administration of the questionnaire 377
24.3.1. Implementation of an Internet survey 378
24.3.2. Reflections on the relevance of these methods and extensions . . 380
24.4. Dissemination of results. 381
24.4.1. Difficulty of disseminating research results. 381

24.4.2. Proposition of a solution for dissemination of results through
the Internet: the development of an automated specific analysis 381
24.4.3. Limitations and expectations of the process 383
24.4.4. Broadcast of research: "Internetized research area".......... 384
24.5. Conclusion 386
24.6. Bibliography 386

Conclusion. 389
Bernard REBER and Claire BROSSAUD

Postscript 397
Roberto BUSA

List of Authors 401

Index 405

Foreword

The New Manufacture of HSS

The picture that is presented in this book is welcome because it deliberately mixes the "social" issues of ICT, and "scientific" issues of humanities and social sciences (HSS). This is indeed a very tangled situation where a "society" radically changes its device reflexivity and where researchers who make it their profession to ensure this reflexivity need to follow all ongoing innovations, with the fear of become disqualified. If some can say that the Internet and digitization are equivalent to what happened in the Renaissance with printing, then we can hardly doubt that it is *another scientific era* that is opening up, especially for HSS.

If Elisabeth Eisenstein [EIS 91] refuted any technological determinism, she was showing that current changes had gained strength through printing, by its power of spreading information to the masses, by its consultation of works authorizing the confrontation of knowledge, and by the durability of printed text not subject to being rewritten. The Internet and digital publishing have changed the scale of available information for the first two criteria, while in contrast, they introduce permanent instability, unlike printing. Surely this possibility of revision [LIV 94] noticed worldwide, allows us to confront heterogenous knowledge and constitutes a profound destabilization of organization of this knowledge itself. Let us specify in what fields:

– Digitization allows and forces *explicitation*. Working with a computer scientist is a radical test for any HSS researcher as it forces him to use very well defined categories, if they are to be manipulated by an algorithm. But all social participants are in the same situation: they need to explain, declare and document what they do, what they say, the people, objects and process, as demanded by quality management. This redocumentarization of the world, which we analyzed in the collective context of Pédauque [PED 06], is still at its beginning. It can certainly usefully help to formalize what was left implicit so that the participants can better coordinate or discuss issues. It can also turn into madness in its generalized form and is a powerful indicator, like some managerial tendencies, including research professionals, of what they want through omniscient ERP (Enterprise Resource Planning). For the HSS, the same risk exists, that of only adopting the paradigms incorporated in their digital tools, in their indicators, at all levels of the production chain of scientific statements: gathering/production, processing, analysis and restitution. The "increased reflexivity" that could be expected through ICT would only lead to a new positivism, calculated certainly, but equally unproductive.

– ICT produce intangibles, we are told, yet the ordinary experience of the computer user makes him feel the very materiality of the machine and of the networks, especially when they break down or do not act as required through the interface commands. Therefore, it becomes possible to say that *we "manufacture" HSS*, only in the sense that, previously, their equipment seemed weak compared to other sciences. The indirect effect of HSS is to rediscover the work, production, and division of labor that constitute every scientific activity. Objects, terminals, sensors, "monstrators" and software can be thought of and adapted to researchers' objectives. HSS can now claim to "use a laboratory" [LIC 96], in the same way as other sciences have done through their equipment. It is also for this reason that we have encouraged all observation platforms to be equipped and shared among disciplines [BOU 04]. It would be unfortunate if HSS copied the separation movement with respect to "society" that constituted the scientific model of contemporaries [LAT 92], having always buried laboratories far from public view. For it is the whole "society" that is coupled to its objects, its ever-more sophisticated devices, such that it no longer knows how to drive them. It is probably not a coincidence that some HSS approaches have rediscovered that the supposed society would not survive long without material objects that compose it in an inextricable lattice.

– If there is a reason to avoid carrying out laboratory experiments as in the good old days of natural science, it is that all the tools used by HSS are in everyone's hands and they gain, through the network, an uncontrollable distribution power. The *Internet destabilizes all authorities* precisely because the notion of copyright is being relativized at high speed. All cultural activities based on copyright are suffering, but neither "hard" sciences nor HSS will be able to avoid it. The asymmetries of knowledge and power can potentially all be destabilized. It is no longer the undisputed laboratory facts that constitute the contemporary truth regime, but rather the proliferation, debate and a certain disorientation created by the "opinion economy" [ORL 99] that governs the Internet.

– Knowledge produced by the new observation tools and computing centers thus made available can hardly stand in disciplinary boundaries. Paradigm circulation is facilitated by means of indexing, emerging from supposedly shared references and we know that creativity is often born of these heterodox circumstances. But we could also believe that here again it is the tools that would give birth to new paradigms, to such an extent that some disciplines might feel endangered by some of their specialties: linguistics with NLP (natural language processing), geography with geomatics and cognitive sciences by cognitics, for example.

The paradox is nevertheless that digital networking enables authors to revisit their ideas of other types of "social" model. Let us make this hypothesis: if Durkheim was the statistical state child and produced a theory on the social structure in line with its history and the thinking tools of that time, then Tarde should have been considered as the father (or ancestor) of networks and digital since he set up all conceptual tools of contagion, imitation and propagation in late 19th Century before these techniques became so visible and handleable today. The question is whether the HSS will only change the data processing software or whether they will take advantage of it to change the "conceptual software" that govern them, in France at least.

– In all cases, socio-technical realities that the Internet and digital have built constitute *land-continents to explore*, requiring specific tools and methods due to their instability. The idea that on the Web, for example, a new library would be created is a typical way of reproducing old frameworks to grasp new phenomena, and reflects this difficult sidestep in the face of what is emerging. It is not in any way a matter of duplication of a "real world", but indeed that of another world – an ultra-world that requires an ultra-reading, as proposed in [GHI 03], just as real as the other. HSS will be the essential surveyors of this new universe.

– One of the advantages of the emerging technical worlds' instability, at their birth, is indeed to give rise to controversies, with exploration possibilities. HSS, when equipped, often depend nevertheless on what is on offer from computer scientists in laboratories or on the market. They are therefore led to make choices that would later appear erroneous or would force them to make unnecessary efforts to adapt. It is the duty of HSS, because of the knowledge of this constant tension between divergence and convergence that animates every social life, to claim a *pluralism of architectures* [LES 99]. Technical fatalities do not exist, any more than a guarantee of winning standards. It is therefore important that the debate – that digitization strengthens – also involves these technical architecture policies, both for researchers and for all collective activities beyond research. There is no constant winning policy for archiving architectures or for those of the publication, to take but two examples.

It is convenient to make choices, but under the condition of organizing debates and preserving pluralism of choice. We have noticed to what extent, over 10 years, the semantic Web model has come to dominate debates by monopolizing all efforts in the field of knowledge engineering, for example. However, its limits, that could be determined from its own theoretical assumptions (the language designed as a label, for simplification) have only recently been allowed, and upset, not so much by researchers as developers who launch applications under Web2.0 everywhere, from the so-called "social" Web. Research then registers this by moving towards a socio-semantic Web and risks preventing the debate. Client-server architectures have been outrageously maintained as unique references in computer laboratories in the same way despite mass spreading of peer-to-peer architectures in general public use.

HSS could at least have the advantage of not suffering from the influence of the social world of computer science: they are then able to ask and require, at least for themselves, tools adapted to emerging phenomena. More generally, they could be those defending the social dynamism of these techniques and therefore their pluralism and controversies. This is not to recite a new version of critical sociology praising the marvels of popular autoproduction and resistance to domination, but to reopen debates that we would want to conclude too quickly, under the cover of supposed technical certainty. When HSS become equipped, they certainly suffer the tensions and uncertainties of the digital universe that are emerging, but they should also be able to assert their experience of these tensions and controversies as an advantage to show their diversity. If this temporarily slows their effectiveness, it is probably positive in the long term because very few can say exactly how effective HSS is.

These debates go on through the very diverse chapters of this work, and allow us to measure the considerable internal conversion work that is being carried out throughout HSS upon the emergence of ICTs – the new knowledge support.

Dominique BOULLIER
Professor of sociology at the University of Rennes 2
Director of UMS CNRS Lutin 2809, Director of LAS, EA 2241[1]

Bibliography

[BOU 04] BOULLIER D., "The platform effect in social sciences", *1st European User Lab Conference*, Lutin, City of Sciences, Paris, 24 November 2004.

[EIS 91] EISENSTEIN E.-L., *La Révolution de l'Imprimé dans l'Europe des Premiers Temps Modernes*, La Découverte, Paris, 1991.

[GHI 03] GHITALLA F., BOULLIER D., GKOUSKOU P., LE DOUARIN L., NEAU A., *L'Outre-lecture. Manipuler, (s') Approprier, Interpréter le Web,* Public Information Library, Centre Pompidou, Paris, 2003.

---

1 http://www.uhb.fr/sc_humaines/las/.

[LAT 92] LATOUR B., *Nous n'avons Jamais été Modernes. Symmetrical Anthropology Test*, La Découverte, Paris, 1992.

[LES 99] LESSIG L., *Code and Other Laws in Cyberspace*, Basic Books, New York, 1999.

[LIC 96] LICOPPE C., *La Formation de la Pratique Scientifique*, La Découverte, Paris, 1996.

[LIV 94] LIVET P., *La communauté virtuelle*, Editions de l'Éclat, Combas, 1994.

[ORL 99] ORLEAN A., *Le Pouvoir de la Finance*, Odile Jacob, Paris, 1999.

[PED 06] PÉDAUQUE R.T., *Le Document à la Lumière du Numérique*, C&F editions, Caen, 2006.

Introduction

"The tool certainly does not make science, but a society that claims to respect sciences should not ignore their tools."

Marc BLOCH, "Preface" in DUBY G., *Apologie pour l'histoire ou Métier d'historien*, Armand Colin, Paris, p. 67, 7th edition, 1974.

It was a long time ago when, in 1949, philosopher Roberto Busa tried with great difficulty to convince the founder of IBM, T.J. Watson (Jr), to develop a piece of software enabling the analysis of a lexicon and navigation through the works of medieval philosophers. The person who agreed to write the Postscript of this book is now recognized as one of the pioneers of computer linguistics and an authority in the field of computers for humanities, with his famous *Index Thomisticus*, linking information and communication sciences (ICTS) and humanities in an original way. Similarly, an expression such as "knowledge society", frequently associated with the emergence of information and communication technologies (ICT), seems to characterize this association today. However, their ability to inform and support communication soon becomes problematic if they are understood in an interdisciplinary way, among researchers from ICTS and researchers in human and social sciences (HSS), as is the case here with the authors convened. Evaluation of technological innovation depends on these technologies, sometimes combined in the same way as the layers of a millefeuille, but also widely in the way HSS captures them. Indeed, many reports[1] are actually careful about the relationships between "ICT and society" and call for a profound consideration. On the one hand, ICT are complex and, on the other hand, how HSS approach them deserves an explanation

---

Introduction written by Claire BROSSAUD and Bernard REBER.
1 See the two recent contributions in particular from the authors [BRO 05, REB 03].

articulating methodological, theoretical and epistemological considerations, far from the two pitfalls of technical and social determinism. The construction of sophisticated tools, both physical and informatics, are concurrent to the construction of variable research objects and often competing in HSS disciplines.

The reader will find in this book analyses polarized on uses and the impact of technologies on society. However, the ambition here is more radical: it tries to examine, both from a theoretical and practical point of view, how HSS concepts or notions important today are processed by the possibilities of these cognitive[2] technologies. We have used eight thematic parts that involve all concepts currently being widely discussed by researchers and through this we open an updated dialog on the role of science and technologies in our societies. The fact of questioning these themes reflects a willingness for formal debate, supported by some conception of philosophy and sociology, to bring together very diverse knowledge and to favor reflexivity. Throughout the chapters, we quickly understand in fact that we are not simply faced with a new type of data, overwhelmed by access to a huge number of documentary resources, or faced with new collaboration possibilities among scientists, but that beyond opportunities, ICT affect the theoretical cores of certain disciplines, their organizations and social and political implications.

Certainly, the authors of these chapters are all strongly rooted and recognized in a discipline, but they are aware of the design, practice and understanding of ICT, and are engaged in this perilous crossing among *a priori* remote sciences. This is the case for those from ICTS, who are subscribed in computer science (with various specializations) and in mathematics, but equally for HSS researchers and teachers, including here sociology, philosophy, cognitive sciences, sociolinguistics, geography, anthropology, management, history, as well as information and communication sciences. Despite inherent risks of dispersion in this type of exercise, exacerbated by the fact that we have voluntarily appealed to researchers often recognized on a broad geographic scale, that of the *Francophonie*[3], this collaboration constitutes a real scientific added value that would not have been achieved without this diversified expertise, avoiding early or outrageously simplistic results. This book aims to make as seamless as possible articles of cultures and scientific languages from disciplines that rarely encounter one another. It therefore concerns neither a classic collective[4] book mainly organized around the same

---

2 We have qualified these technologies as such to better reflect various activities that they support and that concern knowledge.
3 All French-speaking countries.
4 This treatise could furthermore be considered as a way of viewing HSS dictionaries, which rarely try to prove the diversity of their theoretical and methodological resources to answer a question, such as socio-technical interactions, but which often prefer to present great authors, trends or even themes.

discipline, nor an eclectic collaboration. A long and rigorous editorial framework has thus been approved by the authors who carried out this work, supported by an equally great effort by its 40 contributors[5]. The "conversation" among such varied disciplines constitutes a multi- and interdisciplinary challenge, sometimes even transdisciplinary, because certain authors have introduced knowledge of disciplines other than their own. This rare opportunity to get so many disciplines to work with one another is already an achievement in itself, without even mentioning the ICT object, although it does seem to impose itself, as D. Boullier recalls in his preface.

This treatise is in eight parts that involve in a reflective way all the major questions asked by the joint fields of HSS and ICTS around ICT: time, space, networks, text and hypertext, interpretation, cooperation, politics and socio-informatics. The chapters of which the parts are made up have been designed according to an identical coherence: they are marked by the same concern for factual and conceptual exactness, and by a spirit of balance between information and communication technologies sciences (ICTS) and human and social sciences (HSS). From Part 1 to Part 5, Digital Cognitive Technologies involves the epistemological evolution of HSS disciplines in terms of cognitive technologies. We can find their impact on methods and theories. Parts 6 to 8 put this evolution more clearly in a social and political perspective. They further study relationships of technologies between various spheres of society: deliberation and design of scientific documents, new transactional traffic flow coordination arrangements or even collective expertise systems. For concepts and ideas used in every part, the analysis combines three types of input:

– the first type, provided by a researcher generally from ICTS, indicates in what way each idea or concept used is concerned with ICT;

– the second studies how the chosen idea or concept, important for the HSS, is reworked by ICT and their existence in society;

– the third presents experiments showing how ICT can concretely support research in HSS in processing the idea or concept.

Bibliography

[BRO 05] BROSSAUD C., TIC, Sociétés et espaces urbains : bilan et perspectives de la recherche francophone en sciences sociales (1996-2004), Action Concertée Incitative Ville, Maison des Sciences de l'Homme " Villes et territoires ", Tours University, 2005.

---

5 We would like to thank them for their enthusiasm, their patience and their efforts in the "translation" between disciplines. We are also grateful to Claude Henry and Michel Cardon for their help, which was more than just a re-reading, but rather an informed and knowing look over the whole project.

[REB 03] REBER B., Les TIC dans les processus politiques de concertation et de décision. Multiples perspectives ouvertes par les recherches en sciences humaines et sociales, Ministry of ecology and sustainable development, 2003.

Presentation of authors

Editors

Claire Brossaud (claire.brossaud@gmail.com) is a sociologist at the University of Lyon. Her interest is in how sciences and technologies question public spaces in their territorial and deliberative dimensions. After completing a thesis on the "imaginary builder" of a new town, she conducted several research works, in particular for House of Human Sciences "Cities and territories" of the University of Tours and the GRASS-CNRS Paris. She has coordinated workshops and brainstorming sessions on these themes for the last five years within different organizations (French Association of Sociology, VECAM-European and citizen watch on information and multimedia motorways, etc.).

Bernard Reber (bernard.reber@parisdescartes.fr) is a philosopher and researcher at the Research Centre "Meaning, ethics and society" (CERSES-CNRS/University of Paris Descartes). He studies ethics in the context of participatory technological assessment. He is the author of *Les TIC dans les processus politiques de concertation et de decision. Multiples perspective ouvertes par les recherches en sciences humaines et sociales* ("ICT in consultation and decision political processes. Multiple prospects opened by researchers in human and social sciences") for the Ministry of ecology and sustainable development (2003). He co-edited *Le pluralisme* ("Pluralism"), *Archives de philosophie du droit*, vol. 45 (2005) and is the author of *La démocratie génétiquement modifiée. Sociologies éthiques de l'évaluation des technologies controversées* ("Genetically Modified Democracy. Ethical Sociologies of Evaluation of Controversial Technologies") (2010). He is a member of the international group "Eco-ethica" and on the editorial board of *Journal of Agricultural and Environmental Ethics* (see: http://cerses.shs.univ-paris5.fr/spip.php?article113)

Both editors have contributed to different research groups on the relationships between ICTS and HSS, including:

– a research group for computerized textual analysis (ARCATI-IRESCO-CNRS) aiming to compare computerized textual analysis software in HSS research;

– a specific multidisciplinary action of the CNRS "Distributed collective practices and cooperation technologies" (AS PCD/TC).

Co-authors

Henry Bakis is Professor of geography at the University of Montpellier III. He runs (and founded in 1987) the *Netcom* (*Networks and Communication Studies*: http://recherche.univ-montp3.fr/mambo/netcom_labs/) journal, published under the auspices of CNFG and International Geographical Union (IGU) of which he chaired the ICT commission (1992-2000). He is a member of the UMR "Mutations of Territories in Europe" (since January 2007) and the author of numerous works, including *Enterprise, Space and Telecommunications* (1987), *Geopolitics of Information* (1988), *Networks and their Social Issues* (1993).

Luc Bonneville (luc.bonneville@uottawa.ca), after completing a doctorate in sociology (PhD) at the University of Quebec at Montreal and carrying out research as a postdoctoral researcher within the Health Administration Department of the Faculty of Medicine at the University of Montreal, has been a Professor, since 2004 in the Communication Department at the University of Ottawa. In 2004, he was awarded the best doctorate thesis on the issue of "Computer Science and Society" at the annual competition of the Coordination Centre for Research and Education in Information Science and Society (CREIS) in France. In 2005, it was the Institute for Research in Contemporary Economics (IREC) that rewarded him at its annual prize giving ceremony for the best doctorate thesis. His research focuses on the computerization of healthcare organizations, especially in terms of organizational communication. Author of several scientific articles in diverse journals, he is also a researcher at the Interdisciplinary Research Group on Organizational Communication (GRICO) at the University of Ottawa.

Dominique Boullier is Professor of sociology at the University of Rennes 2, Director of the UMS CNRS Lutin 2809, and Director of LAS, EA 2241 (http://www.uhb.fr/sc_humaines/las/).

Grégory Bourguin (http://lil.univ-littoral.fr/~bourguin) is a lecturer in computer science and a member of the ModEL team as well as the Coastal Information Science Laboratory (LIL) at the University of Opale Coast (ULCO).

Stéphan Brunessaux (sbrunessaux@wanadoo.fr) joined Matra in 1990 as Head of the Research Centre for Information Processing of the EADS Defense and Security located in Val-de-Reuil (27). This is the source of the European research project for Internet voting CyberVote, which he ran from 2000 to 2003. He has received the European prize for Information Society Technologies.

Roberto Busa is Professor of philosophy (ontology and epistemology). We owe to him the *Thomisticus Index*, which exists in 56 volumes, available in CD or software (http://www.corpusthomisticum.org/it/index.age). He is one of the founders of the Association for Computers and Humanities (http://www.ach.org/), and of the Alliance of Digital Humanities (http://www.digitalhumanities.org/).

Alberto Cambrosio is Professor in the Department of Social Studies of Medicine at McGill University (Montreal). His work focuses on biomedical innovation, in particular at the clinical-laboratory interface in the field of cancer genomics.

Dominique Cardon (domi.cardon@orange-ftgroup.com) is a sociologist at the Usage Sociology Laboratory of France Telecom R&D.

Jean Clément is a lecturer in Information and Communication Sciences in the "Hypermedias" Department of the University of Paris 8 and head of "Hypertext Writings" team of Paragraphe Laboratory. He has written "Cyberliterature between literary game and video game" in N. Szilas and J.H. Rety (ed.), *Création de récits pour les fictions interactives: simulation et réalisation*, Hermès, Paris, 2006.

Bernard Conein (Bernard.Conein@unice.fr) is Professor of sociology at the University of Nice Sophia-Antipolis and Director of ICT Uses Laboratory in Sophia Antipolos, specializing in cognitive and innovation sociology.

Pascal Cottereau (pascal.cottereau@aguidel.com), a graduate of ESSEC Management, specializes in marketing and ESIEE MS Technological innovation and project management. After 12 years in SSII and consulting firms, he joined the AGUIDEL Company in 2005 and is pursuing in parallel a doctorate in Information Sciences.

François Daoust (daoust.francois@uqam.ca) is a computer scientist attached to ATO, and developer of the SATO software.

Jules Duchastel (duchastel.jules@uqam.ca) is Professor in the Sociology Department at the University of Quebec (Montreal) (UQAM) and holder of the Canada Research Chair in globalization, citizenship and democracy (http://www.chaire-mcd.ca). He works mainly on the analysis of new forms of political regulation in the context of international organizations and development of a deliberative transnational space. He is the author of an abundant methodological production in the analysis of computer-aided speech that he developed at the Centre for Text Analysis by Computer (ATO), which he founded in 1983 http://www.ling.uqam.ca/ato).

Thierry Foucart (thierry.foucart@univ-poitiers.fr) is Associate Professor in Mathematics and a lecturer at the University of Poitiers, authorized to conduct research (http://foucart.thierry.free.fr/). He is interested in epistemological relations between statistics and social sciences.

Anne Goldenberg (goldenberg.anne@courrier.uqam.ca) is a doctorate student in sociology at the University Sofia Antipolis in Nice and in communication at the University of Quebec in Montreal (UQAM). Her thesis focuses on cognitive and political schemes of public wikis.

Sylvie Grosjean (sylvie.grosjean@uottawa.ca) is Professor in the Communication Department of the University of Ottawa. She is particularly interested in communicative interactions during distance collaborative work situations (http://www.uottawa.ca/academic/arts/communication/fra/grosjean.html).

Gaël Gueguen (g.gueguen@esc-toulouse.fr) is Professor of strategic management at Toulouse Business School – University of Toulouse. A researcher within the Management Research Centre, he develops works on the strategic management of SME and firms of Information Technology sector.

Andrea Iacovella is a Deputy Director at ENSIIE and research engineer in the languages and civilizations of ancient worlds, and in computer science. He runs an interdisciplinary research network on the construction of meaning in historical disciplines and is interested in the consequences of digital on their methods and epistemology (http://www.porphyry.org/Members/aiacovella).

Nicolas Larrieu (nicolas.larrieu@enac.fr) is Professor and researcher at the National School of Civil Aviation in Toulouse. He works in the "Electronics and Networks" Department, where he addresses the problems of metrology of Internet traffic and its applications in network security. He is a member of the LAAS-CNRS, 7 Avenue du Colonel Roche, 31077 Toulouse cedex 4, France.

Philippe Laublet (Philippe.Laublet@paris4.sorbonne.fr) is a lecturer in computer science at Sorbonne (University of Paris 4), a former senior scientist at the National Office of Aerospatial Studies and Researches (ONERA) and co-head of the specific action *Web Semantic* of the CNRS (ITCS Department).

After a PhD in Sociology, *Christophe Lejeune* now teaches information and communication at the University of Brussels Interested in mediated interactions, he is the author of a qualitative analysis free software (Cassandre) part of the Hypertopic collaborative platform (http://cassandre-qda.sourceforge.net/).

Arnaud Lewandowski (lewandowski@lil.univ-littoral.fr) is a lecturer in computer science, member of the ModEL team, of the Coastal Information Science Laboratory (LIL), University of Opale Coast (ULCO).

Pierre Maranda (pmaranda@videotron.ca), Professor Emeritus, is research director in the Anthropology Department of Laval University, Quebec. He is the designer of the site http://www.oceanie.org/.

Alain Milon (alain.milon@u-paris10.fr) is Professor of philosophy at the University of Paris Ouest. He is the author of the following works: *Bacon, l'effroyable viande* (Bacon, the Horrifying Meat), Encre Marine, Paris, 2009, *Maurice Blanchot, lecteur de René Char?* (Maurice Blanchot, Rene Chart reader?), Complicites, Paris, 2009; *Levinas, Blanchot: penser la différence* (Levinas, Blanchot: Think the Difference), University Press of Paris Ouest, Paris 2008, *The Book and its Spaces,* University of Paris Ouest, Paris 2007; *La fabrication de l'écriture à l'épreuve du temps* (Writing manufacture facing the test of time), Complicités, Paris, 2006; *L'écriture de soi: ce lointain intérieur - moments d'hospitalité autour d'Antonin Artaud* (Writing for Granted that Distant Inner-Moments Hospitality around Antonin Artaud, literary hospitality moments around d'A. Artaud), Encre marine, La Versanne, 2005; *La réalité virtuelle, Avec ou sans le corps* (Virtual Reality, With or without the body), Autrement, Paris, 2005.

Maxime Morge (morge@di.unipi.it) is a doctor in computer science. He works in the artificial intelligence field, particularly on multi-agent systems. He obtained his doctorate in 2005 at the National Superior School of Mines in Saint-Etienne. After having worked in the fundamental computing laboratory in Lille, followed by a brief postdoctoral stint at the University of Ottawa in Canada, he is currently a member of the Computer Science Department at the University of Pisa.

Andrei Mogoutov is designer and co-founder of the AGUIDEL Company. He is specialized in designing solutions for heterogenous data analysis applied in particular to scientific, technological and economic surveillance, as well as to the strategic management of innovation.

Philippe Owezarski (owe@laas.fr) is a member of LAAS-CNRS, 7 Avenue du Colonel Roche, 31077 Toulouse cedex 4, France.

Stefan Popowycz, after completing studies in sociology of medicine at McGill University, now works as an analyst with the Reports Department on hospitals of the Canadian Institute of Health Information.

Christophe Prieur (prieur@liafa.jussieu.fr) is a lecturer in computer science at University Paris Diderot and researcher at LIAFA, posted in Usage Sociology Laboratory of France Telecom R&D.

Serge Proulx (proulx.serge@uqam.ca) is a sociologist, professor at the University of Quebec in Montreal (UQAM), director of the research group on media uses and cultures (GRM) (http://grm.uqam.ca/) and the laboratory of computer-mediated communication (LabCMO) (http://cmo.uqam.ca).

Richard Rogers (rogers@uva.nl) is director of "new media" at the University of Amsterdam, visiting Professor in the Department of Scientific Studies at the University of Vienna and director of Govcom.org (Amsterdam) Foundation. He is the author of *Technological Landscapes*, Royal College of Art, London, 1999, editor of *Preferred Placement: Knowledge Politics on the Web*, Jan Van Eyck, Maastricht, 2000 and author of *Information Politics on the Web*, MIT Press, Cambridge, USA, 2004 (first prize in ASSIST (*American Society for Information Science and Technology*)).

Eddie Soulier is a research professor in computer science at Charles Delaunay Institute in the University of Technology in Troyes, FRE CNRS 2848, "Cooperation technologies for innovation and organizational change".

William Turner (william.turner@limsi.fr) heads social informatics research in a group working on architectures and models for interaction at the Computer Science Laboratory for Mechanics and Engineering Science (LIMSI) of the French National Research Council (CNRS). He has edited with Geof Bowker, Les Gasser and Manuel Zacklad a special issue for the *Journal of Computer Supported Cooperative Work* entitled "Information Infrastructures for Distributed Collective Practices" (June 2006) and has started to publish extensively on work he coordinated for UNESCO aimed at computers supporting the formation and consolidation of Diaspora Knowledge Networks.

Karl M. Van Meter, doctor in mathematics (ordered algebraic structures), is a sociologist and has directed in particular: *Sociologie*, Larousse, Paris, 1992; *Interrelation between Type of Analysis and Type of Interpretation*, Peter Lang Verlag, Bern, 2003. He has directed the *Bulletin de méthodologie sociologique* (BMS) since its creation in 1983.

Tania Vichnevskaia, cofounder of the AGUIDEL Company, is a consultant in the field of heterogenous data analysis.

Philippe Vidal (philippe.vidal@univ-lehavre.fr) is a lecturer at the University of Havre, and member of the UMR CNRS 6228 IDEES/CIRTAI. He has been, since 1st January 2007, President of the "Information Society" commission of the French National Committee of Geography (CNFG). He is also associate director of *Netcom* journal. Author in 2002 of a thesis entitled "The region facing the information society", his main entrance key focuses on the capacity of local executives to integrate the ICT issue in their economic and social development strategies.

Saïd Yami is Associate Professor in Strategic Management at the University of Montpellier I (ISEM), and Professor at EUROMED Management – Marseille, France. A member of the ERFI, he has published many research articles and several books related to competitive relationships through the topics of rivalry and disruptive strategies, collective strategies and "coopetition". In addition he develops research on entrepreneurship and strategy in high-tech industries.

Manuel Zacklad (zacklad@utt.fr) runs the multidisciplinary laboratory Tech-CICO (http://tech-cico.utt.fr/) at the University of Technology in Troyes (UTT/ICDFRE CNRS 2848).

PART I

Can ICT Tell History?

Chapter 1

Elements for a Digital Historiography

1.1. Introduction

History starts with a document, which is essential evidence from the time it was produced. The traditional document, defined by the inseparable relationship between writing and printing has been succeeded by the digital document, characterized by the introduction of stand-alone writing. This separation alters the operating conditions of most historiography methods when using documents.

Digital technology brings characteristics of a major epistemological break, which is able to affect the continuity and autonomy of changes that mark historiographical evolution.

The mobility and flexibility of digital documents means they can be exploited beyond that of traditional documents, ensuring a link with previous changes. The digital document, in its current rough or future sophisticated version, is a new historical material which requires a new generation of methods for historical analysis and a big change in discipline structures[1].

Chapter written by Andrea IACOVELLA.
1 Conditions of the new writing, from the separation of the traditional document, are explained in the context of a writing theory [DER 67b] and a theory of support for which all knowledge comes from written material from which it is interpreted [BAC 99].

1.1.1. *Epistemological mutations of historiography*

Historiography results from a set of epistemological mutations that are the root of its methodological renewals[2]: critical analysis of the document, use of the hypothesis, conjuncture/structure relationship, sampling and quantitative methods. Problems from history have been enriched by contributions from other sciences, providing an interdisciplinary exchange which has deeply marked historiography, both in its objectives and methods [BLO 74][3]. The actual construction is characterized by the coexistence of various conceptions of history, sometimes radically opposed, creating complex and reactionary debates. This coexistence depends on an activity that is autonomous and separate from methods of analysis, whose use does not alter the processed material. If methodological contributions have had an impact on already existing methods, the changes have remained limited and have not resulted in transformations affecting the operating conditions of each of these methods. Although historical material ensures stability of its objects and methods, at the same time it constitutes a lever for historiography development and its epistemological mutations.

1.1.2. *History and documentation*

Documents have been used throughout history. From the start of history, the document appears as a privileged material with dual status: witness in the information sense and proof in the juridical meaning of evidence [PIC 73][4]. It is in the mid-19th Century that the discipline started to develop a set of rules that controls document analysis, named critical method[5]. Twenty years later, according to Croce the document is the *essential condition to existence of history* [CRO 89][6]. The

---

2 Methods listed in this chapter are not exhaustive and do not constitute a historiography of reference. The study is conducted as an exploration of historiographical methods, by evaluating the impact of digital methods on them.
3 "Science is not defined solely by its purpose. Its limits can be limited, as much, by the nature of its methods", p.50.
4 In 519/518 B.C., Athens and Megara (Greece) fought for the possession of Salamis island which was under the authority of Sparta. To argue in favour of Athenian occupation of the island, Solon called for the funeral tradition to decide the argument. According to this, the Athenians buried a single deceased person oriented to the West per grave, while the Megarians practised multiple burials oriented to the East. To separate the parties, graves were opened and finally it was Athens that won the case, p.46.
5 The critical method was described in a manifesto [LAN 1898].
6 "Talking about a history for which we do not have any documents would be like speaking of anything, which would be missing an essential condition to its existence […] the narration in which it occurs is historical narration only if it is a critical statement of the document" (translation by the author), p.16.

fragment and indirect statement[7] has imposed itself as a tool for which society guarantees throughout archives, libraries and museums [BLO 74][8].

1.2. Historiography facing digital document

1.2.1. *The digital document*

The development of information and communication technology sciences (ICTS) in recent years is explained by the upsurge in human and social uses of the digital document, most of whose characteristics are partly found in the continuity of the traditional document, while others represent a change. The digital document, whose use is inseparable from technologies that produced it, offers two sides – the object itself and methods by which it was created – by which it is convenient to examine the impact on historiography. By its object nature and wide adoption in everyday life, it constitutes a new material for historical studies[9]. By its dependence on scientific methods inherited from ICTS, digital documentation constitutes a supplementary mutation for historiography.

1.2.1.1. Interdisciplinary research on digital documentation

A group of researchers gathered in a multi-disciplinary network dedicated to digital documentation[10] has conducted a collective reflection on ongoing mutations. By combining the expertise of various disciplines and by making the document a research object, these works promote a renewed understanding of human and social activities. The network has published three texts that put together the three analysis levels to which the document was subjected: the document as form, sign and medium. The document as a form [PED 03] with its physical and material nature

---

7 *Ibid.*, p.18-19: Croce mentions the case of Greek painting, history without document, which came in the form of stories or copies.

8 Bloch reaffirms the importance of the critical method and enlarges the scope of the document by keeping to history: "its largest meaning [...] that prohibits no direction of inquiry", p.31. He uses F. Simiand's expression *knowledge by traces*: "what do we mean by documents if not a "trace", that is, the mark [...] left by a phenomenon in itself impossible to grasp?", p.56. Still great news: "The tool certainly does not make science, but a society that claims to respect sciences should not ignore their tools", p.67.

9 As an example, we will consider a youth study, as an indication of economic development conditions affecting flexibility of work, by relying on the anti-CPE (First Employment Contract) movement, which was developed in February and March 2006 in France. Such a study cannot avoid taking into account the new documents that represent electronic mail, websites, blogs and SMS that have played a decisive role in the development of the movement and the new historical participant configuration.

10 Multi-disciplinary thematic network on digital document, RTPDoc: http://rtp-doc.enssib.fr/.

acts as a technical instrument. The document as a sign [PED 05] is analyzed in its dual material and intelligible nature for its interpretation through reading and writing methods. The document as a medium [PED 06], studied as a social object and place of information exchange, comes into the configuration of new forms of human and social transactions [ZAC 05]. It concerns historiography as new participant object of human activities and for whose study methods remain to be explored[11].

1.2.1.2. Instrumentation of a digital document

The traditional document, which is defined by the inseparable relationship between writing support and display support, succeeds the digital document, where the concept of support loses its position as a stable component, due to the proliferation of devices that combine support, access and information coding in complex ways: disks, memories, caches, servers increase, monitors, printers and others. This transformation has the effect of deconstructing the document in a registered resource on one hand, and dynamically calculated views on the other hand [BAC 04c]. The resource is not accessible as it is. The user of a digital document is only confronted with processed documents. As a consequence, the digital document represents the new document, which is accompanied by complex and growing technological interdependence at the same time it is assigned a representation property based on an increase in non-hierarchical views.

1.2.2. *Consequences related to a traditional document*

The dissociation caused by a digital document ruins the conditions belonging to a traditional document:

– authenticity: the objectivity enabled by a traditional document is difficult due to the dynamic reconstruction of the new document. The "reader" does not compare the representation that he has in front of his eyes with the reality of the resource. Where the traditional document relays the set of representations, the digital document reader is deprived of this objectivity which is reduced to a hybrid condition that destroys "the authenticity" of traditional document;

– integrity: readers are confronted with multiple representations of a document without being able to evaluate them for adherence to the original document. The qualities of adaptation to different contexts ruin "the integrity" of the traditional document which has the effect of disorienting the reader. Without any stable space, the reading of the digital document requires a constant reconstruction of the document which is akin to a new form of a new appropriation;

---

11 For now, historians are interested in the Web as a tool for publishing and broadcasting their works [RYG 05] and documentary sources; this overshadows methodological questions linked to email exploitation and other ICT-based supports.

– identity: the growing number of interventions and visualizations, the uninsured character of the precision of the content represented, opacity of supports and potential fragmentation of resources ruin conditions of "identity" of traditional document[12];

– sustainability: technologies interdependence, the use of distributed resources coupled with technological evolutions as both fast and disconcerting, contribute to ruin "sustainability" conditions of traditional document. The sustainability of digital document [CHAB 04] requires a strong instrumentation[13] that results in concentrating the weakness of the device on some points by rendering it vulnerable to a massive and profound alteration of contents.

1.2.3. *Consequences on historiography*

The loss of authenticity, integrity, identity and sustainability with digital documentation ruins the conditions of critical analysis of the document, summarized as: Who produced it? How? When? and Where? It is the unprecedented upheaval in operating conditions when the document is used that results in changes in the historiographic structure's its methods, principles and results. This is why the digital document configures a major epistemological renewal in favor of a digital historiography. We must insist on the scale and complexity of a hypothetical instrumentation of the new document, able to durably restore conditions of traditional historiography to which the discipline appears to be acquired. This instrumentation, which does not allow itself to be reduced to a single technological project, but to a set of mutations often dictated by urgency, does not take into account the historiographic objective: a document is the essential condition for the existence of history. It is mobilized by norms and standards research aimed at a tight mesh of the international community, for strict communication functions.

The second part of the study describes the contribution of ICTS to methods of historiography that fall into two major orientations: the use of computer formalisms to ensure formal control of scientific constructions and semiotization of documentary sources, understood as a modeling and semantic units traceability process. Through its close relationship with the interpretation of sources and historical construction, the nomenclature constitutes the entry point of a comparative study of both technological trends.

---

12 In the example cited earlier, the reality of the anti-CPE movement imposes the new historical actor without it being possible to identify many authors of emails who use pseudonyms, generic IP addresses, trivialized terminals.

13 "Conservation" by a constant updating of materials, "saving" contents by a multiplication of copies and storage areas, "sustainability" of accesses to documents by an action of constant technological migration of formats and encodings.

1.3. ICTS contributions to historiography methods

1.3.1. *Nomenclature and historical semantics*

Nomenclature[14] is a tool for ensuring the link between documentary material and scientific construction, whose consistency it guarantees. It is an interpretation tool that gives a meaning[15] to each term. There is no nomenclature without a document and no history without nomenclature. History is the tool by which the outline is clarified and vocabulary develops from its material. The historian "is forced to substitute a nomenclature to the language of sources, because it restricts the use, distorts the meanings without warning the reader, without ever realizing itself" [BLO 74][16]. The nomenclature aims to overcome the imperfection of documentary evidence and return the deep bonds of facts. Once this has been taken into consideration, each important term becomes a real element of knowledge: it is the method of historical semantic [FUS 82][17].

1.3.1.1. *ICTS models*

Models that have been developed by the ICTS follow the two guidelines introduced by Boch: whether they prefer the stability of contents handled by the categories of the historian or the semantic shift when they are updated, depending on the context of the study. Research covers two fields that are subject to distinct changes in historiography that will be exposed in the study: the use of formalisms and ontological engineering for the formal control of scientific content, the use of documentary engineering for enriching sources and observation of signification modes. The study shows that, far from being mutually exclusive, the two trajectories tend to converge with each other.

1.3.2. *The initiatives of formalisms*

The criticism of formalism redefines artificial intelligence (AI) as "artefacture", in the form of engineering using knowledge emerging from hermeneutic and intersubjectivity [BAC 97]. Knowledge engineering (KE) results from logicism

---

14 The term nomenclature is used in a generic sense in human sciences, as a vocabulary-based tool defined by the discipline. The nomenclature model is that of the taxonomy that relates to natural sciences. It defines two things: classification of elements and science of classification.

15 Signification is a definition standardized by a discipline; [RAS 91]: "By signification we mean the linguistic content considered out of context; by sense, its content in context", p. 74, see in particular Chapters 3 and 4 on semantics and artificial intelligence.

16 *Ibid.*, p. 143.

17 *Ibid.*, p. 31-33 and cited by J.P. Vernant in [GER 82], p. 8.

formalism of the nature of the knowledge (Vienna Circle) and a calculative design of logic following computational formalism (Hilbert and Turing). KE adopts an approach based on a modeling principle that refers to the domain knowledge of the problem and a principle of effectiveness that ensures their operating character[18]. For formalism, knowledge is only knowledge when it is formal and when formal it is operational.

The effective model should reflect reality in its structure and demonstrate a descriptive compliance but in practice there is difference between the modeling principle claimed by formalism and interpretative deadlock of objects coming from it. The difficulty is in defining basic concepts in "primitive"[19] areas. These result from construction based on a particular point of view, but do not define all possible points of view in the field. The consequence is that the predicates of formalism are manipulated as primitives, although in practice their interpretation depends on the context in which they are used. In other words, formalized "primitives" cannot be interpreted as "primitives" in this field [BAC 00].

1.3.2.1. *From nomenclature to instrumentation of interpretation*

Formalization is not a proper modeling approach because it is based on a preliminary modeling. Without a modeling function, computer formalisms become a knowledge support, redefined as intellectual writings and tools [BAC 04a]. They lead to a new instrument of formal control of scientific constructions: ontology is a linguistic and formal representation of the concepts of a domain in an applicative context, which, in the case of history, formalizes the significance of nomenclature. Although the ambitions and hopes raised by formalism are strongly reconsidered, KE has the effect of reviving the instrumentation of semantic models on human sciences.

1.3.2.2. *Works and experiments in historiography*

Archaeology is a fertile disciplinary field in terms of methodology where natural, mathematical and human sciences come together. The indifference of historians towards archaeology is no longer relevant [BLO 74][20]. The development of the

---

18 For a recent status of the works, see [CHA 05].
19 As part of this study, "primitives" do not refer to the lexicon of the "nomenclature" used by the historian, but language entities that are treated as conceptual categories for the validation of historical models. For the difference between components and primitives, see [RAS 91], p. 141-142.
20 "If the most famous theorists of our methods had not shown towards the techniques of archaeology such an astonishing and superb indifference […] undoubtedly we would have seen them less prompt to push us towards an eternally dependent observation", p. 55.

stratigraphical method ensures the study of relationships among objects, the observation of a divided space and a historical reading of facts [SCH 74].

1.3.2.2.1. *Formalization and publication of erudite speech*

Starting from 1974, Gardin presented an analysis of erudite speech [GAR 74] and the project of reformulation of archaeological writing based on a formal approach to "get them out of a practical crisis, due to the proliferation of publications, their high costs and inadequacy of their form, as well as a theoretical crisis" [GAR 75][21]. Gardin proposed to bring the architecture of all archaeological monographs back to a single diagram in order to present an interpretation where archaeological science is developed. The visionary relevance and timeliness of these works show that "[…] all the scientific disciplines face the exponential growth of documentation and development of analytical methods of processing militate in favour of other forms of storage and data broadcasting other than printed literature" [GAR 75][22]. These researches resembled documentary techniques at first, then AI by using expert systems [GAR 87]. It is only recently that the connection with KE and hypertext has initiated a new type of instrumentation capable of meeting the theoretical challenge.

1.3.2.2.2. *The Arkéotek project*

The logicism project took a new turn with the use of CD-ROM publication of research into ethno archaeology [ROU 00]. This is extended in an online journal. *Arkéotek*, on the archaeology of techniques [ROU 04b], was presented as an alternative. "Faced with the scientific publications crisis, combine both electronic edition and new writing practices"[23]. The rewriting of texts, based on the logico-semantic model, uses a formal language – SCD (Scientific Construct and Data) [ROU 04a]. The project uses ontology to describe and structure knowledge gathering [AUS 06].

1.3.2.2.3. *Other extensions: the Palamède project*

Outside the strict publication framework developed by *Arkéotek*, logic formalism was tested as part of a historiography study on the emergence of the first economic and socio-political structures: the *Palamède* project [FRA 89]. This system submits the same archaeological facts to the different theories of emergence of protohistoric urban societies to observe how they differ. The system formalizes doxography[24] in the form of interpretive reasoning of theories, and physiography, which gives a

---

21 p. 6-8.
22 Chapter IV-2: "Radical innovation".
23 www.arkeotek.org.
24 By doxography, we mean the various "theories" related to the same phenomenon.

relevant interpretive value to archaeological facts. On a methodological level, the *Palamède* project represents the first contribution of the comparative method to historiography, which leads to the objectification of theories and to an epistemology of diversity [RAS 02].

1.3.3. *A semiotics of the documentary object*

After formal research on stabilization of content, the other experiment developed by the ICTS aimed to observe the semantic shift of categories at the time of their updating. By focusing on the observation of changes occurring during the study, it takes a close interest in the practices of the historian and aims to identify the modes of signification of the discipline; it is in this way that it relates to semiotics[25]. These works relate to the new modes of reading and ownership of digital documents where handling, owning and interpreting impact on activity: corpus, document and sign [BOU 03]. The autonomy of media, display and writing, introduces breaks in the conception and elaboration of contents [BAC 04b].

1.3.3.1. *New reading methods of the digital document*

On an experimental level, enrichment of sources by annotations is observed as a continuous and dynamic process [BEN 04b] of semiotization that turns them into documentary objects. Semiotization reinstates the knowledge modeling function which, as we have shown[26], is outsidecomputer formalisms. The contextualization of annotations and their hypertextual solidarity with sources ensure the traceability of semantics units linked to the enrichment of the referenced document. It is expressed in formalism of networks of description by the use of directed acyclic graphs[27]. The resulting non-linear reading, between annotations and documents, avoids disorientation[28] of hypertext reading [BAC 99], carried out from document to document, and represents an interpretative path [IAC 02] for expert and specialized use. The new methods involve the reader in an annotation-based ownership of the

---

25 Understood here in its generic meaning of a project studying the life of signs from a social aspect, without focusing on the language and society [GRE 76, PEU 58], but in a position of observation significant in sciences [DER 67a, DER 72].
26 See section 1.3.2.
27 Variant of semantic networks used in computer science to represent knowledge [SOW 84]. Formally, a network is a graph whose vertices are called "nodes", representing concepts, and whose arcs, called "links", show binary relationships among these "concepts". For a critique of semantics networks, see [RAS 91], Chapter IV, p. 121-137.
28 Disorientation remains when going from one textuality to another, unique to each document, known well by the historian: "a text, a search, an image are all different 'speeches', going their own way with their own logic and who must be made to come across each other somewhere" [HAR 82], p. 691.

document, the resulting reformulation of which indicates the distance between content read and reading carried out. The new documentary objectivity [BAC 04c] has two sides: that of the author, who designs the canonical form of multi-support and multi-use presentations, opening up to the multi-structuralism of the document [CHAT 04] and that of the reader-author, who rebuilds a version he or she considers to be the proper reformulation.

1.3.3.2. *The Porphyry project*

Porphyry[29] is a set of methods, based on the semiotization of sources, combined on a single computer platform [BEN 04a]. It is a digital workshop aimed at communities of experts looking to share digitalized documentary corpora: publications and archives (reports, photos, maps, plans, etc.). The system aims to store a hypermedia structure on digitalized documents. It provides tools to assist in the creation, organization, annotation and publication of an enriched document.

The starting point lies in a theoretical questioning of the construction of meaning in historical disciplines [BEN 02] and representation models of the semantics of documents [CAL 03]. In particular, it identifies as a methodological impasse in the use of the only normative[30] definition of categories for historical observation and description of facts [BEN 01]. It delineates production of specialized knowledge and their circulation within a community, as a semantic field to combine needs of the situation expert, exposed to a knowledge update, and the know-how of a reference that characterizes the field [IAC 06].

The system is structured in three interdependent technological levels, experienced in the context of historical disciplines [IAC 05]:

– documentary objects shared by experts;

– common methods of sources enrichment;

– modeling assistants that apply specialized activities of expertise.

---

29 www.porphyry.org. The platform is under open source license. Research activities are conducted in the framework of a transdisciplinary network *Artcadhi-Porphyry* (transdisciplinary research workshop on the construction of meaning in archaeology and in the other historical disciplines).

30 By "normative definition" we mean the signification standardized by the discipline that is subject to formalization described in section 1.3.2. The *Porphyry* model tends to combine the formalization of scientific contents with a hermeneutic approach that introduces knowledge modeling. Formalization and modeling combined form part of a current science observation.

1.3.3.2.1. Generic methods[31] of *Porphyry*

Expertise: the expert invests the space of documentary sources (source, source fragment, subset of sources and/or fragments designated by a file) that he or she enriches by using a system of contextualized annotations (collection). Methods available include naming, clustering, inclusion and the reading step, as well as historic situations that show properties of the enrichment (who did it? what? when?). The guided tour results from a set of annotations where we never pass by the same link twice.

Intersubjectivity: this is a collection of references linked to documentary objects by a user or a group of users. It represents the work of an expert. The system allows the production of several viewpoints from the same objects enabling intradisciplinary and multi-disciplinary[32] comparison.

Diachrony: because each node of the *Porphyry* graph identifies all of the transactions carried out, their author and date, the system ensures traceability of semantic units representing variations that have affected the field of reference and that of its members, showing a set of varied temporalities [ROI 04] that historiography knows well [HAR 82][33].

1.3.3.2.2. Time modeling

Modeling wizards are tools that complete the generic relations of the *Porphyry* platform by using specialization relations, such as temporal relations. Uses of formalism and AI allow us to represent the knowledge of a field reference within the community. The researcher is involved in a dynamic model where he or she can interact with the community [ACC 05]. It concerns the time and rhythms of history on one hand and the conjuncture/structure relation on the other.

Manual temporalization of a corpus involves associating documentary object annotations with a temporal content, probably from knowledge organizations (thesaurus or ontologies[34]). The "temporal" structure derived allows temporalized exploitation of the corpus [ACC 04b]. Modeling times, whether they are multiple or thick, short or long, fast or slow, are presented as a result of a construction [BRA

---

31 By methods, we mean technological tools available in the *Porphyry* system.

32 Current researches focus on the modeling of a language, allowing us to alter the views of Porphyry, such as to compare two or more expertise [GES 06].

33 "[…] text, search, display are themselves and each in its own way, multiple, complex and have known by different temporalities, changes and mutation", p. 691.

34 A thesaurus is a list of terms used in a classification; ontology is a linguistic and formal representation of concepts in a field for an application.

89]³⁵ and decryption "where phenomena are, as the place of their intelligibility" [BLO 74]³⁶. Braudelien historical time is marked by the different rhythms deciphered. A chronology is never a datum but from the result of a combination of several indices. Its validation is continually challenged by the emergence of new data. Archaeology expresses its rhythms by relative dating where events are periods indicated by intervals.

In the *Porphyry* system temporal relations of Allen [ALL 91] that express dependence, inclusion and events simultaneity are explained in the formalism of description networks, where every time modeling done by the expert is represented by a *Porphyry* viewpoint. When the expert relies on the use of temporal relations in the annotation of a document, the system spreads the modeling done in the chronology of reference of the field and reports any inconsistencies, leaving the initiative to change such inconsistencies with the expert [ACC 04a].

Vovelle said of historical conjuncture that it is a consistency of times that can be reformulated by interweaving of different rhythms [VOV 78]. The formal modeling of time provides a unique environment for the construction of "rhythms" of history and observation of the mutual relations that configure forms of consistency. In this context, the conjuncture/structure relation, so problematic in history, opens to an infra- and inter-rhythmic observation. The shift from structure to conjuncture then appears as a change of pace that affects the orientation of the interweaving and shows historical time³⁷.

1.3.3.3. *Traceability and methods of historiography*

Space structured by the writing of an algorithm is a method regulating process in time [BAC 96]. The digital universe is no longer a hierarchical structure organized in classes, but a force deploying time of prevision and possibility of computer algorithm [BAC 04a]. The traceability that ensures the representation as stages crossed during use constitutes a unique property of the document, which modifies the link to historical material and its exploitation. Together with the consequences that ruin the conditions of a traditional document³⁸, traceability enables us to observe and analyze conditions at the heart of what historiography means by internal and

---

35 p. 15-38.
36 p. 36.
37 In the example of anti-CPE movement (see footnotes 9 and 12), the rhythm of institutions of the legislator, although accelerated by the use of procedure 49.3 and despite enactment of the law, was overwhelmed by the stubborn, conscientious and zealous rhythm of the movement, (the configuration, organization and activity) and was conformed by the use of new media. Rhythm interlacing is narrowed in the decision taken by the political authority to not apply a law voted in by parliament.
38 See section 1.2.2.

external criticism of the document. This is also the case when using a hypothesis[39], where traceability of the reconstruction of answers and the verification or refutation of the initial hypothesis inaugurate the systematic exploration of all relationships. Furet recognizes the methodological value of calculation techniques because "there is no "neutral" technique and any statistical calculation raises the question of its compatibility with historical knowledge" [FUR 74][40]. The handling of serial sources requires the historian to consider the implications that the conditions of their organization may have on their quantitative utilization.

In terms of methods, we need to distinguish calculation models, which accentuate or minimize such aspects of a series[41], and the coding of variables that fall under categories for which the historian structures reports, by their uses. Variable coding has an impact on the calculation and "interpretation" of results [IAC 94]. To contain these difficulties, we use a triangulation of methods, where the same data are subject to several algorithms [IAC 97]. The traceability of a digital document introduces a new type of quantitative method, where variable coding, its definition and identification in sources will be explored in a systematic and rational way; in particular, it becomes possible to evaluate the impact of a change in the definition of a variable on the entire series.

1.4. Conclusion

The entire methodological chain of historiography, built since the 19th Century, is affected by the digital document and ICTS in three areas:

– the modalities of the traditional document, which is an essential condition in history, are ruined by the new document and use a strong instrumentation of the latter that have to be updated;

– the specific forms of digital document, such as electronic messages, websites and SMS require the production of new methods to exploit them as historical sources;

---

39 The work by hypothesis challenges the document regarding the source of historical issues, but there is no "witness" role who can validate challenges. Nora wrote: "The whole positivist era believed that existence of sources and possibility of their exhaustive exploitation dictated the historical issue when the reverse is true: it is the problem that creates practically inexhaustible sources", cited in [CAR 81].
40 p. 43-51.
41 In archaeology, statistical methods have been subject to a critical evaluation that takes their context of utilisation into account [DJI 91].

– traditional methods of historiography are upset in their usual configuration when encountering the new methods of reading and writing the new document, and its traceability properties.

Current experiments based on ICTS cover the whole chain of the historian's work, from observation, analysis, criticism and formulation of hypotheses to publication of the results. On a technological level, nothing prevents the two approaches, formal and semiotic, from being expressed within a single digital workshop of historiography that would bring together the formalism of scientific contents and knowledge modeling. Current developments in computing only partially cover traditional methods. Guided by specific constraints of modeling and production from a computer platform, these developments contribute in a way to mixing up references to historiography and do not facilitate multidisciplinary researches.

The digital historiography project shows the characteristics of a major epistemological break capable of affecting the continuity and autonomy of changes that have marked the evolution of traditional historiography. This upheaval affects the conditions of scientific activity and heralds a profound change in structures of the discipline. The conversion factor is concentrated on the writing out, from which came the scientific thinking [LER 65][42]. However, the digital methods affecting the linear development of actual writing, contaminated by hermeneutic of the object and forms of intersubjectivity, lead to questions on the consequences of forms of reasoning that are echoed in works related to "computational reason" [BAC 04a]. Current development of ICTS puts Leroi-Gourhans proposals into practice [LER 65][43] and confirms that relations with historiography are based on the increased mobility in devices for fixing thought in material supports. This a significant feature of human and social evolution [LER 64][44].

1.5. Bibliography

[ACC 04a] ACCARY-BARBIER T., BENEL A., CALABRETTO S., IACOVELLA A., "Confrontation de points de vue sur des corpus documentaires: Le cas de la modélisation du temps archéologique", *Proceedings of the 14th Francophone Congress AFRIF-AFIA of Pattern Recognition and Artificial Intelligence*, Toulouse, p. 197-205, 28-30 January 2004.

---

42 "Writing has constituted by its deployment in single dimension, the analysis tool from which came the philosophical and scientific", p. 261.
43 "Scientific thinking would find a considerable advantage if some process allowed presenting books such that the content of the different chapters open simultaneously in all its implications", p. 262.
44 p. 41.

[ACC 04b] ACCARY-BARBIER T., CALABRETTO S., "La temporalité dans les corpus archéologiques", *Revue Scientifique et Technique*, Digital Document, Special Issue on the theme Time and Document, vol. 8/4, p. 111-124, 2004.

[ACC 05] ACCARY-BARBIER T., BÉNEL A., CALABRETTO S., "Expression and share of temporal knowledge in archaeological digital documentation", *Proceedings of the 5th International Web Archiving Workshop [IWAW'05], held in conjunction with the 8th European Conference on Research and Advanced Technologies for Digital Libraries*, Vienne, Austria, 22-23 September 2005.

[ALL 91] ALLEN J.F., "Time and time again. The many ways to represent time", *International Journal of Intelligent Systems*, vol. 6, no. 4, p. 341-355, 1991.

[AUS 06] AUSSENAC-GILLES N., ROUX V., BLASCO P., *The Arkeotek Project: Structuring Scientific Reasoning and Documents to Manage Scientific Knowledge, Week of Knowledge*, vol. 3, University of Nantes, 2006.

[BAC 96] BACHIMONT B., Herméneutique matérielle et artéfacture : des machines qui pensent aux machines qui donnent à penser. Critique du formalisme en intelligence artificielle, PhD Thesis, Ecole Polytechnique, 24 May 1996.

[BAC 97] BACHIMONT B., "L'artéfacture entre herméneutique de l'objectivité et herméneutique de l'intersubjectivité : un projet pour l'intelligence artificielle", in J.-M. SALANSKIS, F. RASTIER, R. SHEPS (eds.), *Herméneutique: Textes, Sciences*, PUF, Paris, p. 301-330, 1997.

[BAC 99] BACHIMONT B., "De l'hypertexte à l'hypotexte: les parcours de la mémoire documentaire", in C. LENAY and V. HAVELANGE (eds.), *Technologies, Idéologies, Pratique, Mémoire de la Technique et Techniques de la Mémoire*, p.195-225, 1999.

[BAC 00] BACHIMONT B., "L'intelligence artificielle comme écriture dynamique: de la raison graphique à la raison computationnelle", in J. PETITOT and P. FABBRI (eds.), *Au Nom du Sens*, Grasset, Paris, p. 290-319, 2000.

[BAC 04a] BACHIMONT B., Arts et Sciences du Numérique: Ingénierie des Connaissances et critique de la Raison Computationnelle, PhD Thesis, University of Technology of Compiegne, 12 January 2004.

[BAC 04b] BACHIMONT B., CROZAT S., "Réinterroger les structures documentaires: de la numérisation à l'informatisation", *Journal I3*, vol. 4, no. 1, p. 59-74, 2004.

[BAC 04c] BACHIMONT B., CROZAT S., "Instrumentation numérique des documents: pour une séparation fonds/forme", *Journal I3*, vol. 4, no. 1, p. 95-104, 2004.

[BEN 01] BÉNEL A., EGYED-ZSIGMOND E., PRIÉ Y., CALABRETTO S., MILLE A., " Truth in the digital library: from ontological to hermeneutical systems", *Proceedings of the Fifth European Conference on Research and Advanced Technology for Digital Libraries*, Darmstadt, 4-9 September 2001, Lecture Notes in Computer Science, vol. 2 163, Springer-Verlag, Berlin, p. 366-377, 2001.

[BEN 02] BÉNEL A., CALABRETTO S., IACOVELLA A., PINON J.-M., "Porphyry 2001: Semantics for scholarly publications retrieval", *Proceedings of the Thirteenth International Symposium on Methodologies for Intelligent Systems*, Lyon, 26-29 June 2002, Lecture Notes in Computer Science, vol. 2 366, Springer-Verlag, Berlin, p. 351-361, 2002.

[BEN 04a] BÉNEL A., Computer aided consultation of documentation in human sciences: epistemological considerations, operational solutions and applications to archaeology [Online], in F. RASTIER (ed.), Journal-texto.net: site of text semantics (note: PhD thesis in supported computer science at the INSA of Lyon on 12 December 2003), March 2004.

[BEN 04b] BENEL A., CALABRETTO S., "Ontologies…déontologie: réflexion sur le statut des modèles informatiques", in A. BOZZI, L. CIGNONI, J.L. LEBRAVE (eds.), *Digital Technology and Philological Disciplines, Linguistica Computazionale*, vol. 20-21, p. 31-47, 2004.

[BLO 74] BLOCH M., *Apologie pour l'Histoire ou Métier d'Historien*, Armand Colin, Paris, 7th edition, 1974.

[BOU 03] BOULLIER D., GHITALLA F. et al. *L'Outre-lecture. Manipuler, (s')approprier, Interpréter le Web*, Pompidou Centre Bpi, Studies and research, Paris, 2003.

[BRA 89] BRAUDEL F., *Ecrits sur l'Histoire*, Paris, Flammarion, p. 15-38, 1989.

[CAL 03] CALABRETTO S., Modèles de représentation de la sémantique des documents. Applications aux bibliothèques numériques, PhD Thesis, INSA Institution of Lyon, 2003.

[CAR 81] CARBONELL C.-O., *L'Historiographie*, PUF, Paris, 1981.

[CHAB 04] CHABIN M.A. , "Archivage et Pérennisation", *Digital Document Journal*, vol. 8/2, p.12-134, 2004.

[CHAR 05] CHARLET J., TEULIER R., TCHOUNIKINE P. (eds.) *Ingénierie des Connaissances*, L'Harmattan, Paris, 2005.

[CHAT 04] CHATTI N., CALABRETTO S., PINON J.M., "Vers un environnement de gestion de documents à structures multiples", *Proceedings of the Conference Advanced Databases [BDA'2004]*, Montpellier, p. 47-64, 19-22 October 2004.

[CRO 89] CROCE B., *Teoria e Storia della Storiografia*, Laterza, 1917, reprint Adelphi, Milan, 1989.

[DER 67a] DERRIDA J., *L'Écriture et la Différence*, Seuil, Paris, 1967.

[DER 67b] DERRIDA J., *De la Grammatologie*, Minuit, Paris, 1967.

[DER 72] DERRIDA J., *La Dissémination*, Seuil, Paris, 1972.

[DJI 91] DJINDJIAN F., *Méthodes pour l'Archéologie*, A. Colin, Paris, 1991.

[FRA 89] FRANCFORT H.P., LAGRANGE M.S., RENAUD M., *Palamède: application des systèmes experts à l'archéologie de civilisations urbaines protohistoriques*, Working Paper No. 9, 2 volumes, CNRS LISH, Paris, 1989.

[FUR 74] FURET F., "Le quantitatif en histoire", in J. LE GOFF and P. NORA (eds.), *Faire de l'Histoire*, Gallimard, Paris, vol. 1, p. 42-61, 1974.

[FUS 82] FUSTEL DE COULANGES N.D., *La Cité Antique*, Albatros/Valmonde, Paris, p. 31-33, 1st edition in 1864, 1982.

[GAR 74] GARDIN J.-C., *Les Analyses de Discours*, Delachaux and Niestlé, Neuchatel, 1974.

[GAR 75] GARDIN J.-C., LAGRANGE M.S., *Analysis Essays of Archaeological Statements*, Archaeological Research Centre, Valbonne, France, 1975.

[GAR 87] GARDIN J.-C., GUILLAUME O., HERMAN P.Q., HESNARD A. et al., *Systèmes Experts et Sciences Humaines : le Cas de l'Archéologie*, Eyrolles, Paris, 1987.

[GER 82] GERNET L., *Anthropologie de la Grèce Antique*, Flammarion, Paris, 1982.

[GES 06] GESCHE S., CALABRETTO S., CAPLAT G., "Confrontation de points de vue dans le système Porhyry", in G. RITSCHARD and C. DJERABA (eds.), *Proceedings of the sixth day Extraction and Management of Knowledge (EGC'2006)*, Lille, France, 17-20 January 2006, Cépaduès, Toulouse, 2 volumes, p. 725-726, 2006.

[GRE 76] GREIMAS A.J., *Sémiotique et Sciences Sociales*, Seuil, Paris, 1976.

[HAR 82] HARTOG F., "Histoire ancienne et histoire", in F. Hartog and A. Schnapp (eds.), *Le document : éléments critiques* (dossier), *Annales. Economies, Sociétés, Civilisations*, A. Colin, Paris, no. 5-6, p. 687-696, September-December 1982.

[IAC 94] IACOVELLA A., AUDA Y., "Nécropoles de Sicile : étude de l'utilisation des espaces funéraires dans le temps (du IXe au Ier siècle av. J.C)", *Archeologia e Calcolatori*, vol. 5, p. 69-86, 1994.

[IAC 97] IACOVELLA A., "Etudes des proximités dans l'espace funéraire : le cas de la nécropole occidentale de Mégara Hyblaea", *Archeologia e Calcolatori*, vol. 8, p. 67-102, 1997.

[IAC 02] IACOVELLA A., "Modèle opératoire de navigation pour les experts : Appropriation sémantique et délimitation de l'espace documentaire", in F. GHITALLA (ed.), *La Navigation, Les Cahiers du Numérique,* Special issue, vol.3, no. 3, p. 175-190, 2002.

[IAC 05] IACOVELLA A., CALABRETTO S., BENEL A., HELLY B., "Assistance à l'interprétation dans les bibliothèques numériques pour les sciences historiques", in J.-L. LEBRAVE (ed.), *La Société de l'Information et ses Enjeux, Proceedings of Assets of the Interdisciplinary Programme Information Society 2001-2005*, CNRS, Paris, p.167-179, 2005.

[IAC 07] IACOVELLA A., BENEL A., PETARD X., HELLY B., "Corpus scientifiques numérisés: Savoirs de référence et points de vue des experts", in R.T. PEDAUQUE (ed.), *La Redocumentarisation du Monde*, Cépaduès, Toulouse, p. 117-130, 2007.

[LAN 1898] LANGLOIS C.V., SEIGNOBOS C., *Introduction aux Études Historiques,* Hachette, Paris, 1898.

[LER 64] LEROI-GOURHAN A., *Le Geste et la Parole. Technique et Langage*, Albin Michel, Paris, 1964.

[LER 65] LEROI-GOURHAN A., *Le Geste et la Parole. La Mémoire et les Rythmes*, Albin Michel, Paris, 1965.

[PED 03] PEDAUQUE R.T., "Document: forme, signe et médium, les re-formulations du numérique", *RTP CNRS 33 [RTP-Doc], Documents et Contenu: Création, Indexation, Navigation*, CNRS, 2003. (Available at http://archivesic.ccsd.cnrs.fr/docs/00/06/21/99/ HTML/index.html, accessed February 4, 2010.)

[PED 05] PEDAUQUE R.T., "Le texte en jeu, Permanence et transformations du document", *RTP CNRS 33 [RTP-Doc Documents et Contenu: Création, Indexation, Navigation*, , CNRS, 2005. (Available at: http://archivesic.ccsd.cnrs.fr/sic_00001401, accessed February 4, 2010.)

[PED 06] PEDAUQUE R.T., "Document et modernités", *RTP CNRS 33 [RTP-Doc] Documents et Contenu: Création, Indexation, Navigation*, CNRS, 2006. (Available at: http://archivesic.ccsd.cnrs.fr/sic_00001741, accessed February 4, 2010.)

[PEI 58] PEIRCE C.S., *Charles S. Peirce, Selected Writings*, P.P. WIENER (ed.), Dover, New York, 1958.

[PIC 73] PICCIRILLI L., *Gli Arbitrati Interstatali Greci* (338 av. J.C.), Marlin, Pisa, p. 46, 1973.

[RAS 91] RASTIER F., *Sémantique et Recherches Cognitives*, PUF, Paris, 1991.

[RAS 02] RASTIER F., "Anthropologie linguistique et sémiotique des cultures", in F. RASTIER and S. Bouquet (eds.), *Une Introduction aux Sciences de la Culture*, PUF, Paris, p. 242-267, 2002.

[ROI 04] ROISIN C., SEDE F. (ED.), "Temps et document", *Digital Document Journal*, vol. 8/4, p. 7-168, 2004.

[ROU 00] ROUX V. (ed.) *Cornaline de l'Inde: Des Pratiques de Cambay aux Techno-systèmes de l'Indus*, Editions de la Maison des Sciences de l'Homme, Paris, bilingual CD included, 2000.

[ROU 04a] ROUX V., BLASCO P., *Logicisme et format SCD: d'une Epistémologie Pratique à de Nouvelles Pratiques Editoriales*, Hermes, Paris, 2004.

[ROU 04b] ROUX V., GARDIN J.-C., "The Arkeotek project: A European network of knowledge bases in the archaeology of techniques", *Archeologia e Calcolatori*, vol. 15,p. 25-40, 2004.

[RYG 05] RYGIEL P., NOIRET S. (ed.), *Les Historiens, Leurs \revues et Internet (France, Espagne, Italie)*, Publibook, Paris, 2005.

[SCH 74] SCHNAPP A., "L'archéologie", in J. LE GOFF and P. NORA (eds.), *Faire de l'Histoire*, Gallimard, Paris, vol. 2, p. 3-24, 1974.

[SOW 84] SOWA J.F., *Conceptual Structures: Information Processing in Mind and Machine*, Addison-Wesley Longman, Boston, 1984.

[VOV 78] VOVELLE M., "L'histoire et la longue durée", in J. LE GOFF, R. CHARTIER, J. REVEL (eds.), *La Nouvelle Histoire*, CEPL, Paris, p. 316-343, 1978.

[ZAC 05] ZACKLAD M., "Transactions communicationnelles symboliques et communauté d'action: innovation et création de valeur dans les communautés d'action", in P. LORINO and R. Teulier, *Entre la Connaissance et l'Organisation, l'Activité Collective*, Maspéro, Paris, 2005. (Available at: http://archivesic.ccsd.cnrs.fr/sic_00001326, accessed February 4, 2010).

Chapter 2

"In Search of Real Time" or Man Facing the Desire and Duty of Speed

2.1. Introduction

Time and speed are two concerns that have not ceased to occupy the mind of Man. What are they? Are they not only imitations arising from the consciousness of individuals dealing with moving objects? No one can answer these questions regularly posed by philosophers with certainty. Following philosophers' legacies, they are a major source of interest for anthropologists, sociologists, psychologists and historians in modern and contemporary Western societies.

More recently, we have seen geographers, urban planners and managers interested in the concepts of time and speed. Thus these concepts have become the ultimate problem of how society functions as a whole. Sometimes bearers of progress, sometimes causes of so-called diseases of civilisation, time and speed have become essential variables that everyone seeks to own and control. They are therefore precious resources in a context strongly marked by competition, or should we say "hyper-competitiveness". Their control is both a source of freedom and alienation, depending on whether we endure them or we succeed in controlling them. They are the source of freedom when we allow for adapting to rules specific to capitalism based on productivity, described by classical and neoclassical economists as the ability to do as much as possible in the shortest time possible, with the smallest possible resources (capital, work) [BON 03b]. On the other hand, they

---

Chapter written by Luc BONNEVILLE and Sylvie GROSJEAN.

are a source of alienation for the productivist individual enduring denial, voluntary or involuntary, conscious or unconscious, of life itself[1]. Thus, we find individuals who seek to control them because they consider them as symbols of progress, while others seek to avoid them as far as possible because they see them as an important source of alienation.

Information and communication technologies (ICT) are precisely at the heart of these fundamental questions. On the one hand, they allow the completion of activities that were not possible previously (networking, distance exchange in differed and real time, for example), while, on the other hand, they contribute to an acceleration of our pace of life [AUB 03a, AUB 03b] and to an intensification of work[2] [VEN 02a]. Therefore, we consider it necessary to question the close relationship between technology and time, especially in a context where daily use of ICT takes on increasingly important proportions in all spheres of social activity. Thus isn't the technologization of our lifestyles and work (the human-time-speed relationship) a sign of the emergence of a new social temporality? We will answer this question by suggesting a critical reflection based both on a review of various research works and the results of our previous and ongoing research.

2.2. Rate, speed and ICT: emergence of a new social temporality

We can clearly see new spatio-temporal frameworks being drawn which are distinctively based on what is called immediacy, instantaneous or even presentification. It is time brought back to the present moment, substituting sequences, rhythms and cadences that could be considered as specific to temporal modernity, to borrow an expression widely used in several works[3]. This new relation to time is based, among other things, on ICT generalization (computers, cell phones, beepers, etc.) that enables individuals who wish to communicate and exchange a large amount of information remotely and in real time. So the idea of adopting a deterministic vision of ICT is a long way off. But this new temporality leads some individuals to represent time according to the actual moment [BON 03b]. This arises in relation to the possibility of starting and exchanging information at a distance, avoiding movement from a point A to a point B, leading to a certain contraction of the space creating a break with conventional methods of space as a place of

---

1 Which also made Bartoli say that the economy should be at the service of life itself, and not the opposite way around [BAR 96].
2 Intensification of work is characterized by the fact that there is no longer any dead time in the workday and that tasks follow one another, leaving little or no room for contingency management [VEN 02a].
3 For an overview of the work on modern time (or temporal modernity), see [SUE 94] and [PRO 96].

excellent time deployment. In addition writing in a speed logic that is symptomatic of a qualitative reconfiguration of the relationship to time, we can increase productivity and do as much as possible in an increasingly short time.

Manipulation of time took a specific turn from the 18th Century in the wake of the development of increasingly precise and complex measuring instruments. These instruments reflect a permanent quest for speed symbolized by the ability to quantify an increasingly short period of time. In about 1800, we were measuring one-hundredth of a second; around 1850, we were measuring the millisecond; in 1950, we were measuring the microsecond (one millionth of a second, 10^{-6} s); in 1965, the nanosecond (one billionth of a second, 10^{-9} s); in 1970, the picosecond (one thousandth of a billionth of a second, 10^{-12}); and in 1990, the femtosecond (one millionth of a billionth of a second, 10^{-15} s). Today, we are on the threshold of the development of the attosecond (10^{-18} s). This obsession with measuring time turns out to be symptomatic of an ambition to own time, in the most rigorous way possible, as if it was an object that we could seize to better live with, or foil. Why should we be surprised to see that ICT that can overcome distance and, therefore, avoid displacement of certain daily activities, has contributed in developing in some individuals a feeling of omnipotence linked to the idea of a greater individual productivity? This is one of the challenges that every individual wishes take, as it is a symbol of excellence [AUB 91]. Also in the last few decades we have seen an emergence of fast-food restaurants and invasion of our personal space by products which are said to be more and more rapid (automobiles, computers, aerodynamic clothing, etc.). More recently, television channels have started to show news in continuous time, with the weather at the bottom of the screen, stock exchange on the left, and latest news on the right. Behind this phenomenon lies the anthropological dream of controlling time, which implies greater conformity of nature to Descartes' dream of making man the master and owner of nature [DEC 62].

Without doubt it is this persistent quest to control nature that has led to the widespread impatience that is a reflection of a desire to reduce time to its own unique scale, even to question the objective time binding as it was built in modernity. This is consistent with invasion, in all spheres of social activities, of time management. The "passing of time" is now considered as dead time, in a context where time should be managed in the most rigorous way possible. Today, this obsessive control of time is associated with an extreme taste for anticipation and planning [DEG 05]. These are indicators of a society increasingly seeking to manage all, even speed. However, it is precisely this temporal pressure that several psychologists have considered to be the source of emerging temporal pathology [REI 79], or waiting pathology [BON 02], based on fetishism or this cult of generalized urgency [AUB 03a, AUB 03b], that imposes itself on individuals and which requires increasingly large individual ownership of time and, consequently, speed [BON 03b]. This reveals a desire of immediacy [JAU 03a] symptomatic of a willingness to abolish time [AUB 03a, AUB 98] in the case where waiting is perceived as *a priori* "too long" and where, consequently, the individual is always in a hurry [JAU 03b]. The individual thus tries to use ICT in order to carry out the largest possible number of activities in the fastest possible time, simultaneously. Therefore, the sequence of events fades behind a big "now". Time is thus placed in relation to the present moment, to its ability to ensure that a result is presented without delay ("immediately"). Individuals find themselves prisoners in a world of "already", itself based on this obsession of speed and fastness.

2.3. Speed: stigma of a new socio-economic and socio-cultural reality

As we have just mentioned, "changes in speed in the 20th Century (automobile, train, airplane, telecommunications) have lead to an acceleration of time and a narrowing of space" [LEC 99]. Our rhythms and lifestyles today are marked by urgency, as underlined by Aubert: "[…] if the mode of action *urgently* has long remained a relatively exceptional way of dealing with a situation by using a specific device to bypass the blocking of habitual structures (on the medical, legal, economic, political or social level...), it is no longer the case. The last years of the 20th Century seemed in fact marked by the irresistible rise of the urgency reign, which is fast becoming a privileged mode of social regulation and a dominating mode of collective life organization" [AUB 02]. This new temporal reality imposes itself on all types of social organizations (from private enterprise to the family), confronting the individual with both a desire and a duty of speed [GRA 99]. The reactivity of individuals and organizations[4] has become a consequence, or a response to urgency, by contributing in particular to the emergence of new forms of time and work space management. We are witnessing progressively a "networking" of work activity, contributing to the set up of greater compliance and flexibility. But also, as we have seen, contributing to the development of ICT which contributes to the emergence of a new space-time relationship [BON 06a].

2.3.1. Emergence of a "speed economy"

In the context of globalization and competition that exists everywhere, flexibility, response time and qickness of operations become important economic and strategic issues [BUR 98]. Let us consider organizations which compete time-wise with each other in an era of "speed economy" [RUL 00, VEL 02]. "Temporality issue is at the centre of the enterprise: time of release of a new product/service, logistics, time of integration of technologies" [DAU 01]. ICT

---

4 Reactivity here corresponds to the speed at which an individual or organization can react to shocks of conjuncture or solicitations.

allows real-time access to databases, responding rapidly to market demands or to satisfy customers immediately. This exacerbated responsiveness is in an economic context in which a "response model" is opposed to models of standardization or variety [COH 92]. The response model reacts to current economic constraints (economic uncertainties, globalization). The speed of responsiveness to environmental variations is paramount when facing important competition and personalization of products and services [DAU 01, FLI 04]. It then forges within enterprises new forms of temporality, new methods of organization in terms of time. Individuals work in the context of tight flow or of just-in-time [MART 97] that can be summarized by a succession of zeros: zero stock, zero delay, zero defect [VEN 02b]. Furthermore, the temporal variable becomes a characteristic of culture of some organizations which develop their marketing strategy on their ability to satisfy the client at speed (*FedEx, Chronopost International, Photo minute,* etc.).

Several authors have spoken of the emergence of new rules of the game in terms of competition, consistent to this idea of the emergence of a new economy [BON 03a]. This is based on a new process of productivity growth where it is the quality of the service being sold which constitutes the essential value of the enterprise. However, this quality is synonymous with speed. The faster an enterprise offers a service, the more appreciated it is by clients. As highlighted by Vendramin, "ICT plays a role in the extension of opening or accessibility hours of many services, particularly via online services that are accessible 24 hours a day. These introduce [in particular] weekend work and evening work in sectors where this practice was quite exceptional, such as the bank and insurance, marketing services, after-sales services" [VEN 02a]. Services to individuals and enterprises must be accessible any time and in "real time", involving an almost permanent connectivity and availability of teams of workers.

We witness organizational changes[5] to achieve higher responsiveness, as reflected by two heavy trends common to numerous enterprises:

– the explosion of borders of the organization,

– the emergence of transversal management [BLA 00].

The effectiveness of an organizational structure is based on a classic pyramidal scheme structured around hierarchical levels and a bureaucratic type of organization is questioned[6]. The objective of being responsive also requires the delegation of management tools, reconfiguration of decision-making circuits, instrumentation of

---

5 Enterprises therefore evolve towards more compliant and flexible organizational forms which borrow, for the most part, from the network principle [CAS 98].
6 For example, internally, traditional vertical functions are affected by the increase of group projects, matrix structures, and electronic workgroups [BLA 00].

communications and real-time access to data [GRO 06]. Technologies such as enterprise resource planning (ERP) software[7] are integrated into this new configuration of organizations by programming the content and sequence of tasks to be executed, in order to control and to make information flow faster [BON 06a]. Furthermore, to be able to respond quickly and to conquer market shares, organizations are required to operate in networks, specifically in a "global network of interaction" [CAS 98].

New forms of organizations are based on a virtual proximity and new concepts, widely used by the media, such as "business networks" and "virtual organizations", are emerging. This new model of the organization is for Veltz a real "structural turn-over" [VEL 00]. It is the case for virtual[8] organizations, which are geographically dispersed, which rely on the use of ICT and whose duration is temporary, and often tied to an ongoing project [BEL 00].

A virtual organization implements work processes whose unit of place and time does not have any discriminating characteristic. The work place can become an abstraction, no longer determined by a geographically identified space [TAS 03]. Thus, "virtual offices" are appearing within this new organizational environment, challenging both the workplace by remote work [DAV 98] and the classical break between "working time" and "non-working time". These new work spaces are both answers to the acceleration of time in organizations, but they also contribute in intensifying the rhythm, or the pace, of work [VEN 02b].

The space-temporality that once determined the time of work, but also the territory of the worker, and was the place of interactions, tends to disappear in favor of a redefinition of the borders and time of work. With this reconfiguration of geographical constraints, time marked by organizational imperatives remains the only guide for the worker. As highlighted by Hueber and Jacquis, "the virtual space reflects the shift from a spatial logic to a temporal logic" [HUE 00]. These new ways of work organization are instruments of flexibility and act in conjunction with urgency occurring in enterprises [MET 04].

According to Bühler and Ettighoffer [BUH 95], ICT utilization, assuming relatively important control of various information processing devices also makes it possible work from home and carry out work that is often precarious. According to these authors, we might think that this type of work must now be designed as a service to the employer (in exchange for a contractual agreement), which may

---

7 For more details, see [LEMA 02, TOM 03].
8 "Virtual organizations are organizations electronically connected that go beyond the borders of the conventional organization, with links that may exist both within and between organizations" [BUR 98] (unofficial translation).

require the remote execution of certain tasks through a network. This, moreover, encourages individuals to create their own jobs, to become, according to Ettighoffer, "electronic nomads" [ETT 95] who, due to ICT skills, can work anywhere around the world achieve highly by networking with others.

The role of workers consists of organizing the information "already here", to establish, in a network, a service or good entirely developed online, fully carried out remotely [ETT 95]. This assumes that they are able to process information, analyze it, reorganize it, update it, circulate it and, thus, to build an added value to it with utmost speed. This vision is equally shared by Castells [CAS 98], for whom this new type of work is the fruit of an organization based on information technology, which results in a new division of labor characterizing what he considers as a "booming informational paradigm" [CAS 98].This new organization of work upsets the dynamism of the productive-reproductive system on the basis of new principles governing the form of work: "The new information technology redefines the work process and profit of workers, and consequently employment and professional structure" [CAS 98]. In the information and speed economy, ICT expands the working time into the non-working time, so that we can no longer distinguish the latter from the production process. All these structural and organizational modifications create an omnipresence of workers [MOE 96]. The borders of work time are blurred for an increasingly important number of workers, particularly from the fact of the increase in permeability between working time and other social times. It does happen that professional time invades family time [VEN 04].

2.3.2. *Ambivalence of "hypermodern Man"*

Multiple and crossed changes (technological, economic, social and cultural) within society have contributed to the emergence of a new individual who can be qualified as "hypermodern Man" [AUB 04]. He is an individual in a hurry, enslaved to the clock and subjected to the duty of speed, but he is also an ambivalent individual who lives in excess, is always well informed, spurred by urgency, and always efficient. The relationship of the "hypermodern Man" with time is complex and paradoxical, sometimes motivated by a desire for speed, sometimes expressing nostalgia for the past (wanting to take his time or to waste his time doing nothing). But, as highlighted by Aubert [AUB 04], this relationship to time also raises the question of the relationship to speed. "Hypermodern Man" seeks for example immediacy of sensations, conducts brief and interchangeable encounters, building superficial and ephemeral links, with daily routine soon becoming unbearable to him[9].

---

9 This involves more fragile family structures.

The new economy, making us compete on time and allowing "hypermodern Man" to enter the game of "chrono-competition" [STAL 90], has contributed to anchoring workers in a new dimension: hyperwork[10]. Several individuals who are hyperactive at work always do more than they are asked. For these individuals, work overload, over-implication and increased responsibilities become sources of pride and self-exaltation [AUB 04]. However, the impacts of hyperwork are multiple. Aubert highlights in particular the loss of social ties because we no longer take time for dialog and exchange with others [AUB 03a], and we send emails [MIL 03]. The consequences of this hyperwork can be disastrous on the physical and psychological level: causing insomnia, hypertension, digestive disorders, feelings of depression, and heart problems [LEG 04, TRU 04].

Moreover, given in particular the use of ICT, we notice an increasing production of information (notes, memos, flow of messages through email), and some people complain of being "drowned", submerged by all sorts of information they sometimes deem unnecessary [LEJ 06].

Several individuals admit that they receive too many demands from too many people and do not have the time to assess how much we want them to accomplish. They feel they lack time and the result is chronic delays, which often lead to a loss of reason, to a sense of losing control of their own life [LAH 97]. These individuals submitted to speed duty are victims of cognitive overflow syndrome (COS) [LAH 97]. This is explained particularly by the fact that with ICT a proliferation we are seeing cognitive attractors[11] present in working environments [LAH 00] and, more widely, in the everyday life of an individual. However, a working environment too rich in cognitive attractors leads in several cases to "amusement" and to "procrastination". Workers become tired, or exhausted, wanting to carry out a series of urgent tasks and can no longer manage their priorities effectively, continually postponing deadlines [BON 06a].

We can say that technological advances and the introduction of ICT in professional life (but also family life) have upset our daily rhythms. Indeed, with ICT and the reign of the Internet, the slogan "real time" is involved in all economic activities and in our daily life. We can for example access our bank account or our professional emails at any time. Thus, ICT has brought a new form of time control of which, moreover, the family world is no exception. In 2005, a survey from

---

10 It was during a symposium organized by the Institute of Psychodynamism of Work in Quebec in 2002 that the concept of hyperwork was defined thus: "significant exceeding of a reasonable burden of work over a long period of time, this generally without any explicit demand" [LEG 04].

11 Cognitive attractors are tangible and intangible elements that participate in an activity and are simultaneously present in the working environment of the subject [LAH 00].

Statistics Canada, titled 'Internet and our time table'[12], shows that among major users[13], Internet usage is done at the expense of time devoted to family life.

We can also see that contemporary families experience the sense of urgency. Their daily vocabulary is full of expressions that are proof: we are "in a hurry", "overwhelmed", "stuck". Thus, the observation is made [DAL 00] that in many families the shortage of time or, simply, lack of time leads to the perception of b eing subject to a certain tyranny of time. Overloaded schedules of children and parents have made time a major concern in managing multiple family demands in contemporary life. Researches conducted in recent years show that many parents who have jobs struggle to reconcile their professional and family obligations [TRE 04]. Indeed, time organization off work is affected by the changes experienced by the job market, in particular with telecommuting, diversification of types of jobs (temporary job, on call, casual work) and schedules (part time, etc.) [TRE 06]. It seems that families have become poor in time. Far from seeing the entertainment society setting as announced more than 40 years ago by sociologists [DUM 62], we see a culture of overwork setting in [DAL 00].

Consequently, the acceleration of our lifestyles has made time a political issue. Indeed, as time becomes a scarce and precious product, this gives rise to claims based on the right for leisure time, the right to time spent with one's family, the right to the reduction or planning of one's working time. These include particularly the negotiations on reduction of working time in various European countries [HAY 98] or the job-family reconciliation measure in Canada [TRE 06], which have become important political issues. Time is an asset, a scarce resource that must be managed, controlled and protected from abusive exploitation, which gives it a political dimension and allocates it within the dynamics of possession, negotiation and control [DAL 00].

2.4. Conclusion

We have seen that ICT makes its mark on our relationship to time and, as highlighted by Vendramin and Valenduc, "From the medieval clepsydra (water clocks) to the recent methods of time management built into computers and networks, the measurement and control of time have always been an important field of technological innovation" [VEN 04]. Moreover, the observation is made that ICT

---

12 The investigation focused on the links between time spent online and other aspects of daily life of Canadians over a 24-hour period [STAT 06].

13 Major users are the people who have spent more than one hour on the Internet during the reference day.

modifies informational rhythms and places the individual in a live culture, of just-in-time, and of "real time".

Some individuals, qualified as "hypermodern Man", are obsessed with speed, quickness, productivity – symptoms of a society that could be qualified as hyperactive and in a hurry. In line with this new relationship with time, an observation is made: the more ICT helps to "shorten" time by a compression of "duration", the more it is time itself which is beyond the individual. The "hypermodern Man" is an overwhelmed individual with a gift of exacerbated relative ubiquity: he will phone while driving a car, talk on the phone while browsing the Internet for online purchases, for example, attend a meeting by sending emails, write a text while checking the cinema schedule, and consult weather forecasts while reserving a hotel abroad. But he is also an ambivalent man allowing himself to get carried away by the search "for real time" and the desire for speed, while seeking to control time, handling it to avoid becoming enslaved by it.

Time is a scare product. The hunt for lost time, dead time, is open. Moreover, all social activities are now subject to this "chrono syndrome" [EGG 05], and many individuals become "time sick". So we should not be surprised to see emerging, in recent years, forms of resistance which occur not as reluctance to use ICT as such, but rather a refusal to use it for all and everything[14] [LEMO 05]. This phenomenon is also associated with new psychotherapies which aim to rehabilitate the "time sick" to the natural rhythm of things [MARQ 98].

2.5. Bibliography

[AUB 91] AUBERT N., DE GAULEJAC V., *Le coût de l'excellence*, Paris, Seuil, 1991.

[AUB 98] AUBERT N., "Le sens de l'urgence", *Sciences de la société*, 44, p. 29-42, 1998.

[AUB 02] AUBERT N., "Pathologie de l'urgence", *Séminaire organisé par le Centre d'Etude de l'emploi sur le thème intensification du travail*, Noisy-le-Grand, http://www.cee-recherche.fr/fr/ sem_intens/seance16/patho_urgence.pdf, 31 January 2002.

[AUB 03a] AUBERT N., *Le culte de l'urgence, la société malade du temps*, Paris, Flammarion, 2003.

[AUB 03b] AUBERT N., "Urgence et instantanéité : les nouveaux pièges du temps", in F. ASHER and F. GODARD (ed.), *Modernité : la nouvelle carte du temps*, Éditions de l'Aube, Le Moulin du Château, p. 169-185, 2003.

[AUB 04] AUBERT N., *L'individu Hypermoderne*, Erès, Ramonville-Saint-Agne, 2004.

---

14 Research is being completed to better understand certain forms of disconnection of ICT [BON 06b].

[BAR 96] BARTOLI H., *L'économie, service de la vie. Crise du capitalisme. Une politique de civilisation*, Presses universitaires de Grenoble, Grenoble, 1996.

[BEL 00] BELCHEIKH N., SU Z., "Pour une meilleure compréhension de l'organisation virtuelle", *Actes de la 9e conférence de l'AIMS*, Montpellier, 24-26 May 2000.

[BLA 00] BLANCHOT F., ISAAC H., JOSSERAND E., KALIKA M., DE MONTMORILLON B., ROMELAER P., "Organisation : explosion des frontières et transversalité", *Cahier de recherche*, n° 50, EMR Dauphine, CREPA, Paris, 2000.

[BON 02] BONNEVILLE L., "La temporalité du réseau Internet est-elle encore moderne ?", in F. JAUREGUIBERRY and S. PROULX (ed.), *Internet, nouvel espace citoyen ?*, L'Harmattan, Paris, p. 205-222, 2002.

[BON 03a] BONNEVILLE L., La mise en place du virage ambulatoire informatisé comme solution à la crise de productivité du système sociosanitaire au Québec (1975 à 2000), PhD Thesis, University of Quebec, Montreal, 2003.

[BON 03b] BONNEVILLE L., "Informatisation sociale et représentation de la temporalité", in J.-P. DUPUIS (ed.), *Des sociétés en mutation*, Nota Bene, Montreal, p. 77-92, 2003.

[BON 06a] BONNEVILLE, L. GROSJEAN S., "L'Homo-Urgentus dans les organisations : entre expressions et confrontations de logiques d'urgence", *Revue Communication et Organisation*, GRE/CO, Bordeaux, n° 29, 2006.

[BON 06b] BONNEVILLE L., JAURÉGUIBERRY F., Les formes de déconnexion aux technologies de l'information et de la communication (TIC) dans le secteur de la santé, Recherche dans le cadre de l'Initiative de développement de la recherche (IDR) subventionnée par le Conseil de recherche en sciences humaines du Canada (CRSH), 2006-2008.

[BUH 95] BÜHLER N., ETTIGHOFFER D., "L'homme polyactif", in G. BLANC (ed.), *Le travail au XXIe siècle : mutations de l'économie et de la société à l'ère des autoroutes de l'information*, Dunod, Paris, p. 203-218, 1995.

[BUR 98] BURN J.M., "Aligning the On-Line Organisation-with What, How and Why?", *Proceedings of the 8th Business Information Technology Conference*, Manchester, 4-5 November 1998.

[CAS 98] CASTELLS M., *L'Ère de l'information, Vol. 1 : La société en réseau*, Fayard, Paris, 1998.

[COH 92] COHENDET P., LLERENA P., "Flexibilité et évaluation des systèmes de production", *ECOSIP, Gestion industrielle et mesure économique. Approches et applications nouvelles*, Economica, Paris, 1992.

[DAL 00] DALY K., De plus en plus vite : La reconfiguration du temps familial, Institut Vanier de la famille, http://www.vifamily.ca/library/cft/faster_fr.html, Ottawa, 2000.

[DAU 01] DAUTY F., LARRÉ F., "La réactivité industrielle : caractéristiques et outils", *Les Notes du LIRHE*, n° 349, http://lirhe.univ-tlse1.fr/publications/notes/349-01.pdf, 2001.

[DAV 98] DAVENPORT T.H., PEARLSON K., "Two cheers for the virtual office", *Sloan Management Review*, p. 51-65, 1998.

[DEC 62] DESCARTES R., *Discours de la méthode*, Presses de l'imprimerie Bussière, Paris, (original edition in 1637) 1962.

[DEG 05] DE GAULEJAC V., *La société malade de la gestion : idéologie gestionnaire, pouvoir managérial et harcèlement social*, Seuil, Paris, 2005.

[DUM 62] DUMAZEDIER J., *Vers une civilisation du loisir ?*, Éditions du Seuil, Paris, 1962.

[EGG 05] EGGER M.-M., "Pour une économie politique de la vitesse", *Choisir*, n° 547-548, www.fondationdiagonale.org/article.php3?id_article=150, July-August 2005.

[ETT 95] ETTIGHOFFER D., "Networkers, les nomades électroniques", in G. BLANC (ed.), *Le travail au XXIe siècle : mutations de l'économie et de la société à l'ère des autoroutes de l'information*, Dunod, Paris, p. 219-236, 1995.

[FLI 04] FLICHY P., "L'individualisme connecté entre la technique numérique et la société", *Réseaux*, n° 124, 2004.

[GRA 99] GRAS A., "Le désir d'ubiquité de l'homme pressé et le devoir de vitesse", *Revue Quaderni*, n° 39, p. 41-54, 1999.

[GRO 06] GROSJEAN S., BONNEVILLE, L., "TIC, organisation et communication : entre informativité et communicabilité", *Actes du colloque international Pratiques et usages organisationnels des sciences et technologies de l'information et de la communication*, Université de Rennes 2, France, 7-9 September 2006.

[HAY 98] HAYDEN A., Le nouveau mouvement pour la réduction du temps de travail en Europe, Congrès du travail du Canada, Rapport de Recherche, 13, 1998.

[HUE 00] HUEBER O., JACQUIS E., "NTIC et Entreprises Virtuelles", *Actes du 3e Colloque du CRIC*, Montpellier, 30 November 2000.

[JAU 03a] JAUREGUIBERRY F., *Les branchés du portable*, PUF, Paris, 2003.

[JAU 03b] JAURÉGUIBERRY F., "L'homme branché, mobile et pressé", in F. ASHER and F. GODARD (eds.), *Modernité : la nouvelle carte du temps*, Editions de l'Aube, Le Moulin du Château, p. 155-168, 2003.

[LAH 97] LAHLOU S., LENAY C., GUENIFFEY Y., ZACKLAD M., "Le COS, tel que défini par l'ARCo", *Bulletin de l'Association pour la Recherche Cognitive*, n° 42, Annexe au CR du groupe ARCo-industrie sur le Syndrome de saturation cognitive (COS), Compte-rendu de la 152e réunion du CA de l'ARC, November 1997.

[LAH 00] LAHLOU S., "Attracteurs cognitifs et travail de bureau", *Intellectica*, vol. 30, n° 1, p. 75-113, 2000.

[LEC 99] LECLERC G, *La société de communication. Une approche sociologique et critique*, Presses Universitaires de France, Paris, 1999.

[LEG 04] LEGAULT-FAUCHER M., "Hypertravail, quand tu nous tiens...", *Prévention au travail*, Automne, Montréal, vol. 17, n° 4, p. 8-14, 2004.

[LEJ 06] LEJOYEUX M., *Overdose d'info. Guérir des névroses médiatiques*, Seuil, Paris, 2006.

[LEMA 02] LEMAIRE L., Systèmes ERP, emplois et transformation du travail, Fondation Travail-Université, Centre de recherche travail et technologies, http://www.ftu-namur.org/fichiers/SERPETT-r%E9sum%E9-FR.pdf, September 2002.

[LEMO 05] LE MONDE, "Les résistants anti-portable", p. 20-29, 19 February 2005.

[MARQ 98] MARQUIS S., *Bienvenue parmi les humains*, T.O.R.T.U.E., Longueuil, 1998.

[MART 97] MARTY C., *Le juste à temps : produire autrement*, Hermes, Paris, 1997.

[MET 04] METZGER J.-L., CLÉACH O., "Le télétravail des cadres : entre suractivité et apprentissage de nouvelles temporalités", *Sociologie du travail*, vol. 46, n° 4, p. 431-574, 2004.

[MIL 03] MILLERAND F., L'appropriation du courrier électronique en tant que technologie cognitive chez les enseignants chercheurs universitaires. Vers l'émergence d'une culture numérique ?, PhD Thesis, University of Montreal, 2003.

[MOE 96] MOEGLIN P., "La mobilité entre ubiquité et omniprésence", *Actes des séminaires Actions Scientifiques*, n° 6, France Telecom, 1996.

[PRO 96] PRONOVOST G., *Sociologie du temps*, De Boeck-Wesmael, Bruxelles, 1996.

[REI 79] REINBERG A., *L'Homme malade du temps*, Stock, Paris, 1979.

[RUL 00] RULLANI E., "Le capitalisme cognitif : du déjà vu ?", *Multitudes*, mai 2000.

[STAL 90] STALK G., HOUT T.M., *Competing Against Time: How Time-based Competition is Reshaping Global Markets*, Free Press, New York, 1990.

[STAT 06] STATISTIQUE CANADA, "Enquête sociale générale : Internet et notre emploi du temps-2005", *Le Quotidien*, 2 August 2006.

[SUE 94] SUE R., *Temps et ordre social*, Presses Universitaires de France, Paris, 1994.

[TAS 03] TASKIN L., "Télétravail et organisation, les mythes d'une success story. Entre autonomie et contrôle", *Gestion 2000*, vol. 2, p. 113-125, March/April 2003.

[TOM 03] TOMAS J.L., *ERP et progiciels de gestion intégrés : de la décision d'implantation à l'utilisation opérationnelle*, Dunod, Paris, 2003.

[TRE 04] TREMBLAY D.-G., *Conciliation emploi-famille et temps sociaux*, Presses de l'Université du Québec et Octarès, Québec/Toulouse, 2004.

[TRE 06] TREMBLAY D.-G., NAJEM E., PAQUET R., "Articulation emploi-famille et temps de travail : De quelles mesures disposent les travailleurs canadiens et à quoi aspirent-ils ?", *Enfances, Familles, Générations*, n° 4, Spring 2006.

[TRU 04] TRUCHOT D., *Épuisement professionnel et burnout*, Dunod, Paris, 2004.

[VEL 00] VELTZ P., *Le nouveau monde industriel*, Gallimard, Paris, 2000.

[VEL 02] VELTZ P., *Des lieux et des liens*, Editions de L'Aube, La Tour d'Aigues, 2002.

[VEN 02a] VENDRAMIN P., "Les TIC, complices de l'intensification du travail", *Colloque Organisation, intensité du travail, qualité du travail*, Paris, 21-22 November 2002.

[VEN 02b] VENDRAMIN P., VALENDUC G., Technologies de l'information et de la communication, emploi et qualité de travail, Ministry for Employment and Work, Brussels, 2002.

[VEN 04] VENDRAMIN P., "Petits arrangements avec le temps", *Tempos*, n° 1, Institut Chronopost, p. 41-46, January 2004.

Chapter 3

Narrativity Against Temporality: a Computational Model for Story Processing

3.1. Background: problems of temporality representation in social sciences

What is time? Can it be reduced to action and action to text, as thought by Ricoeur in his reflection on narratives [RIC 86]? Varied conceptions of the "narrative" play a role in the sociological usage of the term. We can distinguish three meanings when setting up text analysis software addressing time by narration:

– narrative explanation of social processes;

– narrative theories of action; and

– narrative social ontology.

Some researches focus on understanding an evolution over time. The evolution can concern an institution, a practice, individuals, concepts and variables. This evolution consists of the explaining phenomenon and origin of data to be analyzed:

– stages of a social process (change, bureaucracy, strike, revolution, innovation);

– trajectory of an organization (enterprise, career of an employee);

– strengthening of a conduct (addiction, resilience); and

– life transitions (adolescence, retirement).

Chapter written by Eddie SOULIER.

Stage, cycle, trajectory, conflict and other factors affect the process, change, time and sequencing. We refer in this first case to the narrative explanation of social processes, which constitutes an explanatory empirical approach in social sciences [ABB 95].

The temporal dynamic of the social environment can also be seen as being psychological or cognitive, i.e. unique for the social individual who contributes his or her narrative way of perceiving the world and his or her own action in the form of a sequence of events [BRU 90, RIC 83]. If we were to insist on the difference between the cognitive and procedural aspects of dynamic phenomena, we would say that memories are imperfect records of the way in which we have lived events, not replicas of the events themselves. Regardless of the way of perceiving them and representing or debating operations that they may create, events rightly remain as phenomena [LAH 05]. It is nevertheless legitimate to talk in terms of how they are seen by a human cognitive system and to be able to report on their associated phenomenological data. We gather what we can call narrative theories of action under this label [ABE 03].

There are links between these two realms around the problems of temporality description. These links relate to the ontological structure of the reality itself rather than to the effect of the narrative explanation of social processes, or "narrative", as a mode of cognition embodied in the (social) action of individuals and groups. We are talking about narrative social ontology every time we wonder whether or not the sensitive world consists of an uninterruptible stream of events where permanent change is the norm and stability the exception to be explained. In the first part of this chapter, we present a theoretical framework around these distinctions. In the second part we will focus on the description of textual analysis software created for this framework.

3.2. A theoretical framework for processing temporality

3.2.1. *Narrative theories of action*

Cognitive systems of individuals are a key driver of their actions. From social psychology, sociology, cognitive psychology, philosophy of the mind or artificial intelligence, some key concepts account for the inferences of the individual in the situation to guide his or her action and interpret the events. The concepts of schemas [BAR 32], attribution [HEID 58], frame [GOF 74], scripts [SCHAN 77], sense making [WEI 95] or background expectations [SEA 98], although belonging to different theoretical traditions, present a family trait[1]: they share the idea that agents actively elaborate the situation as "a kind of history" in which they play, will play, could have played or have already played a role.

Some of these models of action have object knowledge structures (scripts), and others situated action (framing, sense making). Some give more weight to social structures, others to rationality or to the situation in the determination of conduct. Some are used to anticipate action, others to motivate or justify it *a posteriori*. Action models also diverge on the way the individual approaches what regulates his or her conduct: rational choice, socialization, situational effect or interpretative construction of the meaning of action. The important aspect here is the role that these action models give to accounts or, in a more phenomenological or cognitive perspective, exploitation of the person's experience in formulating his or her mental schemas. If human action is always a "kind of story", what role does reflection or narration of his or her action play in the formulation of his or her beliefs? How is this experience of action reactivated in the action? Is there recollection of knowledge "stored" in memory or reconstruction of a memory in situation? If everyone reconstructs past events instead of remembering them, the creative part of this reconstruction is important and, moreover, subjected to contextual and emotional factors [ORI 06].

In these action models, the reflexive or narrative activity is triggered when the flow of action is interrupted, damaging the cognitive consistency of reality for the participant. This break causes emotions and a reaction (inquiry at Dewey, repair ritual at Goffman, equilibration at Piaget) to be linked to these interruptions. The idea of controlling the situation – and the cognitive bias [KAH 73] or interactional strategies (manipulation, deception, in the sense of Goffman) that would limit this control – is behind these constructs of actions. The event will generate surprise. It causes the action as well as the meaning.

The emergence of the time of the event invites story telling. We find here the idea that in order to have a story, an unexpected event needs to occur [BRU 02]. The human action is a process during which individuals detect and correct anomalies. Human action can also be seen as a process of encoding routines from past experiences deemed effective. Stories reflect this mix of unexpected elements and

---

1 Other concepts propose figuratives of human action, but they evoke less the means to manipulate processes than systems of categories. These are concepts of attitudes [ALL 35], situation definition [THO 20], cognitive style [SCHU 62], social representation [MOS 61], social constructions of reality [BERG 66], tendencies or *habitus* [BOUR 72], habits [DEW 22], ideologies or beliefs. Boudon proposes to shorten these concepts using the acronym ADACC (*action, décision, attitude, comportement, croyance* [action, decision, attitude, conduct, belief]) to highlight that social phenomena are the product [BOUD 03].

extraordinary aspects in a background of an ocean of ordinary actions. Many action models are based on a narrative design of this action [POL 88].

In contrast to theories of rational choice, narrative theories of action assume that preferences guiding the actions of participants are based on *past experience* rather than a hypothetical deliberation from limited information on the world. The concept of past experience does not carry any functionalist conception of action: it is less a matter of socialization or provision than a practical rationality. The action model that explains this reminder of past experience more normatively in a decisional context is the decision based on cases [GIL 01].

According to Abell [ABE 03], the individual elaborates on the story he or she thinks he or she is dealing with: *What is the story here?* The choice of the decision-maker is directly affected if the current situation reminds him or her of similar stories involving decisions that him-/herself or other decision-makers have encountered in the past (similarity function). With the recollection in memory of the problem that occurred in a similar situation, it is also the memory of the choice made and resulting consequences. A story or a case is therefore a triplet composed of a problem, an action and a result. The decision-maker will often choose to retain the past decision that he/she feels has brought him/her the most useful result (utility function). The similarity between the past and present is more or less important, and the decision to apply, adapted. Elaboration, recollection and adaptation form three stages if the decision is based on the case.

The model tries to capture the way in which experience affects decision-making and suggests that learning consists of the permanent addition of new cases to the memory of the individual. The structure of the case is very much in line with the narrative model introduced by Labov [LAB 97], which is the baseline analysis of oral history: the order in which we represent a past experience to reenact and narrative clauses of an oral story reproducing a story whose events have been lived, would be isomorphic.

3.2.2. *Narrative explanation of social processes*

The second sociological design of time no longer sees the narrative as a cognitive modality (or discursive) incorporated in the action of the individual. Narrativity is a common way for history and social sciences to explain social processes. The temporal structuring of social processes allows an explanatory empirical approach in social sciences: the narrative explanation.

There are many types of sequences of events that are not limited to the chronological and linear dimension. Van de Ven distinguishes five types of event progressions that characterize a process [VAN 92]:

– unit;

– multiple;

– cumulative;

– conjunctive; and

– recurring.

Narrative explanation has difficulties that lie in the identification of events to combine to form sequences, in the nature of the links connecting the events [ABB 03], and in the explanation of causal mechanisms that configure the development of processes [ETI 05]. Narrative analysis carries a debate on the epistemological foundations of explanation in social sciences. The synchronic/diachronic distinction covers two ways of studying an object: by its content or by its process. Research on content focuses on identifying the composition of the object, while research that concerns the process aims at the behavior of the object in time.

Beyond this, there is also the opposition between two explanation modes in social sciences. The first is causal analysis, where the cause of the object resides in the correlation of variables supposed to conceptually represent the object content at a given moment in time. The second is the comprehensive and historical explanation, for which the cause resides in the nature of events making up the process as well as the order and sequence of these events in time. This design also focuses on the meaning to be given to events. In other words, is it the time and singularity of each event that explains the object or is it the relationships among conceptual variables of the model that compose it? In both cases, we are concerned with finding "what leads to what". This means that causality introduces the question of time in a decisive way. We assume that previous events have a certain link with events that follow, although this connection is not too clear. The understanding of causality can also depend on our identification of abstract concepts and our vision of their interaction. Narrative and concept are therefore two different ways to consider causality.

Conditional regularities from processes of production of the effect by the cause reflect, in the case of human sciences, social relationships guided by the intentions of individuals and in social contexts [BERN 05]. These social reports are marked by their historicity. The cause is then understood in a speculative way: it is a "historical" configuration because it is a matter of preserving the complexity of situations that underlie the phenomena that interest us, rather than over-simplifying the spatio-temporal context in which the event occurs. As indicated by [RAG 87], in conventional quantitative research, researches focus on the effect of one variable on another in a large number of different situations. In the narrative analysis, however, the generation of effects by the causes depend on the context in which the event is deployed. In short, it is the conjunction of factors in this situation that is causal, not the individual factors.

There are several methods to elaborate histories of processes [ABE 04] such as *event structure analysis* [HEIS 91], sequential analysis [ABB 95, ABB 00], content analysis applied to narrative data [FRA 03] and so-called longitudinal analysis that use the concept of narrative [POO 00]. Some of these analyses are quantitative, while others are more qualitative. Narrative analysis goes through the collection of accounts of events. It relies on software that assists in collection techniques and modes of data classification. The *Ethno 2* software is a tool for the analysis of sequential structures of events [HEIS 91]. Today sequential methods are relevant techniques for analyzing, simulating or explaining processes where "time counts" [ABB 01]. The event is the basic unit of processes. A first type of research consists of identifying sequences in the studied process and comparing them. Another type involves determination of the order in which the stages of a process occur. Recent developments have involved the comparison of sequences and consist of a set of optimal matching techniques based on those used to sequence the human genome [LES 04].

3.2.3. *Social narrative ontology*

By ontology, we mean the definition of what is the social [RAM 00]. A first difficulty arises from the two social analysis modes that we have just mentioned, that splits the question of complex relationships between the micro- and macro-social. On the one hand, individual (and, indeed, collective) conduct results from the ability of the individual to form judgments based on similarity with a past experience in terms of the cognitive situation in which he or she is[2]. On the other hand, global social processes can be explained by the construction of a narration or a story of processes. In reality, micro-social observations have relationships with the way of accounting global social processes. We notice this at three levels:

---

[2] The narrative action model is partially compatible with the cognitive rationality model proposed by Boudon. Cuin, however, criticizes the latter on two points: it is unlikely that individuals decide the outcome of cognitive processes that are of a great complexity; the cognitive paradigm may explain the belief (in a limited version moreover) of the individual at the time of the decision, but analysis of beliefs is not sufficient to explain *actual* conducts. "What the actor believes or represents does not necessarily explain what he does" [CUI 05], p. 546.

– the entrenchment of beliefs and conducts in processes,

– the status of representation and action discourses,

– the narration as a data source on action.

The first point refers to the recurring debate on the place of the subject in the explanation of social phenomena. We will not dwell on this, except to say that comprehensive and derived sociologies give the utmost importance to the individual and to the symbolic and cognitive dimensions of his or her action in the explanation of the social[3]. In addition, we consider today that this action must be understood in its coupling with the circumstances of the action, i.e. the situation [QUE 06][4]. If action is what is shared between the subject and social process Schapp would be right. Unlike Ricoeur, Schapp sees the human being as "entangled in stories" [SCHAP 92].

The second point concerns the fact that individuals spend time telling stories in and about the action [ATI 06], but also around events, whether they are trivial and personal [LAF 96, VIN 01] or extraordinary and collective [ORF 05]. Narrativity no longer results from just the cognitive dimension of the individual action or of the structuring of the collective action, but also by the participant putting his/her action and events that affect him-/herself or the community to which he/she belongs into words. The narrative discourse is thus part of the temporal continuity of the individual and collective experience.

The complexity of the relationship between language and action increases when this language is a narration on action, as produced simultaneously by the individual. We can consider that y the person putting his/her action into words makes him/her a good theoretician of his/her activity or, inversely, feeds an epistemological suspicion with respect to the words of others [PLA 05].

Is the text of action a component of the action or a material that can be used to better understand the action? Technical debates today focus on the place of words in accessing action, which brings us to leave the field of account *in* the action *as* data source on the action. Antoher question is whether the account of the action requested

---

3 The emphasis on understanding induces the idea that beliefs are part of the explanation of conduct and assume even a concept of action rooted in reflexive self-understanding. The challenge posed to some researchers of access to the reflexive understanding of the participant to reach a "complete" explanation of action illuminates the place given to stories or accounts of life in the methodology of social sciences today [BERT 96, DEM 97].

4 Beyond the answer given by sociological models of situated action in terms of nature of the relationship between the individual, situation and structure, Von Wright is responsible for identifying the *structural interaction* among processes explicable by causes, and actions that are understandable, in terms of intentions [VON 71].

by the observer constitutes a valid trace of the action or an entirely different product that is to be defined. Several researchers believe that the organization of verbal accounts of situations associated with an activity, subject to the strict definition of adapted devices, is a way to construct useful data for analysis of the activity [SOU 06a, THE 06] or the articulation of knowledge of action [BAR 04], whether or not mediated by software.

The fact that narrativity is a mode of existence of social objects points towards the necessity of enlarging the concept of narrative social ontology. Wanting to describe phenomena where "time counts" raises several problems. There is first the risk of equating an event to a variable (in terms of causal analysis)[5], an event being an incident, a major activity or a change [VAN 95]. There is then the question of concepts describing the combination of events and processes in evolution. If the concept of a sequence is often predominant, other concepts exist to express the duration (stages, cycles, phases, conflicts, etc.). Concepts describing the dynamics are also part of this ontology (interruption, delay, acceleration and deceleration, assimilation cycles, backwards, breaking points, rhythms). Forms of event progression can be distinguished (unit, multiple, cumulative, conjunctive, recurrent, etc.). They depend on the structuring mechanisms identified in the literature (dialectical, evolutionist, teleological, etc.).

Abbott mentions other problems related to temporality description [ABB 01, ABB 03]. One problem concerns the *choice of unit of measurement* to express the duration of moments or a time period that constitutes the process being studied. The structure, conjuncture and event are thus the three levels of temporality proposed by Braudel. Another difficulty is *demarcation of the process* studied. Periodization is a way of identifying the upper and lower bounds of the observation period of the process. Periodization therefore allows us to isolate the process from its context as it continues to evolve over time. The third problem concerns the way of describing changes occurring during the time covered by the description. It is difficult not to describe a moment in the past in static form (a *static description*) to represent a period ("thirty glorious years"), even if we know that it is a synchronic fiction built from diachronic facts. There is then *decomposition of the process* into elements that participate in its deployment in time, as we have seen previously with the concepts of sequences of events, dynamic factors or forms of progression. This leads to the question of *description over time*, because the description of a story or development of events lies on hypotheses concerning the observed sequential regularities.

---

5 It is understood that narrative analysis opposes an approach using variables [ABE 04], which implies that the construction of generic categories from events of the original account is essentially an inductive process, close to the approach advocated by the grounded theory [GLA 67].

The social process must be thought of as a succession of historical or narrative links whose sequence is the process. Beyond the necessity to specify the relationship between events integrated in the sequence, narrative analysis raises the ontological question of the list of types of historical links that can be considered to explain the nature of the relationship between the events of the social process[6]. There is equally the fact that the description of social process in the form of narration frequently marks out *privileged moments* rather arbitrarily defining a period of change.

We often unconsciously start from the end (or the object of study, the result) of the analysis period rather than the beginning being, by construction, routine moments in the social processes. The last point concerns the *performance of narrative description*. "The agents themselves describe their actions by doing them" [ABB 03][7]. The first descriptions thus have a character of action. They are efficient in the sense that the description activity is the way by which open and undefined events of the present become fixed and defined events of the past, and by the fact that the descriptions of events sediment over time. Stories thus form pre-discourses [PAV 06] and the events they temporarily set are always subject to complete re-description[8].

These few examples go beyond methodology and epistemology to raise ontological problems. If the social world is a world of events, what needs to be explained is the stability. Concepts of social narrative ontology do not the social composition but the way of assigning a mode of existence to objects. Temporality comes from interaction, not substance. Social structures and participants as social objects are defined as networks of events. They are essentially connectors, relaters or "narrative links" that allow temporal dimensioning, scaling, phasing, sequencing, change, evolution, dynamics, temporal leafage of the social process, and are therefore the foundations of social life. Social life is a movement that connects ingredients that are not of social nature [LAT 06]. If the various software that we will present are far below the ambitions set in this chapter, they are nevertheless in a research program that aims to take into account the narrative in the action, the process and social ontology.

---

6 Some researchers speak of causal mechanisms (mediation, learning, framing, translation, etc.) rather than narrative links to highlight that the account of events in itself cannot always constitute a causal explanation of a social process [STE 00]. This is why some individuals prefer process analysis [GEO 05, HAL 03] to narrative analysis.
7 p. 51
8 As Ferry points out, thus made communicable, recounted experiences simply cease to be lived; they are now transmitted [FER 99].

3.3. A computational model for story processing

The various software programs presented in this chapter are based on precise sociological issues and are designed and developed in reference to specific problems. The area that concerns us is the reflective field emerging from sociology of management [BOUS 05, FLA 02, MAU 06, MIS 06], consultancy [VIL 03] and expertise [TRE 96]. Practices and devices of management are at the heart of intangible and relational activities [DUT 05] and foreshadow contemporary life and work: a sum of activities whose function is to more or less radically change the system of auto-regulation of individuals [GOF 79]. This in turn requires a strong reflexivity of professionals [SCHO 83], the inclusion of "working on others" [DUB 02] in new institutional forms and in the context of technicization of practices [POL 04]. The narrations of practices are a good entry point for understanding. Here we present a series of four programs for analyzing narrations of action: *Hyperstoria*, *Ontostoria*, *Sum It Up* and *MemorExpert*.

3.3.1. *Hyperstoria*

Hyperstoria is a program for indexing and analyzing different representations of a story in a narrative. The collection of stories is based on the technique of narrative interviews, which combines the explicitation interview, critical incidents in the variant of the behavioral event interview proposed by Boyatzis and narrative of personal experiences from Labov. The story told of an action or event is a means of accessing the past action or event in reference to a precise situation (advisory mission, management act or other). The conceptual model of the narration involves three components that exploit different structured representations of the story:

– a model of narrative clauses;

– a model of narrative conversation; and

– a model of story.

The model proposes different graphical representations of this story as graphs [SOU 06b]:

– tree of causes;

– decision tree; and

– tree of events.

Narrative clauses of an account are the basic units of the story. The tool allows different attributes (descriptor that gives a property to a character or object of the story, action, event and other) to be assigned to these clauses. Elaboration of the model of the story goes through the segmentation of the text and labeling of narrative clauses, chronological and causal sequencing of these clauses, and identification of scenes in the scripts. The analyst then proceeds to the elaboration of the tree of causes, which allows us to understand the development of events, and the decision tree, which structures the action of protagonists. The different stories associated with a participant and/or situations are stored in a working folder, which gives a global view. Depending on the needs of the analyst, the tool can support other story models (narrative model, grammars of stories, generalized graphs, modal graphs, causal chains and others). With many tools from this category, the formalization of a sequence in *Hyperstoria* only works for a restricted number of events, which reduces its applicability in the analysis of long and complex processes. It is nevertheless an intuitive way to define the concept of story and to experience the complexity.

3.3.2. *Ontostoria*

The rewriting of narrative interviews is a heavy task for the analyst, who must find the relevant information in the text. A tool for indexing important elements of narrations more easily is necessary for dealing with narrative documents on a larger scale. In these documents the information concerns essentially the description of facts or events related to real or virtual[9] acts of some "characters". These descriptions strive to achieve results. They are involved in situations or undergo tests, handle concrete or abstract materials, receive or deliver messages and establish relationships with other characters [ZAR 06].

Ontostoria offers an analysis of narrative documents from a viewpoint close to the cognitive semantics around a model of a story and the narrative ontology of nearly 150 generic concepts required to index any type of story. In *Ontostoria*, an account is a message or an argument that must be communicated and understood.

A story is a problem that needs to be solved. The story is the way of developing the message. To address these two aspects, we consider that the internal structure of a story includes four intertwined aspects:

– The *character* is the carrier, through his/her character, of the theme developed in the story. A system of characters proposes several contradictory points of view on the theme and offers an argumentative scale to the audience.

---

9 Documents describing the misadventures of the journey of a nuclear submarine (the "character") are narrative documents, as are those telling the successive transformations in the "life" of a commercial product [ZAR 06].

– The *theme* is abstract and represents the axiological/moral concept about which the author of the story speaks around the main event: the problem.

– The *plot* is the order in which the thematic concept is developed in particular situations or scenes and that will allow the audience to feel all the dimensions of it.

– The *gender* is determined by the pragmatic aim of the narrator, who chooses to tell his/her message according to the impact targeted to the public.

Concepts are associated with each aspect. The highest level aspect is the gender, which offers four classes depending whether the problem is internal or external for the agents and whether it is related to state or process (situation, action, belief, and reasoning). These classes are based on works concerning the categorizations of processes in linguistic and cognitive semantics (aspectual categorizations of process, temporal semantics, action mode and other). This hypercategorization then allows the definition of semantic categories in terms of the plot (16 concepts), themes (64 concepts) and characters (64 concepts). Several heuristics allow the four aspects and their concepts to be liked among themselves.

A story is therefore represented as a single hierarchy of concepts some of which are close to the semiotic world allowing the characterization of a character, while others, such as themes, express abstract ideas. If an analyst can manually associate classes to stories from narrative interviews, the number of themes and characters is too important to manually index. *Ontostoria* therefore proposes various formalisms (rhetorical structural theory, centers theory, nervures theory and other) and tools for automatic handling of language (syntax analyzers, parsers of accounts, themes detectors) to integrate in the tool [TOD 06]. *Ontostoria* has been developed with the help of software for the creation of ontologies, the ontologies editor *OntoEdit 2.6*, developed by Ontoprise Gmb H[10], which is based on the ontology web language (OWL) and *RDF Outline* based on XML.

3.3.3. *Sum It Up*

Sum It Up is a study tool that explores the collective construction of meaning through the collaborative activity of conceptualization of an account of events. By what processes do two subjects manage to coordinate to sum up an event through the joint collaboration of an account of it? In order to describe the activity of understanding, we refer to a classical theory in textual linguistics according to which the text summary is achieved through an abstraction process of the macro-structural units of this text, through procedural macro rules. Van Dijk defines four macro rules

---

10 www.ontoprise.de.

that explain how the complex semantic information is processed according to a macro treatment to achieve the gradual elaboration of a semantic macrostructure:

– selection (of the most important information);

– deletion (of details);

– reduction (generalization); and

– construction (of new or missing information in the text).

Sum It Up is a support groupware for the understanding of a story. It implements an activity model explaining the application strategies of inference rules to summarize a story.

This model of the understanding activity is based on two observations. First, rules correspond to different inference levels. Deletion and generalization rules allow us to *cut* through the set of narrative clauses, in order to obtain the micro propositions, while integration and construction rules allow us to *substitute* macro propositions with micro propositions. On this basis, it is possible to propose a representation of the general strategy for the elaboration of a semantic macrostructure with the help of a conceptual metalanguage [LEW 06]. Second, we identify two combinations of cutting and selection rules that define elaboration strategies of a macrostructure:

– an *elimination strategy*, characterized by the use of integration and deletion rules; and

– a *production strategy*, characterized by the use of generalization and construction rules.

The *Sum It Up* model is therefore based on the idea that it is possible to classify each of the four rules according to two criteria: its abstraction level (cutting or substitution) and the strategy it can implement (elimination or production). Fianally, the tool offers a *comment* function that allows subjects to justify their choice in the application of rules. These comments can then be analyzed.

3.3.4. *MemorExpert*

MemorExpert explores the fourth trail of the call to use prior knowledge to solve a new, but similar, situation. *MemorExpert* does not aim to provide an in-depth analysis of each case but, instead, to index "many" stories in a data warehouse appropriately and to retrieve the most useful. The challenge of *MemorExpert* is a matter of indexation, which involves three difficulties:

– the representation of stories in the database;

– the research of stories; and

– adaptation to the context.

Beyond the indexation and research of information, *MemorExpert* is a way to systematically explore stories that can be collected on a field of study, a profession, a project, or an enterprise. The tool has been used to analyze the practices of a team of consultants [SOU 05] and the accounts of experiences in using ICT among craftsmen [FAL 05].

MemorExpert is based on a particular model of indexation. A story is indexed around the intentional behavior of the agent of which the narration speaks and of the anomaly that justifies narration of the story. These attributes, qualified as structural, are completed by contextual attributes that precisely define the nature of the task, the role played by the individual, determinants of the situation, and skills and scenarios of action expected in this context. The tool allows the analyst or participants to directly index and research stories in a field of study. The system proposes statistics on stories, the roles of protagonists and types of actions or events for study purposes.

The study of management acts on a mission to provide advice and here we have been able to develop more sophisticated statistics. The statistics on the stories automatically calculate, according on the number of stories indexed, the percentage of stories per role, per task, per discipline, per situation and per anomaly type. Statistics on the roles calculate the distribution of tasks and disciplines mastered for this role. Statistics on the mission calculate the distribution of roles, disciplines and anomalies per task. The establishment of a warehouse of indexed stories has also lead to a learning tool [SOU 06c].

3.4. Conclusion

Hyperstoria allows the narrative analysis of a story, defined as a process consisting of a sequence of events. The sequence of an event in itself constitutes the almost causal explanation of the event or action. Conversely, *Ontostoria* seeks to "enter" into the story/scenario. Consider it as a structure built according to a holographic, fractal mode to make us feel all the sensations, moral conflicts and hardships that the narrator and – more generally – the protagonists have experienced, and the relationship this may have with semiotization processes.

The second more practical aim of *Ontostoria* is the identification of the structure of narrative documents with natural language processing (NLP). It means offering a tool that identifies elements of the rhetorical structure of text, extraction of relevant information (participants, themes, plots) and proposes to index them using a specific ontology of narrative texts. Thus, the user can easily navigate and retrieve relevant information in a database of narratives, such as stories of experts.

Sum It Up allows us to experiment with two subjects that can agree on the interpretation of an event by summarizing the story in a common account.

Finally, *MemorExpert* enables the creation of a memory of a group's stories in a field of knowledge, and the sociological study of the decision from this case. This software processes text in relation to action and time. The risk is in evacuating account's production modes, which from a sociological point of view should be considered as social activities, as pointed out in [DEM 06]. The text of the "account in interaction" [KER 05] is nevertheless a very complex entity whose elaboration is justified when the research object is the account, but possibly less when the account is an element of the temporal analysis of an action. We have however specified through the KTA (Knowledge Telling Analysis) model a technique of analysis of the narrative interaction in a distributed cognition perspective [SOU 06d, QUA 97] that gives rise to the development new software. This research strategy, consistent for a sociological treatment of time, takes its distance from the usual representation of the language or textual materials that are generally exploited as inputs to specialized text analysis software.

Perhaps is it time to create a new software category to take into account collective prediscursive frameworks [PAV 06] that overlap the text associated with a particular enunciation?[11] For its part, the sequential analysis of social processes so far seems to be satisfied with statistical analysis tools. The *SAS* (Statistical Analysis System) software offers, for example, about 10 classification methods [LES 04]. Some sequence-matching programs have however been specifically developed (for example, *Optimize*, TDA – Transitional Data Analysis, etc.). A wishful evolution would certainly lie in taking into account, using sequential analysis software in a similar way to text analysis software, temporal operators of which we have outlined the list in section 3.2.3. But that is another story...

3.5. Bibliography

[ABB 95] ABBOTT A., "Sequence analysis: new methods for old ideas", *Annual Review of Sociology*, vol. 21, p. 93-113, 1995.

[ABB 00] ABBOTT A., TSAY A., "Sequence analysis and optimal matching technique in sociology," *Sociological Methods and Research*, vol. 29, no. 1, p. 3-33, 2000.

[ABB 01] ABBOTT A., *Time Matters. On Theory and Method*, University of Chicago Press, Chicago, 2001.

---

11 See Chapter 11.

[ABB 03] ABBOTT A., "La description face à la temporalité", in G. Blundo and J.P.O. de Sardan (edd.), *Pratiques de la Description*, EHESS Editions, Paris, 2003.

[ABE 03] ABELL P., "The role of rational choice and narrative action theories in sociological theory: the legacy of Coleman's Foundations", *French Review of Sociology*, vol. 44, no. 2, 2003.

[ABE 04] ABELL P., "Narrative explanation: an alternative to variable-centered explanation?", *Annual Review of Sociology*, vol. 30, p. 287-310, 2004.

[ALL 35] ALLPORT G.-W., "Attitudes", in C.M. Murchison (ed.), *Handbook of Social Psychology*, Clark University Press, Worcester, 1935.

[ATI 06] ATIFI H., MARCOCCIA M., "La narration dans les conversations en situation de travail", in E. Soulier (ed.), *Le Storytelling. Concepts, Outils et Applications*, Hermes Paris, 2006.

[BAR 32] BARTLETT F.C., *Remembering: A Study in Experimental and Social Psychology*, Cambridge University Press, Cambridge, 1932.

[BAR 04] BARBIER J.-M., GALATANU M. (eds), *Les Savoirs d'Action: Une Mise en Mot des Compétences?*, L'Harmattan, Paris, 2004.

[BER 66] BERGER P., LUCKMANN T., *The Social Construction of Reality, A Treatise in the Sociology of Knowledge*, Doubleday and Company, New York, 1966.

[BER 05] BERNARD P., "Les chiffres pour le dire: les nouveaux instruments de l'heuristique causale", in D. Mercure, *L'analyse du Social. Les Modes d'Explication*, Laval University Press, Laval, 2005.

[BER 96] BERTAUX D., *Les Récits de vie: Perspectives Ethnosociologiques*, Nathan, Paris, 1996.

[BOU 03] BOUDON R., *Raison, Bonnes Raisons. La rationalité: Notion Indispensable et Insaisissable*, Paris, PUF, 2003.

[BOU 72] BOURDIEU P., *Esquisse d'une Théorie de la Pratique*, Droz, Geneva, 1972.

[BOU 05] BOUSSARD V. (ed.), *Au Nom de la Norme: les Dispositifs de Gestion Entre Normes Organisationnelles et Normes Professionnelles*, L'Harmattan, Paris, 2005.

[BRO 03] BRONNER J., *L'Empire des Croyances*, PUF, Paris, 2003.

[BRU 90] BRUNER J., *Acts of Meaning*, Harvard University Press, Cambridge, United States, 1990.

[BRU 02] BRUNER J., *Pourquoi nous Racontons-nous des Histoires?*, Editions Retz, Paris, 2002.

[CUI 05] CUIN C.-H., "Le paradigme "cognitif": quelques observations et une suggestion", *French Review of Sociology*, vol. 46-3, p. 559-572, 2005.

[DEM 97] DEMAZIÈRE D., DUBAR C., *Analyser les Entretiens Biographiques – L'Exemple des Récits d'Insertion*, Nathan, Paris, 1997.

[DEM 06] DEMAZIÈRE D., BROSSAUD C., TRABAL P., VAN METER K. (eds.) *Analyse Textuelle en Sociologie. Logiciels, Méthodes, Usages*, Presses Universitaires de Rennes, Rennes, 2006.

[DES 04] DE SAINT POL T., LESNARD L., "Introduction aux méthodes d'appariement optimal (Optimal Matching Analysis)", *Bulletin de Méthodologie Sociologique*, no. 90, pp. 5-25, April 2004.

[DEW 22] DEWEY J., *Human Nature and Conduct*, Henry Holt and Company, New York, 1922.

[DUB 02] DUBET F., *Le Déclin de l'Institution*, Seuil, Paris, 2002.

[DUT 05] DU TERTRE C., "Services immatériels et relationnels: intensité du travail et santé", *@ctivités*, vol. 2, no. 1, p. 37-49, 2005.

[ETI 05] ETIENNE J., "L'effet causal des idées dans l'explication de l'action publique et le problème de l'équifinalité: de l'intérêt de l'analyse de processus (process-tracing)", *Communication at the 8th Congress of the French Association of Political Science*, Lyon, 14-16 September 2005.

[FAL 05] FALLERY B., MARTI C., "Comment partager des récits d'expériences entre des artisans? Un apport des méthodes narratives pour la e-GRH", in P. LOUART, *e-HR: Réalités Managériales*, Vuibert, Paris, 2005.

[FER 99] FERRY J.-M., "Narration, interprétation, argumentation, reconstruction. Les registres du discours et la normativité du monde social", in A. Renaud, *Histoire de la Philosophie Politique*, 5, *Les Philosophies Politiques Contemporaines*, Calmann-Lévy, Paris, p. 231-288, 1999.

[FLA 02] FLAMANT N., *Anthropologie des Managers*, PUF, Paris, 2002.

[FRA 03] FRANZOSI R., *From Words to Numbers*, Cambridge University Press, Cambridge, 2003.

[GEO 05] GEORGE A.L., BENNETT A., *Case Studies and Theory Development in the Social Sciences*, MIT Press, Cambridge, United States, 2005.

[GIL 01] GILBOA J., SCHMEIDLER D., *A Theory of Case-Based Decisions*, Cambridge University Press, Cambridge, 2001.

[GLA 67] GLASER B., STRAUSS A., *The Discovery of Grounded Theory: Strategy for Qualitative Research*, Aldine, Chicago, 1967.

[GOF 74] GOFFMAN E., *Frame Analysis: An Essay of the Organization of Experience*, Harper and Row, New York/Evanston, 1974.

[GOF 79] GOFFMAN E., *Asiles. Etudes sur la Condition Sociale des Malades Mentaux et Autres Reclus*, Editions de Minuit, Paris, 1979.

[GRI 94] GRIFFIN L., RAGIN C., "Some observations on formal methods of qualitative analysis", *Sociological Methods and Research*, vol. 23, p. 4-21, 1994.

[HAL 03] HALL P.A., "Aligning ontology and methodology in comparative research", in J. MAHONEY and D. RUESCHMEYER (eds.), *Comparative Historical Analysis in the Social Sciences*, Cambridge University Press, Cambridge, 2003.

[HEID 58] HEIDER F., *The Psychology of Interpersonal Relations*, Wiley, New York, 1858.

[HEIS 91] HEISE D.R., "Event structure analysis", in N. FIELDING and R. LEE (eds.), *Using Computers in Qualitative Research*, Sage, Newbury Park, p. 89-107, 1991.

[KAH 73] KAHNEMAN D., TVERSKY A., "On the psychology of prediction", *Psychological Review*, vol. 80, p. 237-251, 1973.

[KER 05] KERBRAT-ORECCHIONI C., *Le Discours en Interaction*, Armand Colin, Paris, 2005.

[LAB 97] LABOV W., "Some further steps in narrative analysis", *The Journal of Narrative and Life History*, vol. 7, p. 1-4, 1997.

[LAF 96] LAFOREST M. (ed.) *Autour de la Narration*, Nuit Blanche, Quebec, 1996.

[LAH 05] LAHIRE B., *L'Esprit Sociologique*, La Découverte, Paris, 2005.

[LAT 06] LATOUR B., *Changer la Société ~ Refaire de la Sociologie*, La Découverte, Paris, 2006.

[LES 04] LESNARD L., DE SAINT POL T., "Introduction aux méthodes d'appariement optimal (*Optimal Matching Analysis*)", *INSEE*, Series of Working Papers of CREST, no. 2004-15, 2004.

[LEW 06] LEWKOWICZ M., SOULIER E., GAUDUCHEAU N., "Collecticiels pour la construction collective du sens. Définition, principes de conception, exemple", in E. SOULIER (ed.), *Le Storytelling. Concepts, Outils et Applications*, Hermes, Paris, 2006.

[MAU 06] MAUGERI S. (ed.), *Au Nom du Client. Management Néo-libéral et Dispositifs de Gestion. Approches Sociologiques*, L'Harmattan, Paris, 2006.

[MIS 06] MISPELBLOM BEYER F., *Encadrer, un Métier Impossible?*, Armand Colin, Paris, 2006.

[MOS 61] MOSCOVICI S., *La Psychanalyse, son Image, son Public*, PUF, Paris, 1961.

[ORF 05] ORFALI B., *Société face aux Evènements Extraordinaires. Entre Fascination et Crainte*, Editions Zagros, Paris, 2005.

[ORI 06] ORIGGI G., "Mémoire narrative, mémoire épisodique : la mémoire selon W.G. Sebald", *Les Philosophes Lecteurs, Fabula LHT (Littérature, Histoire, Théorie)*, no. 1, February 2006. (Available at: http://www.fabula.org/lht/1/Origgi.html, accessed February 4, 2010.)

[PAV 06] PAVEAU M.-A., *Les prédiscours. Sens, Mémoire, Cognition*, Presses Sorbonne Nouvelle, Paris, 2006.

[PLA 05] PLAZAOLA GIGIER I., FRIEDRICH J., "Comment l'agent met-il son action en mots?", in L. FILLIETTAZ and J.-P. BRONCKART (eds.), *L'analyse des Actions et des Discours en Situation de Travail. Concepts, Méthodes et Applications*, Peeters, Louvain-la-Neuve, 2005.

[POL 88] POLKINGHORNE D.E., *Narrative Knowing and the Human Sciences*, State University of New York Press, Albany, 1988.

[POL 04] POLKINGHORNE D.E., *Practice and the Human Sciences: The Case for a Judgment-based Practice of Care*, University of New York Press, Albany, 2004.

[POO 00] POOLE M.S., VANDE VEN A.H., DOOLEY K., HOLMES M.E., *Organizational Change and Innovation Processes*, Oxford University Press, Oxford, 2000.

[QUA 97] QUASTHOFF U.M., "An interactive approach to narrative development", in M. BAMBERG (ed.), *Narrative Development: Six Approaches*, Lawrence Erlbaum Publishers, Mahwah, p. 51-83, 1997.

[QUE 06] QUÉRÉ L., "L'environnement comme partenaire", in J.-M. Barbier and M. Durand (eds.), *Sujets, Activités, Environnements. Approches Transverses*, PUF, Paris, 2006.

[RAG 87] RAGIN C.C., *The Comparative Method: Moving Beyond Qualitative and Quantitative Strategies*, University of California Press, Berkeley, 1987.

[RAM 00] RAMOGNINO N., "Epistémologie, ontologie ou théorie de la description?", in P. LIVET and R. OGIEN (eds.), *L'enquête Ontologie. Du mode d'Existence des Objets Sociaux*, Editions of the School of High Studies in Social Sciences, Paris, p. 153-182, 2000.

[RIC 83] RICOEUR P., *Temps et Récit*, Seuil, Paris, Volumes I, II, III, 1983.

[RIC 86] RICOEUR P., *From Text to Action. Hermeneutic Trials*, Seuil, Paris, II, 1986.

[SCH 77] SCHANK R., ABELSON R.P., *Scripts, Plans, Goals and Understanding*, Erlbaum, Hillsdale, 1977.

[SCHAP 92] SCHAPP W., *Empêtrés dans des Intrigues. L'Etre de l'Homme et de la Chose*, Editions of Cerf, Paris, 1992.

[SCHO 83] SCHÖN D., *The reflexive practitioner. How Professionals Think in Action*, Basic Books Inc., New York, 1983.

[SCHU 62] SCHÜTZ A., "On multiple realities", in *Collected Papers, vol. I: The Problem of Social Reality*, Martinus Nijhoff, The Hague, p. 207-259, 1962.

[SEA 98] SEARLE J.R., *La Construction de la Réalité Sociale*, Gallimard, Paris, 1998.

[SOU 05] SOULIER E., "Le système de gestion des connaissances pour soutenir le storytelling dans l'entreprise", *French Management Review*, vol. 31, no. 159, p. 245-264, 2005.

[SOU 06a] SOULIER E. (ed.), *Le Storytelling. Concepts, Outils et Applications*, Hermes, Paris, 2006.

[SOU 06b] SOULIER E., CAUSSANEL J., "La représentation des connaissances d'un texte narratif", in E. SOULIER (ed.), *Le Storytelling. Concepts, Outils et Applications*, Hermes, Paris, 2006.

[SOU 06c] SOULIER E., CAUSSANEL J., "L'apprentissage assisté par le Storytelling : le cas des consultants", in E. SOULIER (ed.), *Le Storytelling. Concepts, Outils et Applications*, Hermes, Paris, 2006.

[SOU 06d] SOULIER E., "L'acquisition de connaissances déférentielles dans l'interaction narrative: théorie de la descriptibilité pour la cognition distribuée", in E. SOULIER (ed.), *Le Storytelling. Concepts, Outils et Applications*, Hermes, Paris, 2006.

[STE 00] STEVENSON W.B., GREENBERG D.N., "Agency and social networks: Strategies of action in a social structure of position, opposition, and opportunity", *Administrative Science Quarterly*, vol. 45, p. 651-678, 2000.

[THE 06] THEUREAU J., *Le Cours d'Action. Méthode Développée*, Octares, Toulouse, 2006.

[THO 20] THOMAS W.I., ZNANIECKI F., *The Polish Peasant in Europe and America*, Gorham, Boston, 1920.

[TOD 06] TODIRASCU A., "Outils TAL pour les textes narratifs", in E. SOULIER (ed.), *Le Storytelling. Concepts, Outils et Applications*, Hermes, Paris, 2006.

[TRE 96] TRÉPOS J.-Y., *La Sociologie de l'Expertise*, PUF, Que sais-je?, Paris, 1996.

[VAN 92] VAN DE VEN A.H., "Suggestions for studying the strategy process: a research note", *Strategic Management Journal*, vol. 13, p.169-188, 1992.

[VAN 95] VAN DE VEN A.H., POOLE M., "Explaining development and change in organizations", *Academy of Management Review*, vol. 20, no. 3, p. 510-540, 1995.

[VIL 03] VILLETTE M., *Sociologie du Conseil en Management*, La Découverte, Paris, 2003.

[VIN 01] VINCENT D., BRÈS J. (eds.), "Pratiques du récit oral", *Quebecois Linguistics Review*, vol. 29, no. 1, University of Quebec, Montreal, 2001.

[VON 71] VON WRIGHT G.H., *Explanation and Understanding*, Routledge and Kegan Paul, London, 1971.

[WEI 95] WEICK K., *Sense Making in Organizations*, Sage, Thousand Oaks, 1995.

[ZAR 06] ZARRI G.P., "Représentation et traitement avancé de documents narratifs complexes au moyen du narrative knowledge representation language (NKRL)", in E. SOULIER (ed.), *Le Storytelling. Concepts, Outils et Applications*, Hermes, Paris, 2006.

PART II

How can we Locate Ourselves within ICT?

Chapter 4

Are Virtual Maps used for Orientation?

4.1. Introduction

Whether travelling on a clearly identified road, such as the National 7, or browsing through hypertext links [BOL 91, LAN 97, SNY 96] in the virtual space of the Web, the question about which map to use and the scale of the map. In his essay, *By reading, by writing*, Gracq points out that in mapping, the unsolvable problem of projections arises from the impossibility of representing a curved surface on a map, without distorting it [GRA 81][1]. In fact, there is always a distortion from reality and the maps, from this point of view, are not cures but merely palliative.

4.2. Orientation context

Either we consider the virtual map as an analogical figure to the terrestrial map and, in this case, hyperdocuments are only duplicates of paper documents, or we consider hyperdocuments and hypermaps as hypermedia, i.e. digital documents whose navigation is not preconstructed beforehand. These hyperdocuments are thus released from traditional forms of reading. In these conditions, the hyperdocument, resulting from a digital writing, cannot be reduced to the digitization of writing. It is thus convenient to find with hyperdocuments, and the hypermaps that accompany them, a unique grammar rather than duplicating duplication of an existing grammar.

Chapter written by Alain MILON.
1 p. 179.

We should note that screen writings of SMS, chat, email, Web type are merely anecdotal declensions of an existing grammar. Just as the radio is not a duplication of the press, the television is not a duplication of the radio, the cinema of the television, the cinema of the video, and the video of hypermedia.

In this perspective, orientation in the space of virtual maps is unnatural because such maps are based on a refusal of orientation (in the sense that orienting means finding the orient and then locating the occident). If orientation in virtual space is to be considered, it does not use the same criteria. Finally, what scale do maps offer? Not only analogical maps, such as road or maritime maps, but especially virtual maps, those we find in the digital space of text or the image of sound generators, for example. For all these maps, is the scale a faithful reproduction of reality, a translation of this reality by or, conversely, of a scale impossible to relate to reality? In these conditions, the map can be more than an addition to records; it can quickly become a process in permanent motion. But why reduce the scale to a measuring instrument of the map? Why not consider it as the means to be released from the *tyranny of analogy*? This chapter will focus on the imaginary that the real or virtual map reveals.

4.3. The flat sphere

If it is clear that the map is used for location, there are also circumstances where location is impossible and contrary to the logic of navigation. Navigation generally questions our displacement space so we do not confuse it with the expanse, territory or the place that we occupy. We will thus define space as part of our *a priori* perception: the expanse as the geometric measure of this space, territory as the expression of space socializations, and place as the way of inhabiting a territory.

Browsing a website can therefore be done in two ways: either the browser allows him- or herself to be guided and follows the site plan proposed by the home page, or the browser travels in an accidental and random way without any prior path built. In this case, browsing is done as if the user is facing a flat sphere defined by the computer screen, which is not only a geometrical limit (15 inches, 17 inches, 20 inches, etc.), but a means of transport with a virtual displacement space. This flat sphere has an extension in two dimensions – length and width – without having any left, right, top and bottom; in other words, it defines a *space oblivious to nature*. The eye does not have the opportunity to locate itself with respect to indicators such as those found on an ordinary map. It is for this reason that the sphere is interesting, since it has no point of origin and there is no means to hierarchically locate points relative to one another. The virtual browser is thus faced with a space that exists physically in the screen size, but also exists metaphorically through the open space bounded by the movement of the browser him- or herself.

This movement that is simultaneously physical and metaphorical is possible due to the unique geometry of the sphere. The spherical shape of the globe is such that all points are identical and without hierarchy. Nevertheless this sphere remains flat because it is physically limited by the size of the screen. On this particular sphere, everything is impossible to locate. Some cartographic search engines or Webmapping, such as *Inventix*, *Kartoo*, use the idea of the flat sphere in a metaphorical sense. They do not have a lot of sites following each other linearly, as if they were hierarchically classified according to relevance criteria, but the query results appear in the form of a star chart with links that move according to the choices of the user. Every time the user makes a selection, the whole Web of links moves and changes as does their relevance. This cartographic construction takes the form of a matrix network which reduces any type of ranking among the different addresses.

If we reduce virtual browsing to a flat sphere, we quickly realize that this navigation is no longer imposed by hyperlinks, but is the result of random processes that cannot be saved. The browser never finds the same thing twice: he or she is in a universe where the trails are not made in advance. This hazardous navigation has the same characteristics as those proposed by encyclopedic navigations, not those carried out through encyclopedias governed by an indexation process, such as the *Universalis, Larousse* or *Encarta* encyclopedia, but governed by the mode of association. These imaginary Borges-type encyclopedias take up the metaphor of matrix connections.

In a matrix, there is no point of departure or arrival. The encyclopedia is like a living body in which the browser travels freely for meetings and random circumstances according to a process of association of ideas. He or she will not look for something specific using a determined query, but he or she will discover exactly what he or she was not looking for. If the indexed encyclopedia imposes definite *paths* and links governed by the one writing them, such as websites where all links are links built by the only designer of the site, the associated encyclopedia is free and invents *paths* that are created on the go, and cannot be saved. It is the difference that we find between *creating a path* and *following a route*. The path is to be built, is not reproducible and cannot be saved, while when the route has already been made, it only needs to be followed. It is exactly what Borges proposed in *The Book of Sand*, through his vision of imaginary inventories:

"The number of pages of this book is exactly infinite. There is no first page, there is no last page. I do not know why they are numbered in this arbitrary way.

Maybe to suggest that the components of an infinite series can be numbered in absolutely any way" [BOR 83][2].

This quote from Borges reflects the difficulty for the browser to locate him- or herself, as much in the *exact space* of the book that has its type and spatial arrangements, like the page, cover, footnotes, pagination, as in the *uncertain space* of the active writing tools determined by algorithmic processes, for example as generators of texts, images or sounds. Behind this type of movement, the whole issue of the mode of navigation established by hyperdocuments is being raised. If we push this reflection on hypermedia to an extreme, however, we realize that hyper navigations are often artificial constructions. The problems posed by the map go far beyond the use of virtual maps, which are only illustrations of complex browsing processes. The better understand this complexity, we should go back to the first categories of orientation: the map and browsing.

4.4. Orientating

To be lost, to lose the North, to be disorientated, wandering, travelling "blindly", losing bearings, etc., all these expressions implicitly ask the same question: what is orientation? To reply to this falsely trivial question, Kant proposes to focus on the term of orientation [KAN 59][3]. Orientating is tantamount to cutting space into quarters to find the rising – the *orient* – and then place the setting – the *occident*. The most surprising in this vision of space is the fact that the principle of objective orientation, arranged by geographical markers, is determined by the subjective desire of differentiation of the browser: the orient in the east, the occident in the west [KAN 59][4].

From this relationship between an objective vision of space geography and a subjective approach to single out these spaces and attribute a name to them, comes the magic of maps and sciences that accompany it: mapping. Are maps only used for orientation? Are they localization instruments whose only use would be to cut the expanse in a geometric way with purely mathematical settings? Nothing is less certain! However, maps remain representation suggestions that oscillate between the descriptions of reality and productions of the imagination. In fact, maps do not leave us indifferent. They generate a particular feeling that combines desire to travel, real or imaginary, with the impression of power and discovery of unknown places. In fact, having a map at hand does not necessarily lead to orientating in space, it also reflects a certain control of space. The link with maps is normally affectionate, not

---

2 p. 141.
3 p. 77.
4 p. 77.

because we like their colors and forms, but because they take us to territories, often imaginary, as if summoning us to possess them. This taste for maps that each of us appreciates in his or her own way often stirs a confused feeling: that of a view from above, the view of the cartographer on the contours he or she has just reproduced and also views from above mixing with cross and side views.

4.5. Nature of the map

Maps are essentially used for moving, and even if there is only one space to describe they are multiple. Staff, route, maritime, air, topographical, geographical, basements, sea, star, thematic, planimetric, hydrographical or virtual, maps are always accompanied by a scale used as measurement and calibration to the surrounding space. Cartographers first defined the map as "the conventional representation, usually plane, in relative positions, of concrete or abstract phenomena, localized in space" [GEO 74][5].

Phantasmagorical, maps can lead to unexpected meetings. While remaining objective representations of a space in two or three dimensions, they can also be seen as metaphorical constructions of an unknown territory. Sort of an imaginary translation to allow the mind to overcome geometrical expanses, these compositions create ambiguities. Designed on a major scale, maps, when claiming any objectivity, seek to encircle the world with almost-perfect contours. In these conditions, summaries or archives of the world, maps are likely to kill the places they describe through the bindings they impose. They surround, suffocate, conserve and eventually turn the territory into a simple place of memory whose fences are definite and irreversible.

At a time when maps are in vogue and the term mapping is some kind of slogan used to signify as much the control of the territory as the political reading of a space, geographers – founding masters of cartography – are faced with other situations in which maps are dispossessed of their object. Is this dispossession the sign of a broadening of the concept of maps or the expression of excessive expansion of a term that ends up meaning nothing? A map of trends, map of jobs, map of social movements, object map or mental map – everything is a pretext for mapping places. Maps are thus taken in the midst of the quest for vital space. Some heads of states even use it as a government program when focusing on geopolitics.

---

5 p.58. Definition proposed by the commission of mapping terminology of the French Cartography Committee. The maps are topographic, relief, hydrographical, bathymetric, planimetric, thematic, geographic, geophysical, route, etc., and their scales are variable: 1/10,000, 1/25,000, 1/100,000, 1/500,000, etc. For an analysis of the definition of the concept of the map, see [JAC 92].

But what is hidden behind this situation: a trend, a fashion, a simple movement of mood or the deep feeling of territorial conquest reflecting a genuine demand of belonging and recognition? Is it a sort of quest of the other, where research makes it a *critical distance* or an *interior distance*, a sort of place of survival? How sad it would be if forced to travel with maps that:

– did not offer us something else;

– would not get us lost;

– would foresee everything: from distance to obstacles;

– would describe the territories covered in minute detail ;

overall would be only the translation of the encountered reality!

It would be much stranger to go on an adventure with maps modeling the spaces encountered, maps inventing territories, maps in fact fighting against the *tyranny of analogy*. However if we keep the term map, it is because of the imagination it conveys and because this concept allows many metaphors. This is why we will define "map" as the expression of a track in perpetual metamorphosis; a track that reveals multiple spaces – painted, physical, random, etc.

4.6. The virtual map

The map is alive and transformable to infinity. Its ruse is here, and we would be overlooking its complexity if we were to reduce it to a faithful representation of the observed reality. Beyond its representation of route, marine, rail, air, space, etc., the map tells us gracefully that faithfulness to representation can be an illusion, the means in fact to make life simpler. It also tells us that, behind this schematic representation is hidden not a reality that the map would try to represent, but a reality beyond itself, taking up Mallame's proposition with force: *the fold behind the fold*. The map, because it is alive, tries to kidnap this reality by all means, not to show that it exists as a whole but to assert its limit loud and clear. By kidnapping reality in this way, the map manages to "eliminate" reality and to emphasize the uncertain and revolted contours of these rebellious tracks.

The map fools us by its ambivalence; it was supposed to reassure and locate us, but it does so in order that we do not lose sight of the fact there is no primary reality and secondary representation. There are thus two places to be that give a meaning to territories surrounding us and that make the alley the principle path. To this end, let us leave the dead and passive maps, preferring transformable and ephemeral maps which, without warning, recompose space as they wish. These maps lead us to an unexplored and ignored land, a land they like to rebuild incessantly. Living only in their most intimate folds, they govern us by their own metamorphosis. These maps are alive so they are foreign to us. They provide adventure and if we try to tame them by clipping, their contours are rebuffed and eventually they explode. Invented, folded, drawn, metamorphosed, maps are rich because they are unknown.

What do we expect from reproductions if they are only truncated layers, too deep to be subject to simple topographical surveys? Nothing, but the secret hope that behind them an unknown map may arise at any time. The unknown map does not convey any reality. Neither servile copy, let alone reproduction, it remains alive and transformable to infinity. In perpetual motion, it changes, and by its mutations reality is inspired, first while territory remains and second it can eventually be modeled but does not care. Whether it is followed or copied, the map is not even concerned by the way the territory tries to mimic it. Ironically, the map knows that it is a copy of itself, and by copying finally smiles at its own avatars. What a strange destiny for a piece of paper which, originally, was to reassure everyone. What would we do without maps, say the navigators? Nothing, if the map is an instrument, because without it we get lost. Nothing, if the map makes reality, because the meaning takes shape through it and we are not sure we understand it. Maps do everything, they who, often, pretend that if they are forgotten reality would exist!

Reflecting here on the silence of these maps in motion and on what they do not explicitly say, this may be the program of virtual maps. If the map speaks, it silently says: "Read me, follow me to unknown places, but trust me at the risk of otherwise no longer being an empty territory!" This is why the map can be both a manipulated space when it reproduces the territory to scale and, at the same time, a place of territory manipulation when it imposes its scale on the territory. The mistake would be to imagine that, whatever happens, the scale – even of "1 to 1" – makes the map a flat copy of a three-dimensional reality.

4.7. Map territory

It is rather in the spirit of invented tracks that cartography is attractive, like those crazy cartographers in search of a map identical to the territory and a territory identical to the map – a map to the "format of the empire" in the words quoted by Borges [BOR 51][6]. In these conditions, the map cannot be treated in the sense of a graphical semiology [BER 67][7], or in that of an inventory of different mental maps [POR 96] or a cartographical approach of art [BUC 96]. It is rather that in the context of a map in search of routes not explicitly conveying:

---

6 p. 130. For an analysis of Borges text, see [MAR 73], p. 291-296.
7 The analysis of Bertin is an interpretation of the different forms and figures that the map has according to its semiological value. In the same way, among mental maps defined as productions of the mind, only those released from the weight of analogy are retained.

– topography – the *place*,

– geography – the *land*,

– cosmography – the *stars*,

– cosmology – the *myths*,

– astronomy – the *celestial bodies*,

– astrology – the *predictions*, or

– geometry – the *shapes*;

since all these sciences call for identification and locations, a map in fact that would juxtapose a series of spaces with respect to one another, each having its own track, none dominating another; thus the external reality may act/interfere like a kind of categorical imperative.

This is not to identify the different geometric shapes of the map with their qualities using point, line, area and volume, but to understand how these shapes invite us to reconsider the principles of displacement. The aim is then less to think about maps than to try capturing the particularity of the tracks they propose – tracks absent of all paths since the path is only a track already executed on maps already written. We encounter many tracks under maps, each with its moods, but we will focus only on those imposing their will, that are the cause of an event of which the browser thinks he or she is the author.

4.8. Program of the map

The first intention of mapping is to investigate not its content or settings but more interpretations that can be extracted from internal maps in the hope of finding the scale of each map. More than it represents, the map is something to see. While showing it defines a track, a kind of wire for the update of reality in another form: that proposed by these maps which, at the wills of their tracks, are created on the go according to a scale without a model. The second intention is to show how the unique nature of the traced map offers a very unique type of scale that we will define, not as a link between a *referent* (the territory the map want to represent) and the *representative* (the map as objective representation of a territory) of a *represented figure* (the subject as a piece of space), but as the expression of a *territory* (the referent) that draws an *unknown figure* from which any representative is absent. The last intention is to consider the map as the mapping of a territory that is rebuilt, according to circumstances, by endless entangled folds. These maps, transformable to infinity, rebuild space. They update territories, as the art critic Fromentin did in his time, not in the way the geographer who interprets contours of a territory, but as a painter who would draw tracks that he would have never seen [FRO 90][8].

The map thus proposes two readings of the world that intersect. It can be a flat copy of what it represents, reproducing the contour of a territory. In this case, it describes the world and accommodates the territory bounded by a domestic and familiar border, that of the *Oikoumene* of Greek tradition, for example[9]. Used as an instrument, it becomes a fact-sheet with graphical representations of expanses circumscribed by natural bounds. The map thus becomes its own register of the limits that restrict it; the topographer carrying out the inventory of these limits by fixing them in a final way. This inventory corresponds moreover to the first function of surveyors describing geometrical tracks and boundaries of a known territory. A unit closed onto itself, the map in fact owns what it represents. The representation of a contour, whatever it is, is sufficient in itself to translate its representation object. A map of France or a postcard of the Eiffel Tower speak for themselves. Their descriptions are included in their representations. But the risk with this type of map that strives to represent reality absolutely is that it leads inevitably to what Foucault calls "the alienation of analogy" [FOU 66][10] in the sense that, for these cartographers, only the objective modeling is worthy of being represented.

The map is also presented as *anamorphous*, however: the map traces a territory that is a new form which, under certain circumstances, ironically mimics reality. The anamorphous of which we are speaking, applied to the map, cannot be analogical. It is an invention that can have an origin, on some occasions, in an interpreted reality, but which in all cases is limited by the imagination of the navigator. This invention, making the object appear and disappear as a fragmented and dilated shape, teaches sight, and us, that nothing can be done without its consent. Such a map is no longer an instrument, but is the strongest expression of an intuition in the first sense of the human mind. Maps correspond to pictorial anamorphous as invented images of the world. These inventions are still maps since they set into motion internal distances and imaginary contours.

4.9. Map instruction

The relationship space that Foucault proposes in *Other Spaces* [FOU 94][11] corresponds to these hypermaps. It is in fact about considering space as a formal relationship organized around networks, both community networks in the traditional

---

8 p. 18, quoted by G. Deleuze in [DEL 93], p. 87.
9 *Oikoumene* is defined for the Greeks as a known territory, a land inhabited by men.
10 p. 63.
11 p. 752-762.

form and virtual community networks, such as exchange networks of Internet-type knowledge. The relationship then allows us to better understand the primary nature of the links among objects. Whether it is a beam or a network, the relationship is built around a non-hierarchical, autonomous and heterogenous mesh. It is remarkable to see that, behind this division of space, Foucault has seen problems posed by the information flow in complex systems, such as expert systems or search engines. This is why "space is given to us in the form of a location relationship" [FOU 94][12]. In this space, the algorithm becomes the grammar of the map and this algorithm is much more than a simple combination of elements that form all possible links, that is non-contradictory, among given primitive terms. This opposition can be translated as in Table 4.1.

Combinatory calculation	Generative process
Infinite set	Finite set
Denumerable set	Non-denumerable operatives
Random calculation	Algorithm as complex operative
Determined navigation	Non-predetermined navigation
Fixed syntax	Dynamic grammar
Integrative operative	Generative process
Author	Without author

Table 4.1. *Comparison table*

The virtual map is heterogenous since it mixes bodies of different natures. It is multiple in the sense that the folds entangle each other because they are one. It is significant in the sense that, having no origin, we can always cut a part of it; this will not spoil the rest of the map. But it is especially non-appropriable. Not demarcating any territory, it can belong to the one seizing it. Hypermaps offer a device that questions the very act of creation by the presence of a complex algorithmic process whose first claim is to make the document exist by itself at the same time that it escapes from the author of the computer program. If the questioning of the author "concept" is not unique to hyperdocuments, the construction of a device entirely subject to an autonomous algorithmic process is specific to hypermedia.

The hypermap is only used to measure the limits of traditional writing, however. It is the conclusion which Queneau, when he laid the foundations of the OULIPO (opener of potential literature), reached: potential writing contains its own limits, and they are only literary experiences that are a long way from literary creation. In

---

12 p. 754.

his interviews with Charbonnier, he noted that "the true new structures will have real interest once used in an original way" [QUE 63]. In these conditions, virtual navigation of hyperdocuments fulfils its function when: it finds the means to fight against analogy and duplication of an already existing path; it prevents the establishment of thought (of not finding the origin of things); and when it loses the common from places and things; moving to the idea of Borges, when it releases itself from non-isotropic spaces that have the same qualities, regardless of where we are; when it forgets utopian spaces that are easily located because they are necessarily out of a possible real; and when it opens on heterotopic spaces, those which worry, those offering an ironical sight of the real. This inventory of possible spaces defined by these crazy maps takes on the harbinger of Foucault in *Les Mots et des Choses* (*Words and Things*) [FOU 66]: this book finds its *raison d'être* in the imaginary inventories of Borges.

4.10. Bibliography

[BER 67] BERTIN J., *Sémiologie Graphique*, Gauthier-Villars, Paris, 1967.

[BOL 91] BOLTER J.D., *Writing Space: The Computer, Hypertext, and the History of Writing*, Lawrence Erlbaum, Hillsdale, 1991.

[BOR 51] BORGES J.-L., *Histoire de l'Infamie, Histoire de l'Eternité*, Le Rocher, Paris, 1951.

[BOR 83] BORGES J.-L., *Le Livre des Sables*, Gallimard, Paris, 1983.

[BUC 96] BUCI-GLUCKSMANN C., *L'Œil Cartographique de l'Art*, Galilée, Paris, 1996.

[DEL 93] DELEUZE G., *Critique et Clinique*, Paris, Editions de Minuit, Paris, 1993.

[FOU 66] FOUCAULT M., *Les Mots et les Choses*, Gallimard, Paris, 1966.

[FOU 94] FOUCAULT M., *Dits et Ecrits*. Des espaces autres, 1980-1988, Volume IV, Gallimard, Paris, 1994.

[FRO 90] FROMENTIN E., *Œuvres. Un Eté dans le Sahara*, Gallimard, La Pléiade, Paris, 1990.

[GEO 74] GEORGE P. (ED.), *Dictionnaire de la Géographie*, PUF, Paris, 1974.

[GRA 81] GRACQ J., *En Lisant, en Ecrivant*, Corti, Paris, 1981.

[JAC 92] JACOB C., *L'Empire des Cartes*, Albin Michel, Paris, 1992.

[KAN 59] KANT E., *Qu'est-ce que s'Orienter dans la Pensée?*, Vrin, Paris, 1959.

[LAN 97] LANDOW G, *Hypertext: The Convergence of Contemporary Critical Theory and Technology*, John Hopkins University Press, Baltimore, 1997.

[MAR 73] MARIN L., *Utopiques: Jeux d'Espaces*, Editions de Minuit, Paris, 1973.

[POR 96] PORTUGALI J. (ed.), *The Construction of Cognitive Maps*, Kluwer Academic Publishers, Dordrecht, 1996.

[QUE 63] QUENEAU R., *Entretiens avec G. Charbonnier*, Gallimard, Paris, 1963.

[SNY 96] SNYDER I., *Hypertext: The Electronic Labyrinth*, Melbourne University Press, Melbourne, 1996.

Chapter 5

Geography of the Information Society

5.1. Introduction

What has happened to the geographical space in the era of electronic communication? This type of interrogation is ongoing, and suggests that answer this question could vary sensibly lot across the decades. Two major trends have made their imprint on the different ways in which the relation between territory/information and communication technologies (ICT) has been considered over the past 30 years[1]. At first, due to lack of sufficient recoil, ICT have raised more hopes than they have brought certainties[2]. It seemed possible, through ICT, to overcome the distance following the reduction or disappearance of travels that are costly in time and energy. This approach is no more dominant now, despite the inclusion of ICT in the international trade or development of an intangible economy, creating the myth of a "global village".

Secondly, just before the bursting of the "Internet bubble" there was a need to give meaning to tools whose artificial graft with society would have led to a number of rejects. The solution therefore goes through a territorial inclusion of ICT. Ultimately, far from eliminating differences between places, ICT development

---

Chapter written by Henry BAKIS and Philippe VIDAL.
1 Researchers' analyses have often fed on steps undertaken by politics and *vice versa*. In a way, science has contributed to the establishment of standards, particularly in this field, which are as volatile as those of ICT [OFF 00, RIP 89, VID 02]. The analyses have finally evolved beyond simple trend projections.
2 These observers are mostly prospectivists or futurologists, such as Toffler [TOF 80], Rheingold [RHE 94], who had a big influence among decision-makers and politicians.

increases the territorial disparities within industrialized countries, between North and South, and even within those of the South. At the same time, states and territorial authorities have became more strongly involved: now, the provision of infrastructures and services are a fundamental action of the development of territories[3].

Today, there seems to be a trivialization of ICT in industrialized countries and in the biggest capitals of the planet. After 30 years of ICT presence in these territories, attributes of the geographical space have been enhanced and modified significantly and unequally:

– real time;

– the possibility of rich (large volumes of data possible) and regular communications, even at a very long distance;

– development of wireless links;

– pricing of some communications regardless of distance, etc.

This is why the new term geocyberspace is now being usefully employed in social sciences to describe this symbiosis between space and ICT.

The work that follows was therefore undertaken in three areas:

– the first refers to the way the technological determinism has undergone a reality check (section 5.2);

– the second emphasizes the territorial character of ICT applications (section 5.3);

– the third advocates the idea of a certain trivialization of ICT in the territories of industrialized countries and considers the consequences (section 5.4).

5.2. Technological determinism of the facts

The first period expresses faith in technological determinism. It is politically marked by an opposition between a territory seen as binding and ICT enabling us to overcome it. Researchers and academics have, however, rapidly combined the prospective visions of engineers, beliefs of politicians (French cable plan of 1983) and predictions of futurologists [RHE 94, TOF 80], all announcing the imminence of radical changes.

---

3 The genesis of "information society" is done with guesswork and readjustments, technical blabbering and success stories, commercial failures and popular unexpected successes.

5.2.1. *Avoidance of travel and space contraction: two ideas promoted by policy makers and industrials*

In the early days of network computing, the argument favored by policy makers and their advisors to encourage public investment was based on a simple idea: the avoidance of motorized trips to reduce energy bills. The oil crisis was then a major element justifying the computerization of French society [NOR 78], as the problem of carbon dioxide emissions in the atmosphere and its environmental consequences is today.

The hypothesis of the introduction of distance-working in organizations developed and had the advantage of reintroducing economic activity in rural territories. Palliative (dealing with energy difficulties) and curative (revitalizing the rural world) steps led to the establishment of a strategy that has had great successes (Minitel) and some notable failures (cable plan, distance-working). ICT were considered during this first phase to be prosthetic and prophetic tools, offsetting the difficulties experienced in some territories and announcing major changes in the relationship between man with his environment. Various offers of distance-services, such as the start of e-commerce, participatory democracy via networks, e-government, etc., appeared that it was felt would finally allow man to escape territorial constraints.

5.2.2. *The research world: a frank rebuttal*

Researchers opposed the reading of facts [BAK 83, CLA 97, JAU 93, LEF 98, OFF 00, SAV 98] to the speeches and prospective visions of engineers and politicians (French cable plan of 1983). Quite rapidly, they pointed out the false trails left by these speeches that, implicitly, conveyed the myth of the neutrality of the territory. It became clear that the emergence of a new technique and its uses did not suddenly reorganize the social space [BAK 94]. The issue of travel avoidance is being undermined by numerous geography specialists in transport, especially where, more broadly, there are innovations such as Grübler's[4] (he provides objective indications on a relationship between distance communication and transport[5]).

The graph in Figure 5.1, based on the French case, shows that physical displacements have never been more important than today. The two curves tend to form a single one with the approach of the year 2000, leading its author to consider that the relationship between ICT and transport is not so much one of substitution as of complementarities.

---

4 The graph in Figure 5.1 and its interpretation come from [VID 04].
5 This, despite a deterministic approach.

74   Digital Cognitive Technologies

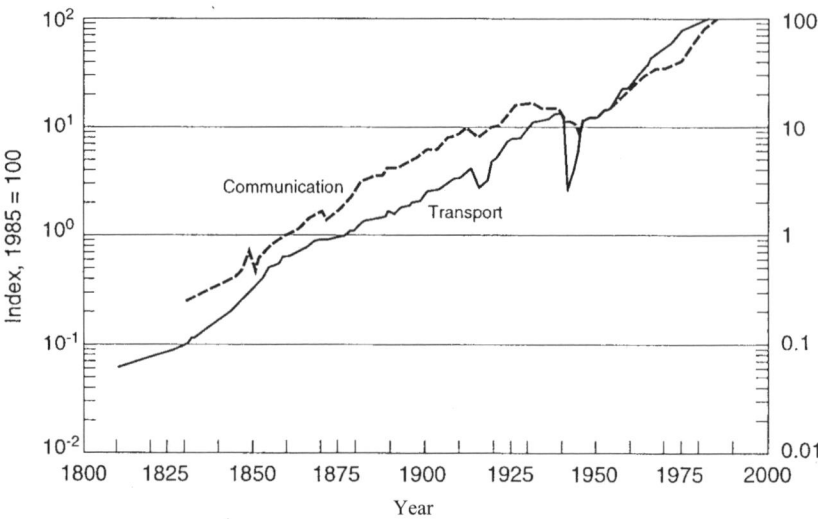

Figure 5.1. *Putting the rate of increase of motorized travel and increase in distance exchanges into perspective [GRU 90, GRU 98]*

5.3. From the "end of geography" to the "territorialization of ICT"

This offer of infrastructures and services has provided arguments for the assertion of a controversial thesis: "the end of geography" [BAT 95, CAI 97, OHM 02, PAT 04]. With telecommunications, even if physical distance does not prevent relationships in "real time" anymore and there is a reduction in the social space-time [ALL 95, BAK 83], geographical space does not disappear as such. Without a hypothetical "end of geography", its renewal is ahead with the ever-greater inclusion of ICT within territories [DUP 05, LAS 00, LEF 98, PIG 04]. Far from eliminating the differences between places, the development of ICT, which is not the mere presence of infrastructures, makes the territorial disparities even more serious. Since then, this regionalization of ICT has contradicted the thesis of "the end of geography". Increasing numbers of works have shifted the focal point from public and private offer of a telecommunications infrastructure in territories towards their ownership through uses and users. This will result in moving the relationship between ICT and society from the macro to the micro level with a real "call to the local" [JAU 93].

What is at stake now is the idea that the spreading of ICT depends on their capacity to be appropriate, and on a scale that had previously been lacking in France: general public fax project (late 1970s) [BRE 93], cable plan (1983), distance-working (early 1990s). Social pull is replaced by technological pull. The principle is simple: the innovation process follows the needs faced by users within territories. It is about proposing a technical answer to a social demand where the user is seen as the co-designer of technologies [EVE 98a, IRI 96]. These evolutions fall into three dominant trends:

– urban participants take over the ICT;

– ICT become mobility and proximity tools;

– finally, that ICT become additional assets in competition between territories[6].

5.3.1. *The appropriation of ICT by urban participants*

If, in the early 1980s, computer network seemed likely to come to the rescue of rural areas, by the mid-1990s the city was subject to experiments, confirming the idea of an ever-stronger reciprocal link between territory and ICT. The city appears to be the natural partner of innovation[7] [GRA 96] primarily because there are a large number of potential consumers that are of interest to industries in the sector and institutions wishing to initiate innovative policies. This is the case with e-commerce "which is more of interest to metropolitan zones already well served by classic trade than less accessible peripheries to whom it might be useful" [DUP 02]. It is also the case for the mobile phone and its growing role in the search for merchants who can meet the expectations of more urban mobile consumers [MAR 05].

Municipalities have, for their part, mainly invested in offering ever-more sophisticated technological services to the population, both in terms of access (the Internet via digital public spaces), content (municipal portals) and 3G mobile communication radios[8]. The appropriation of ICT by urban investors is a phenomenon that occurs in different ways in other types of spaces. For example, the ability that the Mouride brotherhood (a socio-religious movement of West Africa based in Touba) had in harnessing ICT to enhance the "mouridism" within the second town of Senegal illustrates this natural congruence between urban areas and ICT [GUE 03]. This allows us to address another very meaningful aspect when it

---

6 The issue of network planning is therefore raised, for territorial participants, who have long-identified with the sole responsibility of operators (before liberalization).
7 With a few exceptions, covering mostly experiments that are difficult to replicate on other rural territories: digital village *Chooz* launched in 1997, and the European center of IT in Rural Areas (CETIR) was inaugurated in 1999.
8 Other variants of the urban characteristics of these experiments are, for example, the testing of third generation (3G) mobile radio communication systems in European cities, to offer, sometimes in geo-localization condition, a wide range of services including voice, data and video to users.

5.3.2. ICT, tools of mobility and proximity

In recent years ICT have been viewed in relation to physical mobility, while the dominating speech of industry and politicians has long been founded on displacement avoidance. Three key concepts (mobility, proximity and localization) tend to replace in the mind of promoters of the information society those of "space neutrality" and "global village". Such evocative terms were, however, justified in the 1980s and 1990s when many political and economical choices were made in terms of ICT. This premium on territory can be interpreted as a relative failure of political and economical choices prevailing in the 1990s, both in terms of certain policies of land use (the best example is probably the inconsistency of distance-working[9]) and many traditional enterprises (Click and Mortar) who saw cyberspace as a new digital Eldorado.

If e-commerce starts to generate real profits it favors travelers. It is symptomatic that of the first 10 cyber-sellers in France, five of the companies were working in the fields of travel or transport: if ICT permit us to avoid certain displacements, they provoke much more. This mobility is expressed by displacements, often short and numerous, in the context of complex circuits [DES 01]. ICT have become tools controlled by the user before or during his or her displacements. Over the past 30 years, the relationship between ICT and territories of proximity has asserted itself [VID 04].

The field of mobile telephony illustrates this squeezing from global to local. The first generation (1970s and 1980s) allowed a few vehicles belonging to the political or economic elite to communicate on national, international or local scopes. The second generation global system for mobile communications (GSM, 1990s) has installed the individual mobile wireless phone in everyday life, the phenomenon having a major growth from 2000. "Where are you calling me from?" is a common question during conversations. It is a symptom of a return, if not to a local level, to a territory level, from the user to the communicating object [JAU 93]. The third generation appears to have finally inscribed the "local" characteristic of technological evolution. LBS services make the idea that ICT would deny the territory obsolete: by LBS services, we mean territorial multimedia services for consumers/people in the city. The proximity of the service has become the main

---

9 In 1994, DATAR launched one of its first calls for proposals on ICT, entitled "Teleworking for employment and wining back of territories". In it published a book entitled *Teleworking, Tele-economy. A Chance for Employment and Attractiveness of Territories* (see [ROZ 95]).

added value of communicating objects: the territory becomes the *raison d'être* [VID 04]. In an experiment conducted in Rennes, this approach reflects the municipal will to raise the voice of citizens from their place of living. A priority that is reminiscent of the resolutions of the Mayor of Parthenay: "We have conducted a democratic experiment: we want our citizens to value their own information and become content creators". For promoters of the information society, territory is now a resource and not a constraint. It is no longer just a matter of providing services from a distance, but also offering city inhabitants close, interactive and personalized services. This new period clearly establishes reintroduction of the territorial dimension in debates between ICT and society.

5.3.3. *ICT, instruments of competition of territories*

From the close relationship that has developed between city, proximity and ICT, follows the fact that telecommunications are becoming major instruments in the competition between territories. Certain authors (Innis, Mac Luhan, Castells) stated that making the planet a small village would lead to on-going dichotomous reading of phenomena, which will feed the idea of the existence of a digital gap between territories in information society and others.

This dominating/dominated opposition extends beyond the aspects of economics and world scale. The mechanism is equally reflected in the scale of cities in the same country. In the past five years, French municipality websites have been evaluated and awarded the Internet city label [10]. This allows the establishment of a hierarchy between the most advanced cities, in terms of provision of services to the population, and others. In terms of infrastructure, each year ORTEL[11] publishes a scorecard of the state of network coverage of the national territory (a digital atlas of territories). Gradually a hierarchical relationship will form between the most and least advanced; the latter being stigmatized as being behind the times.

In France, regional (AEC, ARDESI, ARANTIS, ARTESI, MARSOUIN, SOURIR, POSI, etc.)[12], departmental (SUSI)[13], technopolitan (DigiPort)[14] and

---

10 The Internet city label has existed for six years. It gives one to five @ to local communities wishing to compete on the evaluation of their communal website.

11 ORTEL: regional observatory of telecommunications.

12 AEC: Aquitaine Europe Communication; ARDESI: Regional Agency for the Information Society in Midi-Pyrénées; ARANTIS: Regional Agency Poitou-Charentes of ICT; ARTESI: Regional Agency of Information Society Technologies of Ile de France; SOURIR: Synergy of Observatories of Regional uses of Internet and Networks; MARSOUIN: Armorial pier of Research on the Information Society and Internet uses; POSI: Observation Pole of the Information Society at Reunion and in the India Oceanic zone.

13 SUSI: General Internet Uses Agency.

national (OTV created in 1990, becoming OTEN in 2005) observatories[15] have produced a series of monographs on "territories and ICT". Other associations (AVICCA, ECOTER and FING)[16] carry out extensive monitoring and counsel territorial communities. At the same time, more academic works have appeared than in the past[17], produced by research laboratories specializing in these questions. A normative and prescriptive discourse will follow, largely driven by participants such as the Caisse des Dépôts et des Consignations, which in the 2000s promoted action for the digital development of territories. This discourse will finally bear fruit as the territories establish blueprints, become equipped, develop effective strategies and exchange "best practices"[18].

In addition, we start to incorporate the results of experiments. Some cities are now investing in an ICT strategy, developing nationally-recognized communal portals and public multimedia spaces. Today, ADSL[19] technology allows the transformation of an ordinary telephone line (analog) to a high speed digital data transmission line. The subscriber may have his or her traditional telephone service and broadband Internet on the same line. This technology is better than the failed cable plan of 1983 or the theory report of 1994, which envisaged that by 2015 the whole French territory would be connected through fiber optics.

5.4. The trivialization of ICT in territories in industrialized countries

The technology on offer is constantly improving, the supply of contents becoming more responsive to the demands of the population, public organizations are incorporating ICT in a transversal way in their territorial development programs[VID 02], and users are appropriating the technologies. In terms of technological offers, we not only see a spatial extension of technological coverage throughout the France, but also a diversification of connection possibilities in major cities. These observations are explained by a double phenomenon [DAN 05, RAL 05]. On the one hand, the liberalization of telecoms mandated by Europe has enabled this abundance of supply in urban areas. On the other hand, article L. 1425-

---

14 *DigiPort*: Science Park Lille Metropolis.
15 OTV: Observatory of telecommunications in the city; OTEN: Observatories of digital territories.
16 AVICCA (ex-*Avicam*): Association of Cities and Communities for Electronic Communications and Audiovisuals; ECOTER: Mission ECOTER is a new network of technologies in the public sector; FING: Internet Foundation New Generation.
17 See in particular the geography magazines *Netcom* and *Flux* and theses in the discipline.
18 Through structures such as CREATIF or regional agencies such as ARTESI, ARDESI, which will publish guides.
19 ADSL (asymmetric digital subscriber line) offers a speed of several megabits per second.

1 of the general code of territorial communities, which came into effect in 2004, has provided the opportunity for territorial communities and their organizations to intervene extensively in broadband in case private initiative was insufficient. The second phenomenon has more particularly benefited rural and suburban communities, while some sparsely populated zones, and therefore less profitable areas, are not involved in these local initiatives.

It is therefore an early stage to speak about universal access [DAN 05]. These evolutions augur well for the appearance of a new period: that of an integrated approach of the space and ICT relationship that sees the concept of geographical space become mutated in geocyberspace.

5.4.1. *The geographical space of the 21st Century*

This concept aims to describe geographical space in the era of electronic information networks. It is distinct from "geo-space" (which is the space of kilometers, physical distance and regions). It is also fundamentally different from the concept of "cyberspace", which was created by a science fiction author to describe a virtual world in which inhabitants of hyper-computerized megalopolis dive [GIB 84]. Engineers and researchers in social sciences have adopted this word to refer to computer infrastructures and the mapped visualization of traffic flow on the Web. The works of Batty and Dodge are representative of this [BAT 93, BAT 94, DOD 01].

The word cyberspace is also used, for lack of anything better, to designate the new space of electronic communication. To avoid any confusion, the meanings being very different, the neologism "geocyberspace" was coined [BAK 97c, BAK 01][20] in order to designate a geographical space with new properties. These properties include the possibility of regular long-distance almost instantaneous high quality relations, and sharing such complex data that they have considered remote surgeries, such as "the Lindbergh operation", first transatlantic surgery on humans conducted in September 2001.

There are different rates in the emergence of an information society (depending on places and societies). An ICT standardization process has been carried out. ICT are no longer fantasized about, but are considered additional assets for a territory, in the same way as water, gas or electricity. Testimonies of this standardization are now appearing with a rate and appropriation of forms that show how much the

---

20 Neologism used in the context of the commissions "communication and telecommunication networks", then "geography of the information society", of the International Geographical Union [BAK 97c, BAK 01, PAR 03].

virtual will feed the real, the global will feed the local, the spatial will refer to a-spatial, and vice versa.

However, places and routes remain unavoidable objective dimensions for the human being who lives in a more complex geographical space. The space of electronic communication neither replaces, nor passively overlaps the traditional geo-space: it closely joins with it at all scales. The space of electronic communication is twice that of geo-space: it reflects that the geographical space is structurally changed and has radically new attributes.

5.4.2. *An integrated approach between space and ICT*

To illustrate this idea of an integrated approach between space and ICT, it is convenient to start from the uses of these technologies to demonstrate that the space and ICT relationship is plural – not only hyper-contextualized, nor uniquely virtual, but that the space and ICT relationship tends to vary depending on needs and circumstances. The time seems ripe for the synthesis between global and local approaches to ICT. If we consider, for example, the issue of 3G mobile phones, it becomes clear that a single user will be able to use his or her device for uses that are sometimes territorial and sometimes totally devoid of territorial intentions. Applications called "geolocalized services"[21] cover a range of multimedia services in connection with the environment of the individual. The information delivered may be non-commercial (events organized by the municipality, pharmacies and medical care, etc.) or commercial (local brands aimed at its clients). The information is broadcast by technical devices combined with the GPS system. Here, the link to local territory is obvious, but the user will also be able to use his or her device to escape this territorial context.

5.4.3. *A more complex geographical and social space*

By the second half of the 20th Century, man was living in a geographical and social space that owes much to the telephone [ABL 77, ABL 01, GOT 77]. This trend continued as a result of technical innovations allowing some kind of telepresence, and, more importantly, with the spread of ICT, the Internet especially. The internet has enabled new forms social contact [PIO 91], even unusual complex configurations, such as virtual communities existing only through the Internet [RHE 94] whose effects are as much social, cultural and political [BAK 95, FRO 05, LEV 97] as economical and spatial.

---

21 The provision of LBS services considered here is around territorial services to consumers/users.

At a macro scale, Castells defends the idea of a "new spatial logic" that he calls the "space of traffic flow". This logic is based on a strong economical dimension being the dominant dimension of global society. The author proposed "a spatial model of the network society" that takes shape in what he calls the "global society". The information age is reflected by a relationship of domination of space thorough which traffic flows, on the space of places with economically dominant territories, and others cut off from the new technological system [CAS 89, CAS 96, VEL 05].

On a micro scale, distance-working is finding a new interest, following technological and regulatory evolutions in access to broadband networks, as well as by the nature of the public targeted: it means maintaining the local active population. The example of SOHO (small office home office) illustrates the synthesis between village land and globalism: welcoming independent distance-workers from the tertiary sector in a rural environment and in particular responding to the growing demand for installation by the population wishing to "live and work in the country". The link between space and ICT once more refers to both a local environment for production and a global dimension for the customer.

While, for more than a decade, access to the Internet was through dedicated (cybercafés, multimedia kiosks) or individual places, access also can be via wireless technologies (Wi-Fi, WiMax) in towns and villages. This grid, still has to be improved and is currently unavailable everywhere. Its spread has been facilitated by the reduction in the cost of access. Territorial websites were first showcase websites without providing a service to the population. They then became involved in the local democracy and improvement of public services (Parthenay Digital City, Issy-les-Moulineaux). Today, they provide both a quality local public services (dematerialization of administrative procedures) and a promote the local territory on a global scale.

5.5. Conclusion

While by the end of the 20th Century, having a good quality communal website was rare and represented an indisputable territorial marketing tool [VID 99], there is now no such great advantage. Now, having an intranet no longer makes a business a vanguard structure. The uses are generalized and trivialized, at least in industrialized countries. In developing countries the spread of the Internet has been remarkable and speedy, for example in South Africa. In this country there has been an exponential increase in mobile telephony [CHE 01, LOU 05]. In West Africa, satellite links have become important in providing access to broadband Internet [BER 03].

In short, many territories have these essential infrastructures, which are now seen as basic infrastructures (just as networks of drinking water, electricity, gas and transport are). These infrastructures encourage the emergence of more favorable territorial dynamics. There is a certain "trivialization" of these infrastructures, although this phenomenon does not lead to the disappearance of comparative advantages among territories with digital equipment.

Techniques are changing and requirements in terms of bandwidth or computer power, for example, are still important. The next challenge will be for the responsible organization in territories to offer access to "mobile broadband" to allow the transmission of more data at a faster rate. The supported presence of ICT in territories shows that we can no longer consider ICT to be in opposition or juxtaposed to the territory. New perspectives have come to light in terms of integration and standardization of the space and ICT relationship. They will probably lead to the validation of a new territorial paradigm: geocyberspace.

5.6. Bibliography

[ABL 77] ABLER R., "The telephone and the evolution of the American metropolitan system", in I.D.S. POOL (ed.), *The Social Impact of the Telephone*, The MIT Press, Cambridge, United States, p. 318-341, 1977.

[ABL 01] ABLER R., "Mastering inner space: Telecommunications technology and geography in the 20th Century", *Netcom*, vol. 15, p. 17-28, 2001.

[ALL 95] ALLEN J., HAMNETT C. (ed.), *The Shape of the World: Explorations in Human Geography 2: A Shrinking World? Global Unevenness and Inequality*, Oxford University Press, Oxford, 1995.

[ASC 95] ASCHER F., *Métapolis ou l'Avenir des Villes*, Odile Jacob, Paris, p. 59, 1995.

[BAK 83] BAKIS H., Télécommunications et organisation de l'espace, Thesis on the state of arts and social sciences, University of Paris I, 1983.

[BAK 84] BAKIS H., "Transports et télécommunications – substitution et complémentarité", *Workshop Geography of Telecommunications and of Communication, 25th International Congress of Geography*, Paris, 1984.

[BAK 94] BAKIS H., "Territoire et télécommunications. Evolution de la problématique et perspectives: de l'effet structurant aux potentialités d'interactions au service du développement urbain et régional", *Netcom*, p. 367-400, 1994. [BAK 95] BAKIS H., (ed.), "Communication and political geography in a changing world", *International Political Science Review*, vol. 16, no. 3, p. 219-311, 1995.

[BAK 97a] BAKIS H., "Approche spatiale des technologies de l'information", *Revue Géographique de l'Est*, vol. 37, n° 4, p. 255-262, 1997.

[BAK 97b] BAKIS H., GRASLAND L., "Les réseaux et l'intégration des territoires", *Netcom*, vol. 11, no. 2, p. 421-430, 1997.

[BAK 97c] BAKIS H., ROCHE E., "Cyberspace – the emerging nervous system of global society and its Spatial Functions", in E.M. ROCHE and H. BAKIS (eds.), *Developments in Telecommunications*, Avebury, p. 1-12, 1997.

[BAK 01] BAKIS H., "Understanding the geocyberspace: a major task for geographers and planners in the next decade", *Netcom*, vol. 15, no. 1-2, p. 9-16, 2001.

[BATE 95] BATES S., "The end of geography", *Theories and Metaphors of Cyberspace – Abstracts,* 1995, available online at: http://pespmc1.vub.ac.be/Cybspasy/SBates.html, accessed February 5, 2010.

[BATT 93] BATTY M., "The geography of cyberspace", *Environment and Planning B*, vol. 20 no. 6, p. 615-616, 1993.

[BATT 94] BATTY M., BARR B., "The electronic frontier: exploring and mapping cyberspace", *Futures*, vol. 26, no. 7, p. 699-712, 1994.

[BATT 97] BATTY M., "Virtual geography", *Futures*, vol. 29, no. 4/5, p. 337-352, 1997.

[BER 03] BERNARD E., Le déploiement des infrastructures Internet en Afrique de l'ouest, Thesis in geography, University of Montpellier III, 2003.

[BRE 93] BRETHES J.-C., "Le télécopie: 150 ans d'histoire (1843-1993), *Réseaux*, no. 59, pp.19-131, 1993.

[BRO 04] BROSSAUD C., "La fracture numérique dans la cité: Etat des "lieux" de la recherche francophone", *Proceedings of the International Conference on ICT and Inequalities: Digital Division,* Carré des Sciences, Paris, 18-19 November 2004.

[CAI 97] CAIRNCROSS F., *The Death of Distance: How the Communications Revolution will Change our Lives*, Harvard Business School Press, London, 1997.

[CAS 89] CASTELLS, M., *The Informational City: Information Technology, Economic Restructuring, and the Urban-regional Process*, Basil Blackwell, Oxford, 1989.

[CAS 96] CASTELLS, M., *The Information Age: Economy, Society and Culture – The Rise of the Network Society*, Blackwell, Malden, 1996.

[CEN 03] CENTRE FOR THE STUDY AND OBSERVATION OF THE DIGITAL CITY, *InfoCity, a Citizen Network Portal, Territorial Eecompositions and ICT*, CSODC, 2003.

[CHE 01] CHENEAU-LOQUAY A., "Les territoires de la téléphonie mobile en Afrique", *Netcom*, vol. 15, no. 1/2, p. 1-11, 2001.

[CLA 83] CLAISSE G., *Transports ou Télécommunications, les Ambiguïtés de l'Ubiquité*, PUL, Lyon, 1983.

[CLA 97] CLAISSE G., *L'Abbaye des Télémythes, Techniques, Communication et Société*, Aléas, Lyon, 1997.

[DAN 05] DANG NGUYEN G., VICENTE J., "Quelques considérations sur l'aménagement numérique des territoires : le rôle des collectivités locales dans le déploiement des infrastructures de l'économie numérique", *Proceedings of the Forty-first Symposium of ASRDLF*, Dijon 5, 6-7 September 2005.

[DES 01] DESSE R.-P., *Le nouveau Commerce Urbain, Dynamiques Spatiales et Stratégies des Acteurs*, PUR, Rennes, 2001.

[DOD 01] DODGE M., KITCHIN R., *Atlas of Cyberspace*, Addison-Wesley, London, 2001.

[DUF 03] DUFEAL M., GRASLAND L., "La plantification des réseaux à l'épreuve de la matérialité des TIC et de l'hétérogénéite de territoires", *Flux*, no. 54, pp. 49-69, 2003/4.

[DUP 88] DUPUY G. (ed.), *Réseaux Territoriaux*, Paradigms, Caen, 1988.

[DUP 02] DUPUY G., *Internet, Géographie d'un Réseau*, Ellipses, Paris, 2002.

[DUP 05] DUPUY G., "Réseaux et frontières: Internet aux marges", *Annales de Géographie*, vol. 114, no. 645, pp. 465-576, 2005.

[EVE 98a] EVENO E., *Les Pouvoirs Urbains Face aux Technologies de l'Information et de la Communication*, PUF, 1998.

[EVE 98b] EVENO E., IRIBARNE A.D., "Les utilisateurs comme co-concepteurs de services multimédia interactifs: le projet "Ville numérisée" à Parthenay", *Proceedings of the Symposium Think of the Uses*, Bordeaux, p. 319-333, 27-29 May 1998.

[EVE 04] EVENO E., "Le paradigme territorial de la société de l'information", *Netcom*, no. 1/2, p. 89-134, 2004.

[FRO 05] FROMENT B., BAKIS H., "Migrations, télécommunications et lien social : de nouveaux rapports aux territoires? L'exemple de la communauté réunionnaise", *Annals of Geography*, no. 645, p. 564-574, 2005.

[GAR 03] GARCIA D., Désenclavement et Technologies d'Information et de Communication en moyenne montagne française : l'exemple du Massif Central et de ses bordures, Thesis of the University of Montpellier III, 2003.

[GIB 84] GIBSON W., *Neuromancer*, City Lights Books, New York, 1984.

[GOT 77] GOTTMAN J., "Megalopolis and antipolis: The telephone and the structure of the city", in I.D.S. POOL (ed.), *The Social Impact of the Telephone*, The MIT Press, Cambridge, United States, p. 303-317, 1977.

[GRAH 96] GRAHAM, S., MARVIN, S., *Telecommunications and the City: Electronic Spaces, Urban Places*, Routledge, London, 1996.

[GRU 90] GRÜBLER A., *The rise and fall of Infrastructures, Dynamics of Evolution and Technological Change in Transport*, Physica-Verlag, Heidelberg, 1990.

[GRU 98] GRÜBLER A., *Technology and Global Change*, Cambridge University Press, IIASA, Laxenburg, 1998.

[GUE 03] GUEYE C., "Enjeux et rôle des nouvelles technologies de l'information et de la communication dans les mutations urbaines, le cas de Touba (Sénégal)", Technology, Business and Society, United Nations Research Institute for Social Development, Program Document, no. 8, May 2003.

[INN 50] INNIS H.A., *Empire and Communication*, Toronto University Press, Toronto, 1950.

[IRI 96] IRIBARNE A.D., "Pour une approche socio-culturelle des autoroutes de l'information", *Multimédia et Communication à Visage Humain*, File coordinated by A. HIS, Transversals Science/Culture (file for a debate no. 56), p.134-141, 1996.

[JAU 93] JAUREGUIBERRY F., "De l'appel au local comme effet inattendu de l'ubiquité médiatique", *Espaces et Sociétés*, no. 74-75, l'Harmattan, p. 119-133, 1993.

[KEL 02] KELLERMAN A., *Internet on Earth: A Geography of Information*, John Wiley and Sons, New York, 2002.

[LAS 00] LASSERE F., "Internet: la fin de la géographie?", *Cybergeo*, no. 141 2000, available at: www.cybergeo.presse.fr/, accessed on February 5, 2010.

[LAT 01] LATOUR B., *L'espoir de Pandore, pour une Version Réaliste de l'Activité Scientifique*, La Découverte, Paris, 2001.

[LEF 98] LEFEBVRE A., Tremblay G., *Autoroutes de l'Information et Dynamiques Territoriales*, PUM and PUQ, Toulouse, 1998.

[LEV 97] LEVY P., *Cyberculture*, Odile Jacob, Paris, 1997.

[LOU 05] LOUKOU P.A., Télécommunications et développement en Afrique, le cas de la Côte d'Ivoire, Thesis in geography and planning, Montpellier III University 2005.

[MAR 05] MARZLOFF B., *Mobilités Trajectoires Fluides*, Library of territories, L'Aube essai, La Tour d'Aigues, 2005.

[MOR 05] MORIZET B., BONNET N., "La géographie des centres d'appels en France", *Annals of Geography*, vol. 113, no. 641, p.49-72, 2005.

[MUH 05] MUHEIM F., "Voyage dans l'"haut-delà". Ethnogéographie de la communauté virtuelle Suisse-Québec : pour une critique du cyberespace", in H. BAKIS (ed.), *Technologies de l'Information: des Infrastructures Matérielles aux Communautés Virtuelles*, Netcom, vol. 19, no. 3-4, p. 101-114, 2005.

[MUS 95] MUSSO P., RALLET A., *Stratégies de Communication et Territoires*, Collective IRIS-TS book with the assistance of DATAR, L'Harmattan, Paris, 1995.

[NOR 78] NORA S., MINC A., *L'Informatisation de la Société*, Rapport à M. le Président de la République, La Documentation Française, Paris, 1978.

[OBR 92] O'BRIEN R., *Global Financial Integration: The End of Geography*, Pinter, London, 1992.

[OFF 93] OFFNER J.-M., "Les "effets structurants" du transport: mythe politique, mystification scientifique", *L' Espace Géographique*, vol. 22, no. 3, p. 233-242, 1993.

[OFF 96] OFFNER J.-M., PUMAIN D., *Réseaux et Territoires, Significations Croisées*, L'Aube, La Tour d'Aigues, 1996.

[OFF 00] OFFNER J.-M., "Télécommunications et collectivités locales. Des cyber-territoires en développement "virtuel"", *Troisième entretien de la Caisse des depôts sur le développement local*, p. 14-169, 2000.

[OHM 02] OHMAE K., *The Borderless World – Power and Strategy in the Interlinked Economy*, HarperCollins, New York, 2002.

[PAR 03] PARADISO M. (ED.), "Geocyberspace dynamics in an interconnected world", *Netcom*, vol. 17, no. 3-4, p. 129-190, 2003.

[PAT 04] PATE E., *Competition and the End of Geography*, The Progress & Freedom Foundation Aspen, Colorado, 2004, available at: http://www.usdoj.gov/atr/public/speeches/205153.htm, accessed February 5, 2010.

[PIG 04] PIGUET E., La fin de la géographie? 2004 http://www2.unine.ch/webdav/site/manifsacademiques/shared/documents/li03-04_piguet.pdf, (accessed 3 March 2010).

[PIO 91] PIOLLE X., "Proximité géographique et lien social, de nouvelles formes de territorialité?", *L'Espace Géographique*, no. 4, p. 349-358, 1991.

[RAL 01] RALLET A., "Commerce électronique et localisation urbaine des activités commerciales", *Revue économique*, vol. 52, no. H-S, pp. 267-288, 2001.

[RAL 05] RALLET A., ULLMANN C., "Le "Haut débit", nouveau défi du développement local: approches croisées de l'économie et de la géographie", *Proceedings of the Forty-first symposium of the ASRDLF*, Dijon, 5-7 September 2005.

[RHE 94] RHEINGOLD H., *The Virtual Community: Homesteading on the Electronic Frontier*, Addison-Wesley, Reading, 1994.

[ROZ 95] ROZENHOLC A., FANTON B., VEYRET A., *Télétravail, Télé-économie. Une Chance pour l'Emploi et pour l'Attractivité des Territoires*, DATAR/IDATE, Montpellier, 1995.

[SAV 98] SAVY M., "TIC et territoires le paradoxe de localisation", *Les Cahiers Scientifiques du Transport*, no. 33, pp. 129-146, 1998.

[TOF 80] TOFFLER A., *The Third Wave*, Bantam Books, London, 1980.

[TRE 08] TREMBLAY G., "De Marshall McLuhan à Harold Innis ou du village global à l'empire mondial", Revue tic&société, De TIS à tic&société: dix ans après, available at: http://revues.mshparisnord.org/lodel/ticsociete/index.php?id=222, accessed 27 May 2008.

[ULL 06] ULLMANN C., Les politiques régionales à l'épreuve du développement numérique : enjeux, stratégies, impacts, Thesis of geography and planning, Pantheon Sorbonne, Paris, 2006.

[VEL 05] VELTZ P., M*ondialisation, Villes et Territoires*, PUF, Paris, new edition, 2005.

[VID 99] VIDAL P., "Les sites *web* de quatre municipalités françaises au crible de la géographie des représentations", *The Canadian Geographer*, vol. 43, no. 2, pp. 191-197, 1999.

[VID 02] VIDAL P., La Région face à la société de l'information, Le cas de Midi-Pyrénées et de Poitou-Charentes, Thesis in geography and planning, University of Toulouse-Le Mirail, 2002.

[VID 04] VIDAL P., "Les services géolocalisés dans les métropoles européennes : les opérateurs à la recherche d'un nouveau modèle économique", *Proceedings of the Fourteenth RESER Conference*, Castres, available at: http://www.reser.net/file/4282,/ September 2004.

[WIL 00] WILSON M., COREY K., *Information Tectonics, Space, Place and Technology in an Information Age*, John Wiley and Sons, Chichester, 2000.

Chapter 6

Mapping Public Web Space with the Issuecrawler

6.1. Introduction

Changes in the conceptualization of Web space may be read from the kinds of visualizations made over the past decade. This chapter concerns itself with the visualization modules made for one Web "mapping" device, the Issuecrawler network location software. It briefly periodizes understandings of Web space by examining the contexts in which Issuecrawler mapping modules were conceived and built: the site inlink lists (in the "Web as hyperspace" period), the circle map or virtual roundtable (in the period of the "Web as neo-pluralistic space"), the cluster or issue network map (in the "Web as network" period) as well as the geographical map or the distributed geography of an issue (in the current "locative" period). The focus moves from the metaphysics of software-made space (sphere, network) to the specific info-political geographies that can be charted with the aid of the Issuecrawler and allied tools.

6.2. The death of cyberspace

The symbolic end of "cyberspace" may be located in the Yahoo lawsuit in May 2000, brought before a French court by two French non-governmental organizations, the French Union of Jewish Students and the League Against Racism and Anti-Semitism. The suit ultimately led to the ruling in November 2000 that called for

---

Chapter written by Richard ROGERS.

software to block Yahoo's Nazi paraphernalia pages from Web users located in France.

Web software now routinely knows a user's geographical location, and acts upon the knowledge. You are reminded of the geographical awareness of the Web when you type into the browser, google.com, and are redirected to google.fr. Whilst it may be viewed as a practical and commercial effort to connect users with languages and local advertisements, the search engine's IP-geo-location handling also may be viewed as the demise of cyberspace as place-less space [MIL 01].[1] With location-aware Web devices, cyberspace becomes less an experience in displacement than one of re-placement – you are sent home by default.

What might be called "death of cyberspace", or the revenge of geography, has consequences for any theorizing of the history of Web space. The question posed here concerns how Web space is conceptualized by devices that have sought to "map" the Web, especially without employing conventional politico-geographical cartography or borrowing from geological metaphors, such as thematic islands, peaks or valleys [DOD 01]. The following discusses one device in particular. The Issuecrawler"s sense of Web space is explored through a brief history of the visualization modules created for the software – a history that also seeks to periodize understandings of Web space. It does so through a reflection on how the visualizations provided commentary on contemporaneous Web thought.

The Issuecrawler is server-side Web network location software. Input URLs into the Issuecrawler, and the software crawls the URLs, captures page/site outlinks, performs co-link analysis, and outputs the results in lists as well as visualizations. The software was conceived in the mid-1990s at the Department of Science and Technology Dynamics, University of Amsterdam [ROG 96], and has a forerunner in the Netlocator, also known as the De-pluralising Engine, built in Maastricht during the Jan van Eyck Design and Media Research Fellowship, 1999-2000.

6.3. Tethering websites in hyperspace through inlinks

The Netlocator began with the insight that Websites (or Webmasters) link selectively as opposed to capriciously. There is a certain optionality in link-making. Making a link to another site, not making a link, or removing a link, may be viewed as acts of association, non-association or disassociation, respectively. Later, we learned through a Georgia Tech study and our own observations and interviews, that

---

[1] Since 2000 ethnographers of the web have concluded that geography is more significant than media. It seems that the local and national dimensions are more powerful than practices specific to media of so-called non-localized space.

hyperlinks are matters of organizational policy, especially for corporations and government [KEH 99], [GOV 99]. Selective link-making could create space – when we conceive of space as that demarcated by limited acts of association. The demarcationist approach performs an important break with cyberspace by suggesting that hyperlinking behaviors dismantle the "open-ended-ness" of cyberspace, which formed the idea of "placelessness".

What types of associations are on display in hyperlinks ("reading between the links"), and what could be the shapes of spaces demarcated by link associations? In the late 1990s and early 2000s the leading visualizations we discussed were the Plumb Design's ThinkMap Visual Thesaurus as well as I/O/D's Web Stalker, followed shortly thereafter by TouchGraph's Google Browser as well as Theyrule.net by Josh On.[2] All are non-directed graphs, without arrowheads, which is to say that the elements (synonyms, site pages, board members and companies) are associated (and lines are drawn between them), without specifying a uni- or bi-directional association. Undirected graphs, arguably, derive from a path model of the Web, also built into browsers (with the forward and backward arrows), and lead to ideas about every link being a two-way link [NEL 99, BER 99].

Seeing the Web in terms of paths is not far-fetched, since one may surf from page to page, and use the browser buttons, or the browser history, to retrace one's steps and also move "forward" again. However, on the Web, two-way links, it may be observed, are less frequent than one-way links. Viewing any hyperlink as a bi-directional association, we learned at the time, also has its infamous cases, whereby for example a German ministerial site was accused of being linked to a call boy network [MAR 00]. The Bundesministerium für Frauen und Jugend linked to a women's issues info site, and that info site linked to a call boy network. To the German newspaper *Bild Zeitung* this Web path implicated government.

To stand on the shoulders of Vannevar Bush, Theodore Nelson and other path and hypertext model pioneers would view the Web as pathway space (for the surfer) [BUS 45, MAR 99]. The Netlocator (and later the Issuecrawler), however, strove to distance itself from the Web as pathway space, and instead concentrated on the Web as selective associational space (made by Webmaster linking). How does one view associations? As is well-known, a site's outlinks, most readily in the form of one or more link or resource lists, are viewable to a site visitor. To gain a sense of a site's inlinks, however, requires the use of the advanced search of an engine, or access to the referrer logs of a site. Until the creation of "trackback", a feature implemented in the Movable Type blogging software in 2002 that shows backlinks to a posting, inlinks in the late 1990s were not an everyday concern. Only ranking algorithm

---

2 The *ThinkMap Visual Thesaurus* by Plumb Design is accessible at http://www.thinkmap.com/, and *I/O/D 4: The Web Stalker* at http://bak.spc.org/iod/.

makers, most notably Google with the PageRank system, made use of them. Nowadays, on the Web as well as in the blogosphere and in online news, devices recommend pages routinely by counting inlinks, e.g. "most blogged" stories at the *New York Times* and the *Washington Post*. In all, concern with inlinks as a marker of page relevance or reputation marked a major shift in the underpinnings of Web space.

Counting inlinks addressed the site authority problem. Previously, in the mid-1990s the foremost issue concerning search engine developers related to how to separate the "real name" from the borrowers of the name, e.g. to return Harvard University at the top of the list when Harvard is queried. In leading search engine results (AltaVista's), the "eminent scientist and the isolated crackpot [stood] side by side", as one leading author put it [RHE94]. In their ranking logics, AltaVista granted site owners the authority to describe the content of their sites (in metatags) and their descriptions became the basis for the engine returns. The Web became a space displaying "side-by-side-ness", fitting with contemporaneous ideas about its pluralizing potential [BAR 96]. Google, conversely, granted other sites that authority (hyperlinks and link pointer text). Counting inlinks and having other sites grant authority through linking (and naming their links well) form the basis of most search engine algorithms these days, including Yahoo's as well as MSN's. Once a major competitor to automated search engines, the directory has declined.

6.4. The depluralization of the Web

Which links should be counted? The observation we made in the late 1990s was that search engines' "population" for link-counting was the entire Web (or the percentage of it they were able to index). Instead of focusing on what the "most influential" [social network metric] are calling the carriers of the term, on the record on the Web (which is how I would summarize the dominant search engine ranking algorithms), we preferred to look for what could be called "organizational networks". Insert a set of URLs of organizations working in the same area, and return those organizations (or URLs) which have received at least two links from the starting points.

Thus, like Google's algorithm for the entire Web, the Netlocator's (and, later the Issuecrawler's) algorithm for a portion of the Web crucially sought to take into account sites' inlinks. Once the crawling and co-link analytical procedure of the Netlocator completed, a list of sites in the network (the results) were displayed, color-coded as governmental (.gov), commercial (.com), non-governmental (.org), and scientific (.edu), including country-specific sub-level domains. (.ac.uk, for example, would count as scientific). When an actor was clicked, the links it *received* were highlighted. It was not just inlink counts, but types of inlinks, that concerned us, however. When showing an actor's inlink types, the Web could be made into an actor reputational space by showing which links a site received.

How and why do sites link [PAR 05]? The Netlocator-related link language for outlinks and inlinks provided a schematic for linking behavior generally, according to domains as well as further qualitative characterizations (see Figure 6.1). In one of our first extended case studies, on genetically modified food, inlinks and outlinks provided actor profiles according to types of links received and given. For example, three corporate sites were compared; the sites' respective standings differ according to the types of links received, and sites' respective display of awareness according to types of links given. One corporation has a different standing by virtue of receiving links from NGOs and government, as opposed to from other corporations only.

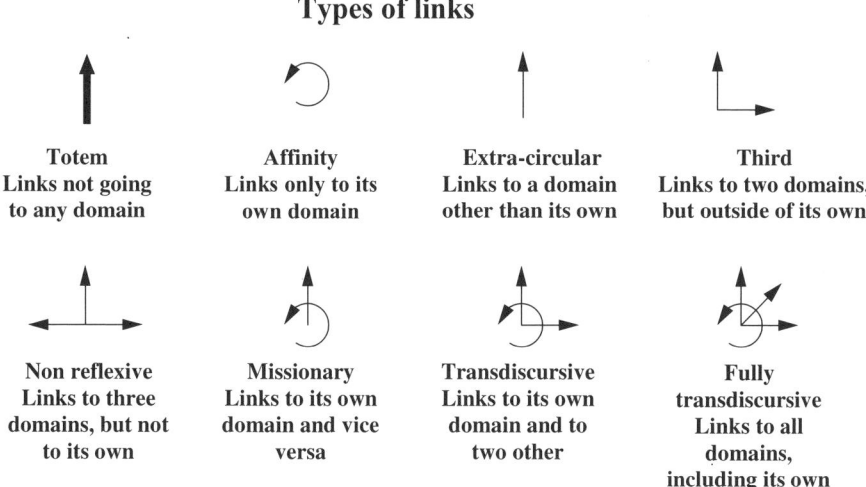

Figure 6.1. *Hypertext link language of actors, Govcom.org, Design and Media Research Fellowship, Jan van Eyck Academy, Maastricht, 1999*

Links are also classified qualitatively: cordial, critical or aspirational. Cordial links are the most common – to project partners and affiliates and other friendly or respected information sources. Critical links, largely an NGO undertaking, have faded in practice, and aspirational links are made normally by smaller organizations to establishment actors, often by organizations desiring funding or affiliation. In the case study on genetically modified food products, the "inspirational" links are as follows: Novartis provides a link to Greenpeace. Greenpeace does not provide a link back. Greenpeace and Novartis provide a link to the government. The government does not provide a link back.

In 1998, before the existence of Netlocator, the shape given to the visualization of associational linking in Web space was based, initially, on astronomical charts. Generally, thinking in terms of the Web as a universe (to be charted) coincided with early ideas of the Web as a hyperspace, where one would jump from one site to another at a great distance. Google's "I'm feeling lucky" button also played upon the trope of hyperspace and the famed hyperspace button (from the Asteroids arcade game by Atari, released in 1999). In the period of starry night site backdrops and random site generators, Websites, arguably, appeared untethered, individual stars whose relationships could be charted (and constellations or configurations perhaps named). The circular maps that we have created from data output by Netlocator (lists indicating links between pages and sites) also evoked the "sphere" of public sphere theory. The idea that the Web was or could be made into a pluralizing space, where familiar hierarchies of credibility may be challenged, became the focus of our visualization work.

With the GM Food (1999) and Russian HIV-AIDS maps (2000), we sought to show interlinkings between sites in a kind of virtual roundtable [ROG 00]. What if the Web, according to network inlink counts, were to determine who would sit at the table, instead of more familiar agenda-setters? Significantly, however, not all the actors have the same standing at the table – some receive more links than others and thus grow larger in size. The links between the actors are considered to be entanglements. Are the linked actors all on the same side? Would only the largest nodes speak, and the smaller ones keep still? Thus our roundtable was not flat; it had complications, which we sought to capture in the notion of the 'De-pluralizing Engine,' the other name for the Netlocator. In fact, the Web should not be seen as a pluralizing space by itself, as it creates hierarchies through inlink counts generally, and through inlink counts from the most influential actors (the basis of Google's PageRank).

In the circle maps, especially those auto-generated in what came to be known as the vanilla version of the Issuecrawler (2001), the de-pluralizing spirit continued with built-in notions of a core network and a periphery, where the latter, called the "waiting room", comprises those actors (or sites/pages) not quite receiving enough links to sit at the table. A variation on aspirational linking was in evidence in the visualization, as it showed only links from periphery to core, and not from core to periphery. Thus the peripheral link showed a desired belonging to the core, as of yet unachieved owing to lack of sufficient inlinks. Referring to the mapping practice, Noortje Marres prefaces her PhD dissertation with the following remark: "When we [took] to the Web to study public debates on controversial science and technology, we [found] issue networks instead" [MAR 05]. Notions of the Web as debate space, with the virtual roundtable construct (however much we strove to complicate it), did not fit with the empirical findings.

Even when we endeavored to *make* the Web into a debate space, by harvesting text from organizations' specific, issue-related deep pages, we found only statement juxtapositions – comments by organizations on a particular statement, but scant inter-organizational exchange. Organizations would release views on an issue on their Websites (which we would capture), but forums and other dialog spaces were not used by what could be construed as the parties to a debate. The Web could not stand in for a building – or an event where debating parties could gather. (Certain authors also began to discuss our work as evidence that the Web (or Net) should not be construed as a public sphere [DEA 02].) The Web as neo-pluralistic space had come to an end.

6.5. The Web as (issue) network space

Not only with the circle maps but also with the cluster mapping module made by Andrei Mogoutov[3], the final version of Issuecrawler (2003), with its instructions of use, is to be described as an "issue network" location and visualization machine. The understanding of the web as a network space, as opposed to a virtual space or online community space, is initially linked to the distinction between a multi-site analysis and a monosite analysis. ("Online communities" these days are still generally geographically concentrated and located on one site: for example, Hyves, in the Netherlands, Facebook in the United States, Orkut in Brazil, Cyworld in South Korea, and Lunarstorm in Sweden.).

When performing multiple site analysis, with the Issuecrawler, the crawling and co-link analysis return the sociable and the under-socialized, so to speak, in the same space. (Thus we achieved a new form of "side-by-side-ness"). In terms of types of associations (found in Web space and network mapping more specifically), issue networks may be distinguished from popular understandings of networks, and social networking, in that the individuals or organizations in the network neither need be on the same side of an issue, nor be acquainted with each other (or desire acquaintance) [MAR 06]. Actors may be antagonistic, oppositional, adversarial, unfriendly, estranged.

Additionally, unlike social networks, issue networks do not privilege individuals and groups, as the networks also may be made up of a news story, a document, a leak, a database, an image or other such items, found on individual pages of Websites. (Thus the Issuecrawler considers "deep pages" as significant for the study of issue networks.) Taken together these actors and "argument objects" serve as a means to interrogate the state of an issue either in snapshots or over time. "Issue

---

3 See Chapter 9.

states" may be gauged, initially, by taking note of the network's actor composition [ROG 04].

For example, we compared in 2004 queries in Google for "climate change" and RFID (radio frequency identification). If we look at the actors who appear most often in the results, for climate change, we have scientists from the United Nations, governmental agencies, and other actors forming part of the establishment. In the results for RFID, we find the trade press, corporations, lone activists and electronics hackers. If we compare the composition of the network in terms of actors (from the results found by the search engine based on the count of links), it clearly appears that climate change is a more mature issue than RFID.

It is important to emphasize the Web's capacity (with the Issuecrawler) to display configured, professional and publicized culture. The networks or lists that are located rely on public displays of connection (hyperlinks), rather than informal, quiet or old-boy relationships [HOB 03]. Indeed, the goal of network mapping is often to make things visible, to reveal non-public relationships, even to dig for dirt. A 2002 search engine query resulted in the newspaper headline: "UN weapons inspector is leader of S&M sex ring" [REN 02].

Understandings of the Web as network space, together with the return of the informality of the Web (particular through the blogosphere), have given rise to an investigative outlook. The impulse relates to the Web's street proximity, its closeness to the ground, including the "fact-checking", evidential spirit of the political blogosphere. The Issuecrawler takes into account a sense of a public "real" – evinced in the making and displaying of a hyperlink. Thus, importantly, the Issuecrawler does not map what is commonly understood as "virtual space" (as an online game environment). A map of a virtual space would be to a computer or video game what a "site map" is to a Website, showing the world (or the pages) that has been built and how one may navigate it.

Since 2005 the Issuecrawler has been considered a mapping device for issue professionals and researchers working with that made public. In an effort to make the Issuecrawler's sense of the "real" even less virtual, the latest visualization module, the Issuegeographer, strives to ground networks (see Figure 6.2). We placed issue network actors on a geographical map to show the proximity (or lack thereof) of the places of actors to the places of issues.

We also show that the climate change issue has happened in certain places according to the network actors on the web.

Figure 6.2. *Localization of an issue. The results of Issuecrawler on climate change prepared by Issuegeographer, 2005*

The focus of the visualization work began to consider actor mobility, whether networked actors move from issue to issue (or whether issues move from network to network). The provocative question read: Do networks form around issues, or are there networks in place that assume issues as they arise? Previously, in social movement research, the idea was mooted that there is "free-floating movement potential", in the sense of a given collection of publics which are able to form a movement, with particular conditions. That is, movements are not spontaneous uprisings, but rather more structural phenomena. May the same be said of networks? Are networks simply there, like Websites under construction, waiting for content?

In particular, global issues may have typical discursive homes, as at (recurring) conferences, summits and other gatherings. Thus, we asked, is there a difference between where an issue is happening, and where it is currently based? The notion of the "base of an issue" takes professional circulation as its point of departure, which results in people asking each other, not where you are from, but where you are currently based. With the Issuecrawler in tandem with the Issuegeographer, the Web becomes a space where we can locate where an issue is based [GOV 05].

6.6. Conclusion

Issuecrawler visualizations have evolved with conceptualizations of Web space – from hyperspace and cyberspace over public sphere and debate space to network and locative media. In each case the visualizations sought to engage with specific notions of Web space. In the hyperspace period the Netlocator tethered sites by showing inlinks. The Issuecrawler broke with the alleged open-ended-ness of cyberspace by showing how hyperlinks demarcate associational space. It also engaged with public sphere theory (Web as debate space) by unflattening the virtual roundtable, showing over-sized nodes and entangling links. The cluster map module organized actors into a particular kind of network, the issue network, where, with the Issuegeographer, one is able to map the distance between where an issue is happening (e.g., on the ground), and where an issue is currently based (e.g. in a summit network). Recent concrete research projects with the Issuecrawler engage with the current locative media period, where the Web, with tools, may be made to show information politics in specific geographical settings [ROG 06].[5]

6.7. Bibliography

[BAR 96] BARBROOK R., CAMERON A., "The Californian Ideology", *Science as Culture*, vol. 6, no. 1, p. 44-72, 1996.

[BER 99] BERNERS-LEE T., *Weaving the Web: The Past, Present and Future of the World Wide Web by its Inventor*, Orion, London, 1999.

[BUS 45] BUSH V., "As we may think", *Atlantic Monthly*, p. 101-108, July 1945.

[DEA 02] DEAN J., *Publicity's Secret: How Technoculture Capitalizes on Democracy*, Cornell University Press, Ithaca, New York, 2002.

[DOD 01] DODGE M., KITCHIN R., *Mapping Cyberspace*, Routledge, London, 2001.

[GOV 99] GOVCOM.ORG, *The Rogue and the Rogued: Amongst the Web Tacticians*, Video document, Jan van Eyck Design and Media Research Fellowship, Maastricht, 1999.

[GOV 05] GOVCOM.ORG, "The places of issues: The Issuecrawler back-end movie", in B. Latour and P. Weibel (eds.), *Making Things Public: Atmospheres of Democracy*, Film, ZKM, Karlsruhe, 2005.

[HOB 03] HOBBS R., LOBARDI M., *Global Networks*, Independent Curators International, New York, 2003.

---

5 The author wishes to thank Noortje Marres for her critique, as well as Anat Ben-David, Erik Borra, Marieke van Dijk, Koen Martens and Auke Touwslager for their maps and programming. For more information on the software presented here: http://movies.issuecrawler.net/.

[KEH 99] KEHOE C., PITKOW J., SUTTON K., AGGARWAL G., ROGERS J., *GVU's Tenth World Wide Web User Survey, Graphics Visualization and Usability Centre*, College of Computing, Georgia Institute of Technology, Atlanta, 1999.

[MAR 99] MARRES N., ROGERS R., "To trace or to rub: screening the Web navigation debate", *Mediamatic*, vol. 9-10, p. 117-120, 1999.

[MAR 00] MARRES N., Somewhere you've got to draw the line: de politiek van selectie op het Web, Master's Thesis (MSc.), University of Amsterdam, August 2000.

[MAR 05] MARRES N., No issue, no public: democratic deficits after the displacement of politics, PhD dissertation, University of Amsterdam, 2005.

[MAR 06] MARRES N., "Net-work is format work: issue networks and the sites of civil society politics", in J. DEAN, J. ASHERSON, G. LOVINK, (eds.), *Reformatting Politics: Networked Communications and Global Civil Society*, Routledge, New York, 2006.

[MIL 01] MILLER D., SLATER D., *The Internet: An Ethnographic Approach*, Berg, Oxford, 2001.

[NEL 99] NELSON T., "Xanalogical structure, needed now more than ever: Parallel documents, deep links to content, deep versioning, and deep re-use", *ACM Computing Surveys*, vol. 31, no. 4, December 1999.

[PAR 05] PARK H., THEWALL M., "The network approach to Web hyperlink research and its utility for science communication", in C. HINE (ed.), *Virtual Methods: Issues in Social Research on the Internet*, Berg, Oxford, p. 171-181, 2005.

[REN 02] Rennie D. "UN weapons inspector is leader of S&M sex ring", THE DAILY TELEGRAPH, 30 November 2002.

[RHE 94] RHEINGOLD H., *The Millennium Whole Earth Catalogue*, Harper, San Francisco, 1994.

[ROG 96] ROGERS R., "The future of science and technology studies on the Web", *EASST Review*, vol. 15, no. 2, p. 25-27, June 1996.

[ROG 00] ROGERS R. (ed.), *Preferred Placement: Knowledge Politics on the Web*, Jan van Eyck, Maastricht, 2000.

[ROG 04] ROGERS R., "The departure of science from the RFID issue space: Mapping the state of an emerging sociotechnological controversy with the Web", *The EASST Conference*, Paris, 28 August 2004.

[ROG 06] ROGERS R., Mapping Web space with the Issuecrawler, govcom.org, 2006, available at: http://www.govcom.org, accessed February 5, 2010.

PART III

ICT: a World of Networks?

Chapter 7

Metrology of Internet Networks[1]

7.1. Introduction

The Internet is changing in terms of its uses. From a monoservice network to transport binary or textual files 20 years ago, the Internet must today be a multiservice network for the transport of diverse and various data, such as audio and video (movies, video on demand, telephony, etc.). Indeed, there is a need to operate a technological mutation of the network to make it capable of transporting different types of information offered by all of the applications using the Internet, with adequate and multiple quality of service (QoS). However, all attempts to ensure the QoS of the Internet have failed, due particularly to a complete ignorance of Internet traffic and the reasons for this complexity. In the end, as it exists today, nobody has control of it, or even a complete knowledge of the network, which goes against the implementation of multiple services of communication of guaranteed quality.

The metrology of Internet networks – literally "the science of measurements" applied to the Internet and its traffic – must provide an answer to these problems. First, to provide services with predetermined qualities, it must be able to measure these qualities. Second, metrology must answer the questions about the model(s) of Internet traffic that are now missing. As it is recent science (it appeared in the early 2000s), it changes the whole process of research and engineering of Internet networks and becomes the cornerstone.

Chapter written by Nicolas LARRIEU and Philippe OWEZARSKI.
1 This chapter is taken essentially from the article by the same authors: "Metrology techniques and tools for Internet and its traffic", *Techniques of the Engineer*, Treatise "Measurements and Control", reference R 1090, 2006, with authorization from the editors.

The metrology of the Internet is divided into two distinct tasks: the first consists of measuring physical parameters of QoS offered by the network or the traffic. In a network of the size and complexity of the Internet, it is already a complex task, as we will see. This activity of measurement and observation makes it possible to highlight visible phenomena but, in the case of networks, it is even more important to deduce the causes, i.e. to determine the components and/or mechanisms of protocol that generate them. We find ourselves in fact facing the same problem as Plato in the allegory of the cave [PAL 02].

In this cave where a wood fire was crackling, he saw only the shadow of the men wandering in the cave and the shadows projected on the walls of the cave were huge, giving the impression they were those of giants. Metrology – measurement of the QoS or simple analysis of traffic – confronts us with a Platonic problem. It only shows us the effects of the mechanics of networks, while what intrigues us is the second part of metrology: by highlighting the causes of the Internet's shortcomings, we pave the way for research to change the mechanisms, architectures and protocols of the Internet.

This chapter therefore introduces the problems associated with the implementation of global measurements in computer networks such as the Internet, which will be the target of all the works described below (section 7.2). It shows the different active and passive measurement techniques, their needs, qualities and faults (section 7.3). Then, from measurements or observations on the traffic, this chapter highlights the causes of the current limitations of traffic growth, thus showing the importance of metrology for research and network engineering (section 7.4). Section 7.5 concludes this chapter by listing a number of network domains that derive large profits from using metrology.

7.2. Problems associated with Internet measurement

When a computer network specialist talks about Internet measurement this can, depending on the case, be related to three types of metric measurements:

– the size of the Internet, which is expressed in the number of machines linked to the Internet;

– the traffic carried on the links (wired or wireless) of the Internet; and

– the measurement of Internet QoS.

The term QoS can have different connotations depending on who uses it, however. To simplify the problem, we can consider that there are two main viewpoints:

– *applications* (or of users): each application has needs in terms of "delay, throughput, loss", the famous triptych for new generation computer networks. These needs are obviously different from one application to another, and each application would wish to have a communication service specifically tailored to its needs;

– *operators*, whose objectives are to optimize the use of the resources of the communication infrastructure (and thus maximize their earnings), to limit losses and delays, and to fairly and consistently charge the services provided to users.

Another characteristic of the measurement of physical parameters of QoS (also called "simple") is related to the size of the Internet, consisting of a multitude of interconnected networks, themselves consisting of numerous nodes and points of presence in different cities around the world. Thus, it is important to break the measurements into two new classes:

– *End-to-end measurements*, concerning the triptych "delay, throughout, loss" between two user terminals. These are measurements that join the aforementioned applications' point of view.

– *Step-by-step measurements*, concerning the triptych "delay, throughput, loss" between intermediate network nodes (routers, switches, gateways, etc.). These are measurements that join the operators' point of view.

7.2.1. *Geographical and administrative dimension*

Clearly, the major inherent difficulties in the creation of traffic or QoS measurements are related to the size of the network and its geographical range. Thus, it is essential to have many measurement points in the network. To watch the traffic at a single point at a given time is quite insignificant because Internet communications are often made over very long distances. However, the Internet is an interconnection of networks and, as such, multiple operators are owners and in charge of the management and operation of one of these networks that make up the Internet. Today it is impossible to position measurement tools at certain points of the network against the will of the operator who owns and manages this point. To carry out metrology of the network, we should be able to place measurement points at all network nodes, but the construction of the Internet prohibits this by default. Furthermore, "to monitor" all Internet nodes is a titanic task.

To circumvent this difficulty, work has been underway for three to four years on sampling problems in both space and time. The idea is to find techniques that, from some measurement points that would work only during short periods of time, would accurately estimate traffic on the entire network over 24 hours. Today, despite the efforts made, the results on sampling are at a standstill and, indeed, there are no significant positive results in this field.

Furthermore, the global scope of the Internet leads to daily variation in activity. It thus makes little sense to watch only one point of the network because on one continent where night has fallen, for example, an intercontinental connection can be faced with potentially reduced traffic (which will therefore pose few problems with compliance to QoS constraints). Then, a few milliseconds later, the network can cross a continent during the peak of the day, and thus meet significant traffic and possible congestion phenomena that is problematic for QoS communication. In such a context, it is thus essential to have measurement points on different continents because local traffic can have significant consequences on the intercontinental connection we were considering. We find the problem mentioned above when making end-to-end or step-by-step measurements. We clearly see from this example that end-to-end measurement gives an overall result of actual QoS for this connection, whereas step-by-step information helps locate the problem, causing service limitations.

7.2.2. *Distribution problems*

Another set of problems in performing measurements on the Internet are related to distributed systems. These problems are even more difficult to solve as the Internet is vast and, more importantly, uncontrolled a single entity. Thus, one of the main issues to be resolved is the possibility of using a unique time reference. Indeed, if the clocks of "probe machines" involved in delay measurement, for example, are out of synchronization, the result will have no value and will not be exploitable.

There are many network clock synchronization protocols, the most well known and used is NTP (Network Time Protocol) [MIL 96], but performance evaluations of this protocol have shown that even if it succeeded in giving fairly satisfactory results for local networks, it is totally inadequate on expanded long-distance networks, such as the Internet. In the absence of clock synchronization protocols on long-distance networks that are better than NTP, research needs to be carried out on large-scale and geographically expanded distributed systems. Today, many engineers and researchers in networks use GPS (global positioning system) which, in addition to giving the current position on the Earth's surface, also carries the pulses of reference atomic clocks.

Another problem encountered in distributed systems – and even more pronounced in the Internet because of its size – is related to the location of measurements that are not necessarily made at the point at which they will be exploited. These measurements should therefore be repatriated to their place of exploitation. This repatriation generates at least two other problems.

The first is related to the amount of traffic that this can generate on the network. For example, even if sometimes, for a delay measurement only a scalar value is transmitted by the measuring probes and therefore does not generate an excess of significant traffic, it is possible that the information to be repatriated is a complete trace of packets that can represent several megabytes, or even gigabytes. In such a case, measurement information signaling is far from being transparent to the network and its traffic.

The second problem is when we wish to exploit the measurements performed in real time. Obviously, the repatriation of data to their place of exploitation takes more or less time depending on the case, but enough time to question the time validity of this measurement. In a high-speed network – which is the case for the Internet today – whose traffic varies a lot and very quickly, it is legitimate to wonder whether the value of the measurement received is still valid.

This is in fact a problem encountered in distributed networks and systems where control or signaling messages in the network use the same transmission medium as the communications data of users. In the case of the Internet, signaling messages and user data are carried in packets that follow the same links and go through the same routers, etc. By making an analogy with road traffic, it would mean for example that in order to determine the time that it would take on a Saturday to go from Paris to Marseille, a vehicle is sent on the same journey on the night before. Obviously, the value obtained on Friday night may not be very different from that on Saturday, but, if the Saturday in question is the first Saturday of August at the start of the summer holidays, it is clear that the value obtained on Friday night will not be connected to the journey the vehicle will take on Saturday. The sudden increase in car traffic between Paris and Marseille on the first Saturday of August, with the creation of bottlenecks (similar to network congestion), makes the measurement performed the night before worthless. In networks, however, there is no parallel medium to convey urgent control or signaling messages, which will make the performances of network management mechanisms or measurement signaling sensitive to data traffic existing on the network. This is also a major problem that must be fixed to be able to implement and deploy measurement systems in the Internet and effectively exploit the results.

7.2.3. *Measurements, estimations and analysis*

This problem leads naturally to the approach concerning the dependence of measurement techniques or their "calibration" on the nature of traffic. Thus, it seems clear that we would not be able to use the same solutions to measure 2 Mbps ADSL traffic and the traffic of an operator at the heart of a network at tens of Gbps. In the same way, the granularity of observation may not be the same in both cases. This granularity will also need to be adapted to other traffic characteristics, for example, if the average size of transmitted traffic flow varies on different links, it will be wise to adapt the granularity of observation to the size of traffic flow carried.

Finally, and this is the main problem of network measurement and metrology, it is almost impossible to work on the multitude of mechanisms and protocols included in the Internet architecture. This general problem discussed in the introduction of the chapter makes it such that measurement alone does not provide information on the network and its behavior. Only seeing the effects of mechanisms and protocols is far from satisfactory. Measurement should therefore be followed by a phase during which metrology methods and techniques, for example based on traffic characterization or analysis and/or QoS measurements, will help explain the causes of the phenomena observed. To do so, we will now explain different techniques that have been designed and developed in recent years to measure the basic physical parameters in networks or to collect traffic traces.

7.3. Measurement techniques

7.3.1. *Active measurements*

The principle of active measurements is to generate traffic in the network to be studied and observe the effects of components and protocols – networks and transport – on the traffic: loss rate, delay, RTT[2], etc. This first approach has the advantage of a user-oriented positioning. Active measurements remain the only way for a user to measure the parameters of the service that he or she may receive. One of the major drawbacks of the network with active measurements is the disturbance introduced by traffic measurement, which can change the network state and, thus, distort the measurement.

Indeed, the measurement result provides information on the state of the network carrying normal data of users, signaling of the network control plan and all the "probe packets". However, we would wish to have information that matches normal traffic only, without "probe packets", which inevitably have an impact on a network's performances. We should therefore be able to estimate the impact of these packets on a network's performances or be sure that they would have a minimal impact, if possible almost zero. It is this last simpler proposition that requires the most research effort. We talk about non-intrusive measurement traffic. Thus, many current works address this problem by trying to find profiles of measurement traffic

---

2 RTT (round trip time): the time taken by a data packet transmitted from a source to a receiver to come back to the source. It is a key performance parameter in a computer network.

that minimize the effects of additional traffic on network state [ALM 99a, ALM 99b, ALM 99c, PAX 98].

Another question raised by such measurements is whether the convergence speed of measurements provide reliably good results. Indeed, in order to asses some parameters, it is sometimes necessary to implement a whole complex process to approximate the solution. For example, to measure the rate available on a path between a source and a destination, some tools transmit "probe packets" rates that increase at each attempt until losses appear. The losses are treated as congestion phenomena. The value of the generated rate is thus the value retuned by the measurement tool as the available rate on the path. However, the process can be lengthy and, where the transported traffic on the path is very variable, the result is unreliable. In some circumstances the tool does not converge to produce a result. Rapid convergence is therefore essential in order to determine the changes in traffic on the path.

Finally, accuracy is crucial. If the measurements have significant confidence intervals, the results are of no interest to researches, engineers or network administrators. For example, in the case of the measurement tool of rate available on a link, accuracy is closely related to convergence speed of the measurement process and to the step increase of the "probe traffic" at each iteration change. In other cases, such as delayed measurement, accuracy can only be related to the quality of time synchronization between the source and destination of the "probe packets".

7.3.2. Passive measurements

Passive measurement projects have appeared much more recently than those of active measurements because they require systems for capturing or analyzing traffic in transit. This is an area that has developed relatively recently – even if some simple software with limited capacity existed previously, e.g. TSTAT, NTOP, LIBCAP, TCPdump, TCPtrace, etc. They nevertheless highlighted that supervision tools, working with a passive approach, were likely to solve many problems with the design, engineering, operation and management of Internet networks.

More powerful hardware is in fact the basis of the current boom in the field of passive metrology. Hardware has paved the way to microscopic passive metrology (defined below). The principle of passive measurements is to watch traffic and study its properties at one or several points of the network. The advantage of passive measurements is that they are not intrusive and do not alter the network state when using dedicated hardware solutions (for example on the basis of *Dag* maps [DAG 01]). They allow very advanced analysis. It is, however, very difficult to determine the service that may be offered to a customer based on information obtained by passive metrology. It is better, in this case, to use active techniques.

Passive metrology systems can also be differentiated according to the method of analysis of traces. Thus, the system can carry out an analysis online or offline. For an online analysis, the whole analysis must be undertaken in the timeframe corresponding to the passage of the packet in the measurement probe. Such a "real time" approach performs analysis over very long periods and produces significant statistics. The maximum complexity of these analyses remains very limited, however, because of the low computation time allowed (getting lower when the network speed increases).

In contrast, offline analysis requires the probe to save a traffic record for future examination. Such an approach requires considerable resources and represents a limitation for traces over a very long duration of time. An offline analysis allows complete and difficult analyses capable of studying non-trivial traffic properties. Moreover, as the traces are saved, it is possible to perform several analyses on them then correlate the results obtained on the trace or different traces for a better understanding of the complex network mechanisms.

The main constraint of installing measurement probes is that the network whose traffic we wish to analyze is almost always an operational network (with the exception of a few experimentation networks in research laboratories), and that, despite the presence of probes, this network should continue to function without any service degradation:

– The primary requirement for the measurement system is total transparency for the network and its traffic. This means it needs to be non-intrusive, the equipment should not cause failures, transmission errors or introduce delays so that the traffic profile and network performances remain unchanged.

– The second requirement in the selection of passive measurement probes is the accuracy and validity of traces it will produce. Thus, it is essential not to "miss" packets transiting on a network, and to have precise information on the transition of these packets, especially in time. The system should therefore be well designed and provide a precise clock that does not deviate.

– The third requirement focuses on the possibility of correlating the events of several traces, for example to track a packet at several points in the network, or to analyze the passage of packets and their acknowledgements, etc. in an overlapped way. In order to finely analyze such events occurring at geographically distant points and at distinct instants but slightly apart in time, it is necessary to have a common and universal time basis for all of the probes.

Finally, there are other requirements that may have secondary importance when designing a measurement tool, but that should not be overlooked. The first concerns the problem of metrology of a complete network and the repatriation of data collected from different metrology points. It is important to find a way to bring these data back to the analysis machines efficiently and without affecting the network and its load. The second concerns the nature of the information to be collected, especially the size of recordings made on each packet. Indeed, collecting all the packet data, with all of the "applicative" information, is *a priori* contrary to the rules laid down by the CNIL (French Commission for Privacy) on the right to privacy. We should therefore carefully consider the data to be collected on each packet in relation to the analysis we wish to do.

7.4. Characteristics of Internet traffic

Let us begin this section by describing the evolution of traffic distribution per application measured in the Internet in recent years. Figure 7.1 illustrates the distribution measured in May 2000 on the Sprint network. The large proportion shown by http traffic (more than 75% of Internet traffic) is remarkable. We also note that the main standard applications are represented: Web, secure Web, email, ftp or news. Newly emerging applications in 2000 are present: traffic flow of multimedia streaming (such as MediaPlayer or RealAudio) and network-distributed games (like Quake). Nevertheless, the most important feature of this traffic remains its elasticity and its QoS time constraints, which are not important (for the vast majority of its applications).

Three months later, in August 2000, the distribution was almost the same with the exception of a new application, Napster, which in the span of a few weeks became one of the major applications of the Internet. Napster was termed a "killer application" because, within three months, it represented between 20 and 30% of the traffic. Over time, this type of P2P (Peer-to-Peer) application has known a very important success among users, thus representing an increasingly important part of Internet traffic. Although Napster had some problems with the American justice system, it has paved the way for a whole set of P2P applications, such as Gnutella, E-donkey, Morpheus and others. Indeed, three years later, P2P traffic was steadily increasing, and on some links of the Renater network, it represents the same proportion as http traffic (see Figure 7.2). Of course, Napster has been replaced by Kazaa and E-donkey. Such an increase of P2P traffic has completely altered the characteristics of global traffic because of the nature the files exchanged (mostly music or films), which are comparatively much longer than Web traffic flow, which was the majority of traffic a few years ago.

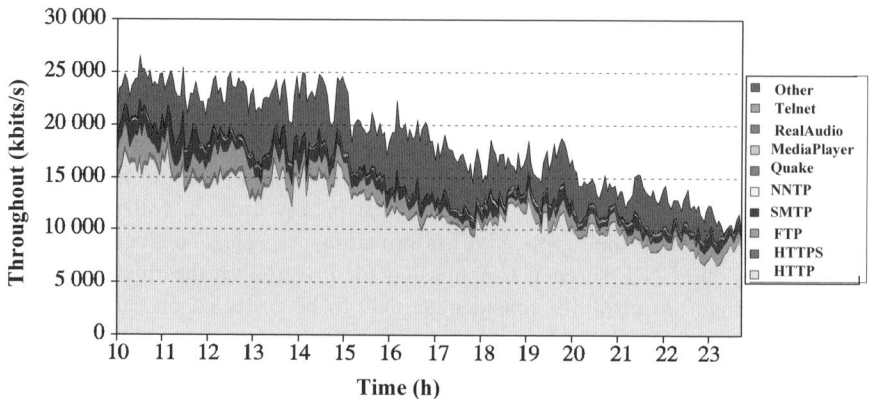

Figure 7.1. *Throughput of main applications (Sprint). Distribution of traffic on the Sprint network (May 2000)*[3]

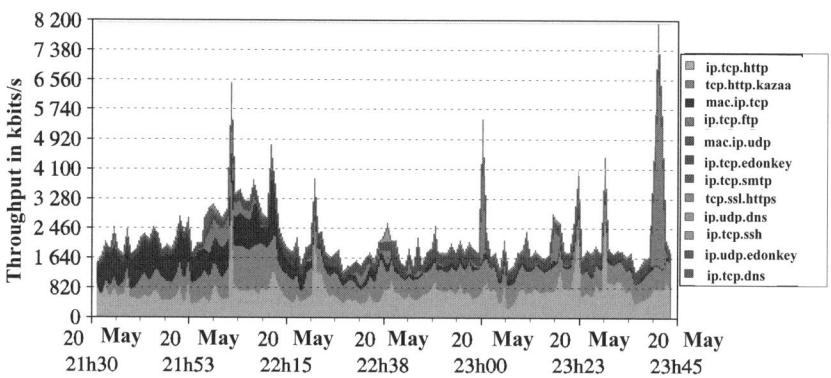

Figure 7.2. *Throughputs of main Internet applications. Traffic distribution on the ReNatER network (May 2003)*[4]

In fact, the increase in P2P traffic coupled with the presence of traditional traffic induces the following characteristics:

– Internet traffic is always composed of thousands of small flows called mice (mainly due to Web traffic as well as P2P control traffic);

– a number of increasing elephant flows.

---

3 The applications are classified in the same order in the legend and on the graph.
4 The applications are classified in reverse order in the legend and on the graph.

This is so much so that distribution of the size of traffic flows in the Internet has changed significantly. This phenomenon has been analyzed since the early 2000s and the results are shown in Figure 7.3. The exponential distribution (the one with the least cumbersome tail) serves as a reference because it is close to the Poisson model[5]. We can see in this figure that the proportion of very long flows has increased dramatically since 2000. If in 2000 this distribution was not very far from an exponential, it is no longer the case at present. On the contrary, it has a heavy tail and is very far from exponential distribution. This is a major achievement of metrology research in recent years; until then, we considered that the Poisson model or slightly more complex Markovian models – those that had been validated for telephone network traffic – applied to Internet traffic. It is clear today that this is not true and that Internet traffic is much more complex and diverse.

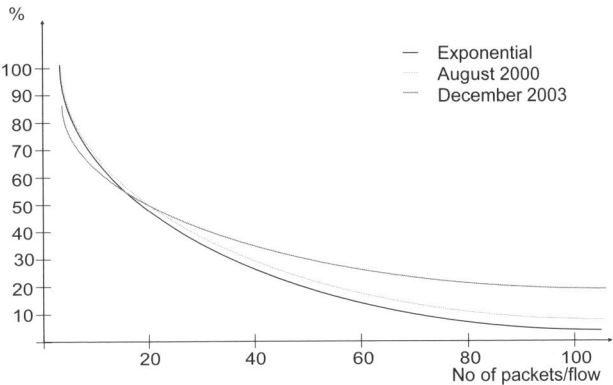

Figure 7.3. *Evolution of the distribution of traffic flow sizes in the Internet between 2000 and 2003*

Returning to this major change in Internet traffic, which consists of an increasing number of long traffic flows, Figure 7.4 illustrates the changes that we can observe. To do so, it compares current Internet traffic with a traffic following the Poisson model. These two streams traffic are observed at different granularities (0.01, 0.1 and 1 second), and it is easy to see that Internet traffic does not smooth out as fast as the Poisson traffic. The analysis shows that this result is totally due to elephants present in the Internet traffic. Indeed, elephant transmission creates the arrival of a great wave of data in the traffic that has to last a relatively longer time (more than

---

5 This model is used as a reference in most cases when it comes to performing simulations or evaluations of network performance. It was considered to be a model correctly representing Internet traffic.

one second[6]). That is why we observe this difference between the two traffic types: elephant transmission induces persistent oscillations in the current traffic.

In addition, the TCP connections used to transmit larger elephant flows last longer and the dependence existing between the packets of the same connection thus spreads over longer time scales. This phenomenon is traditionally called LRD (long range distance). It is attributed to several causes, the main one due to the congestion control mechanisms of TCP (the dominant protocol of the Internet). Among all TCP mechanisms, it is clear that the one based on a closed loop control introduces dependence in the short term, since the acknowledgements result from the arrival of a packet and the emission of all packets following the connection is determined by this acknowledgement.

In the same way, the two TCP mechanisms (slow-start and congestion avoidance) introduce dependence on a longer term between packets of different congestion windows. Thus, by generalizing these observations, it is clear that the TCP packets of a connection are dependent on each other. In addition, the increase in Internet links capacity, by allowing the transmission of increasingly long flows, and increases the LRD phenomenon. This is why we observe the persistence of an oscillatory behavior in Internet traffic in Figure 7.4, even with a significant granularity of observation (1 second).

Since the dependence phenomenon of TCP spreads in the traffic through the flows (i.e. the TCP connections) [VER 00], the increase in flow sizes induces an increase in the extent of dependence, which can reach very significant scales. Thus, an oscillation at time *t* then induces other oscillations at other moments that can potentially be a long way from *t*. It is clear, moreover, that the elephants, because of their significant lifetime in the network and large network capacity (most of the time the links being oversized), have time to reach large values for their congestion control window. Thus, for the traffic flow that experiences it, a loss induces a significant decrease, followed by a significant increase in throughput. The increase in size of flows therefore favors oscillations with high amplitude and a phenomenon of long-term dependency. Of course, the oscillations are very harmful for optimal use of the global resources of the network, given that the capacity released by a flow undergoing loss cannot immediately be used by another (due to the slow-start phase in particular). This means there is a waste of resources and implies a decrease in global QoS of the network. In fact, the more the traffic oscillates, the less significant the performances are [PAR 97].

---

6 Web traffic flows are transmitted in less than one second in the current Internet.

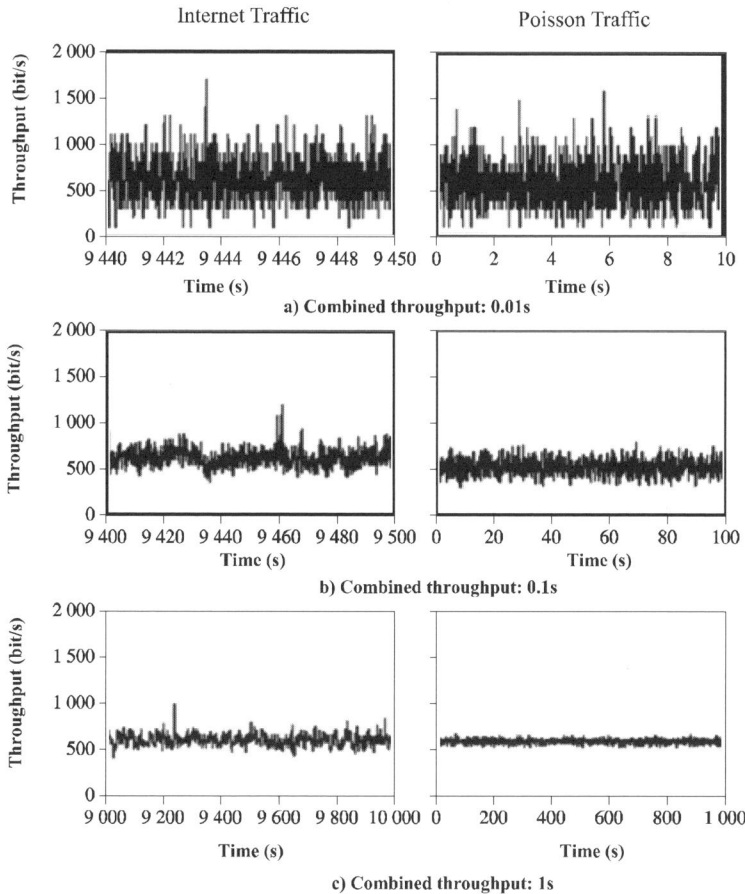

Figure 7.4. *Comparison between the oscillations observed in Internet traffic and Poisson traffic*[7]

7.5. Conclusion

With the advent of the Internet network and the problems posed today by its phenomenal growth, metrology has become the cornerstone of many activities, both in network research and engineering. In particular, the Internet and its mechanisms, change in demands of users, etc., give great complexity to the structure of the IP traffic, one of whose manifestations is for example the phenomenon of long-term

---

7 This study, conducted in the context of *Metropolis*, is based on traffic on an ADSL plate of *France Telecom*.

dependence or self-similarity. Its knowledge needs to be monitored (if not controlled) more or less continuously in terms of its upgradeability. The need for traffic measurements and analyses is justified especially by the following needs:

– to assess the demand of user traffic, particularly in the context of the supply to come from differentiated service classes;

– the size network resources: processing capacity of routers; transmission rate of links; size of "buffers" at interfaces. Adapting the operational management of these resources to the time scalability of traffic demand. This includes in particular all aspects that related to routing in a network;

– control of QoS offered by the network: rate of packet loss, transfer delay and jitter of end-to-end packets for applications with "real-time" constraints, useful transport throughput of data traffic flows;

– testing of the adequacy of performance models developed using analytical calculations or simulations, both in terms of the validation of considered hypotheses and the relevance of results.

Finally, metrology is used for many functions in networks, a non-exhaustive list of that might start with:

– measurement, implementation and management of QoS;

– management and administration of networks;

– pricing and billing;

– congestion and admission control;

– routing;

– security (intrusion detection, network protection through tools such as firewalls, etc.);

– etc.

It follows from this that metrology is not an easy science. It is only used by computer network professionals: engineers, researchers, operators, administrators, etc. Although this science is very young in the field of networks, its capabilities have made its use obligatory for all these professionals and metrology has even started to be taught in tertiary education (universities and schools of engineers). Much remains to be done, however, so that the metrology tool is able to answer all of the questions asked about the Internet and its traffic. One of the major difficulties comes from its multidisciplinary side, requiring technical expertise in the field of computers but also statistics or signal processing for the analysis of captured traffic traces.

Metrology is now also of interest for human sciences, but also the media, advertising, polling organizations, and many more.

7.6. Bibliography

[ALM 99a] ALMES G., KALIDINDI S., ZEKAUSKAS M., *A One-way Delay Metric for IPPM*, RFC 2 679, September 1999.

[ALM 99b] ALMES G., KALIDINDI S., ZEKAUSKAS M., *A One-way Packet Loss Metric for IPPM*, RFC 2 680, September 1999.

[ALM 99c] ALMES G., KALIDINDI S., ZEKAUSKAS M., *A Round-trip Delay Metric for IPPM*, RFC 2 681, September 1999.[DAG 01] Network Research Group. The DAG Project Dag 4 SONET network interface, available at: http://dag.cs.waikato.ac.nz/, accessed on February 5, 2010.

[MIL 96] MILLS D., *Simple Network Time Protocol (SNTP) Version 4 for IPv4, IPv6 and OSI, Request for Comments 2030*, available online at: http://www.ietf.org/rfc/rfc2030.txt, October 1996.

[PAR 97] PARK K., KIM G., CROVELLA M., "On the effect of traffic self-similarity on network performance", *SPIE International Conference on Performance and Control of Network Systems*, Dallas, United States, November 1997.

[PAX 98] PAXSON V., ALMES G., MAHDAVI J., MATHIS M., *Framework for IP Performance Metrics, RFC 2 330*, available online at: http://www.ietf.org/rfc/rfc2330.txt, May 1998.

[PLA 02] PLATON, *L'Allégorie de la Caverne*, La République, Book VII, 4th Century B.C., Flammarion, Paris, 2002.

[VER 00] VERES A., KENESI Z., MOLNAR S., VATTAY G., "On the propagation of long-range dependence in the Internet", *Proceedings of the SIGCOMM'2000 Conference*, Stockholm, September 2000.

Chapter 8

Online Social Networks: A Research Object for Computer Science and Social Sciences

8.1. Introduction

Studies on the uses of new technologies have greatly benefited from input from works on sociability and social networks. With its special features, fixed or mobile telephony has thus been able to enter the directory of ways of meeting people and enriching social works [CAR 05]. The proliferation and diversification of exchanges on the Internet, designed as a relational instrument and not only as an information research tool, calls to extend and expand these opportunities, but also to take into account the specific properties of social relationships on the Internet. Indeed, the network of networks today provides support to large-scale cooperative activities, organized in massively interactive communities such as *Wikipedia*, the groups of freeware developers, blogging platforms, network players or activists of international civil society[1].

The growth of these large groups is accompanied by original forms of regulation where the principles of *self-organization* hold an important place. Cognitive technologies (or intellectual technologies) of the Internet play a decisive role, both in the definition of the formal properties of these communities (geographical dispersion, heterogeneity of forms of commitment, exchange volumes, history of

---

Chapter written by Dominique CARDON and Christophe PRIEUR.
1 To mention but a few. These communities are the subject of ongoing researches within the *Autograph* project: http://overcrowded.anoptique.org/ProjetAutograph (accessed February 5, 2010).

contributions, etc.) and in the cooperation regulation modes that prevail (decentralization, governance by procedures, management of collective learning and group memory, conflicts settlement to consensus, etc.). Also, these cooperative groups on the Internet are relational systems and organized forms. They require the development of an interdisciplinary approach, bringing together sociologists and computer scientists in order to describe their relational and organizational properties. For this, algorithmic graphs offer original tools and open up promising avenues of research. Indeed, if the theory of graphs was used when mainstream social network analysis was set up, it is mainly statistical analysis tools that rapidly imposed themselves. After more than 30 years of research, the algorithms on the graphs have reached a level of sophistication that makes them extremely efficient on very large databases, to which new opportunities arising from the study of the specific properties of large interaction networks are added.

The purpose of this chapter is to explain, through a brief review of existing literature, some of the issues in the analysis of relational forms in large interaction networks. We will focus on two central issues in sociology and algorithmic of graphs, which are the identification of profiles and recognition of communities. Let us first clarify the different types of "social" links that can be retrieved from the World Wide Web and the way in which the application field of the theory of graphs has progressively opened to such data types.

8.2. A massively relational Internet

The network of networks provides a heterogenous set of data that can be extracted and used as material for a relational approach:

– computer scientists have studied the relationships between machines and Internet (servers, routers, IP addresses, etc.);

– retrieval specialists have started work on links between sites [EFE 00];

– linguistics is interested in the semantic proximities between queries or certain descriptive categories of Web pages [HAS 02]; and.

– sociologists seek to extract "social" links between individuals using the Web.

It is on this last category of links that we will focus here, although works in this field are less numerous and successful. It must however be noted that the boundaries between these different categories of links are sometimes far from being obvious as the bodies connected by and on the network of networks are so heterogenous. This is why it is necessary to decompose the different ways of linking people on the Internet.

The Internet first records the interpersonal links of communication tools (email, discussion list and instant messaging). Their "social" character stands out immediately but, on the other hand, to reach and manipulate individual exchange data poses access and confidentiality problems. Studies have, however, been devoted to the analysis of email exchanges, particularly in organizations, when, as in the case of Enron, the data were made public by the American justice system [DIE 05]. Another set of relational materials can be extracted from public Internet sites, particularly those in favor of the development of Web 2.0: discussion lists to public archives, blogs and, more generally, all social media (such as MySpace), and sites of content exchange and tagging (Flickr, del.icio.us, etc.). The dynamism that has developed around these sites is based on the contributions of users who generate content and links between their centers of interest. They thus weave a vast, disorderly landscape of personal expressions, comments and attachments of all kinds. Therefore, it is often difficult to browse and have a "superior" view of these flat and unorganized spaces. Behind their apparent disorder, we assume that it is possible to bring out structured sets according to identifiable social logics.

8.3. Four properties of the social link on the Internet

The links that can be extracted from online groups are primarily characterized by their *public character*. In contrast to interpersonal exchanges through email, chat or instant messaging, they are explicitly displayed on public websites. If they are therefore more easily accessible to researchers, their public character is a structuring property that prohibits us from considering them to be equivalent to personal exchanges. Indeed, in the sociability approaches developed by social sciences, the definition of the social link includes a strong interpersonal dimension, associating two individuals by a set of territorial, situational and intersubjective variables of proximity.

The social relationship is organized around dyadic exchange in which each one addresses the other in the first person (me, I) and in the second person (you). The social links on Internet that we are talking about here, however, have a multi-addressed expression: they are presented as a triangulated communicational exchange always associating a third, the public, as third person (him, he, her, she, them, they) to the relationship between two people. Although the public is forgotten or simply implied, its presence necessarily intervenes in the social relationship that people build on Internet. The exchanges are subject to particular advertising constraints that vary significantly depending on the devices and exchange modes[2].

---

2 See Chapter 5 of this book.

Secondly, the link is often associated with *published content*, so that the relationship between two people goes through an *intermediate informational object*. It is therefore essential to understand the range of contents that serve to link people:

– a post that we comment on a blog;

– an article to which we contribute on *Wikipedia*; or

– a "tagged" photo on Flickr;

– etc.

These variations are the principle of different ways of considering the existence of community forms on the Internet, whereby some sites offer to link people to eachother by the production of common contents without establishing any addressed relationship between them – this model is qualified as "blackboard" by Michel Gensollen [GEN 03]. Other sites are explicitly dedicated to the production of a personalized meeting, as in social networking sites of the MySpace or Friendster type. The link between people thus varies greatly depending on the position taken by the contents in the coordination of exchanges. In this matter, the case of the blog is exemplary. If it seems to definitely be a publishing tool, it is above all a communication tool allowing varied and original modalities to come into contact. We cannot understand the "posts" on a blog without paying equal attention to the comments they generate. Everything happens as if the individuals providing comments have expressed in various forms some traits of their identity in order to put this production at the service of the selection, maintenance and enrichment of their contact list [CAR 06].

Thirdly, social relationships on the Internet often have *a unique signification to the exchange system* incorporated into the meeting system. Indeed, these technologies do not only connect people, they also define *the way* they connect. Boyd shows, for example, how Friendster users are encouraged to multiply and expand their statements to friends as much as possible, since this meeting system among local networks only brings them into contact with the friends of their friends at a distance of four degrees. Also, users tend to considerably widen their circle of "declared friends" to casual acquaintances and those with very indirect relationships.

This extension phenomenon contributes to making the definition of friendship indecisive and in mingling social links that have been built in extremely different relational contexts (school, work, family, leisure, etc.) without distinction, while reducing the trust and reliability of links thus brought together. This is also why users have created a new category to designate relations produced by their electronic practices, the *friendsters*, which they separate from usual *friends* [BOY 03]. Paradoxically, electronic devices asking users to publicly declare their "strong links" thus often contribute to them expanding their circle of "weak links", particularly when these statements are part of a strategic aim:

– to increase opportunities for meetings;

– to increase the chances of finding a job;

– to increase his/her reputation within a network of blogs;

– etc.

Generally, the formation of circles enlarged by weak links is a structuring property of relational forms on the Internet.

Finally, a large number of interpersonal links on the Internet are established around "almost persons", informational contents of a public nature (movies, music, stars) or a private nature (texts, multimedia auto-production) to which people identify themselves before connecting to one another through this intermediary. This characteristic has found an exemplary illustration on Friendster, with the proliferation of *fakesters*, a phenomenon not anticipated by the developers of the site and against which they have unsuccessfully tried to fight. Very early, users started to create fictional characters that they associate with their networks of contacts to enrich the definition of their personal identity.

These *fakesters*, explains Boyd, can be:

– entities of the public culture (God, Homer Simpson, George W Bush, LSD, etc.);

– cultural clusters defined by an ethnic, social or cultural characteristic (Brown University, Burning Man, black lesbians of San Francisco, etc.);

– actual fake real people (for example friends who refuse to participate in the site)[3].

Besides the fact that it reflects the creativity of users and thus allows us to describe by projection some of their personality traits, the production of *fakesters* promotes a large-scale aggregation of individuals who do not know each other. These fictional characters have a much greater relational power than the simple individuals: by registering them in their networks of friends, Friendster users produce massive aggregates that deform the traditional forms of friendly sociability. Friends, tastes, simple acquaintances, music or cinema stars, locations, personal preferences, political opinions or fictional individuals are thus recorded in the same

---

[3] When the hosts of the site began to destroy *fakesters* they judged unnecessary and they thought were distorting the system, users responded reporting a "fakester *genocide*" and concerting their effort on maintaining this diversion of functionality on the site [BOY 04].

area of the relational directory of people, as a sort of catalog of signs that may define what the individual is most attached to. The example of Friendster shows that it is necessary to take this extension of the relational domain to a series of heterogenous beings on the Internet as a research topic in itself, and calls to integrate the sociology of cultural practices to that of sociability [CAR 03, PAS 05].

The study of the social links of the Internet can thus be done according to several scales that complement each other: the fine observation of the properties of these links and motivations of participants who form them helps us to understand the various ways in which these links are arranged. Similarly, this understanding enables us to tackle the analysis of the complex assembly of all these links differently, the analysis for which the technical tools are very sharp, but can lose their relevance without a deep knowledge of the field.

8.4. The network as a mathematical object

Attempting to capture and *measure* the various relational forms that structure the complex assemblies of links that we find on the Internet requires a formalization allowed by graph theory. It is moreover from the application of this theory to anthropology and social psychology that the analysis of social networks was born in the 1950s[4]. However, with the advent of the Internet and the considerable number of interactions that go through it, not content to serve as a tool for the study of networks, graph theory was in turn amended by the analysis of this new field of relations.

8.4.1. *Graphs*

The mathematical object "graph" is defined in an extremely simple way as a *set of binary relations between elements*. More precisely, we call a *graph* an object consisting of a set of elements called *vertices* and a set of pairs of vertices, called *edges* or *arcs*, depending whether the relation is symmetrical or not (in the latter case, we say that the graph is *directed*). When the model is applied to networks, the mathematical terms *vertices*, *arcs* and *edges* coming from geometry are often replaced by *nodes* and *links*, which are more evocative[5]. This mathematical formalism is particularly good for modeling the links of the relational Internet. For example, in the case of blogs, it allows us to model the links between the blogs from relationships among blogs as listed in the blogroll (the list of favorite blogs offered by the blogger to his or her readers) or from the blogs of people who comment on

---

4 For reference books on the analysis of social networks, see [SCO 00, WAS 94].
5 See [BER 67] for more details.

the posts of the blogger. It is therefore possible to draw several types of relational networks from the same communication device [LIN 06].

Since social relationships on the Internet are often built upon informational content (text, topic, photo and video), the graphs may be *bipartite* because the set of vertices can be split into two categories: people and contents[6].

8.4.2. *Complex networks and "small worlds"*

Over the past decade, works on the structure of the graph of the World Wide Web and of other interaction networks from various disciplinary fields (social, computer, biological and semantic[7]) have shown that these networks share properties that make them a very special class of graphs. This justifies the emergence of a new field of study, combining statistical physics and computer science, as well as disciplines in which this type of network appears. We often speak of "small worlds", in reference to the famous experience of Milgram, who showed in 1967 that, within a social network, a *very short distance* could separate any two individuals taken at random[8]. The study of these networks requires us to be able to determine whether the observed properties on a given graph (such as the average shortest distance between two nodes) are "natural" or unexpected. To do this, a common way is to use *random graphs* [ERD 59]. The principle of their use is as follows: we generate a large number of graphs, each with the same number of vertices and edges as the studied graph. If the observed property of the latter is shared by most of the generated graphs, it means that it is probably not characteristic of a particular class of graphs.

Thus, the following properties are observed on networks of blogs, or Web pages in general, as well as on-air transport networks or even networks of lexical co-occurrences in a text[9], and they are not observed in random graphs:

---

[6] On the same model, the analysis of social networks devotes a major part to *affiliation networks*, in which we study the links between individuals and institutions, events or organizations.

[7] See in particular [ALB 02, NEW 03, WAT 99].

[8] A series of letters were distributed to individuals at random with the instruction to pass them to people they know who would do the same, each choosing the next person according to his or her supposed closeness to the final destination, known by his or her name, address and profession. The average length of the chains actually reaching their destination was six [MIL 67].

[9] To cite only a few examples of "complex networks" or "small worlds" discussed in the literature.

126 Digital Cognitive Technologies

– *Low density*. The *density* of a graph is the proportion of existing links compared to the number of possible links. It is easy to see that in a social network, a given individual is linked to only a limited number of people. Significantly increasing the size of the population considered will only add a few links to this individual.

– *Heterogeneous distribution of the number of neighbors*. The *distribution* of the number of neighbors of a graph is the function indicating, for each positive integer d, the number of vertices of the graph having exactly d neighbors[10]. In random graphs, this distribution generally shows an average value around which a great part of the vertices are concentrated, and of which none departs disproportionately. This is a *homogeneous* distribution. In the case of complex networks, it shows instead a considerable gap between the maximum and minimum values and creates an exponential decay[11] (see Figure 8.1). These distributions, similar to those observed for over a century on demographic characteristics such as income or property [PAR 1896, ZIP 49], feed the debate around the concept of *social capital* [BOU 80, BUR 01, COL 88]. Let us emphasize the fact that this phenomenon is observed on large volumes of data and for well-defined relations, for example a network of mail or telephone exchanges over a given period, without prejudging the other types of links that the individuals observed can have in other networks.

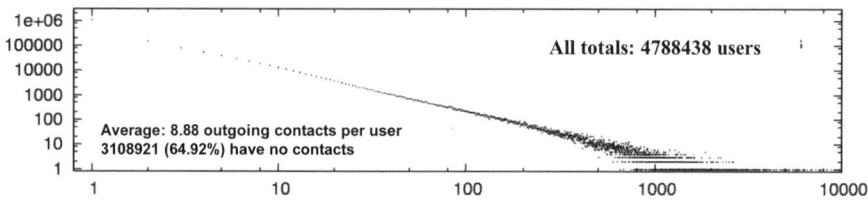

Figure 8.1. *Distribution of contacts on Flickr.com*

– *High clustering coefficient*. The *clustering coefficient* can be defined as the probability that two vertices connected to a third are themselves interconnected. For random large graphs, this amount is generally about zero. More precisely, the larger the graph, the lower this probability is. In contrast, the population size of a complex network has little influence on the clustering coefficient, which remains high[12]: in a

---

10 Two vertices are *neighbors* if there is an arc or an edge between them.
11 On networks of webpages, for example, most pages are not frequently listed, while other very popular websites have a large number of links.
12 Let us be precise here and point out that "high" does not mean 1/3 or 1/4, but rather in the order of 1/100.

social network, for example, whether it is on or off the Web, the links of an individual are located in relation to sociability circles.

It should be noted that as opposed to these properties, which are characteristic of complex networks, the one observed by Milgram is in fact very common to random graphs, and the term "small worlds" would be more justified by the importance of the clustering coefficient[13].

8.4.3. *Switching scale: working on large networks*

Since the advent of the Internet and the sharing of large databases, traditional tools of analysis of social networks have faced a problem, particularly due to the manipulation of graphs in the form of *matrices*. The adjacency matrix of a graph is a table having as many rows and columns as the graph has vertices, and where the element located on row i and column j is zero or one depending whether or not there is an arc between the vertices numbered i and j.

The main advantage of the matrix is that it allows the use of many mathematical tools that are not specific to graphs. However, this mode of representation induces a strong limitation on the amount of manipulable data because the number of values stored in the adjacency matrix of an n-vertices graph is n^2 (n squared so for 100,000 vertices, we will need to store 10 million values), while the low density of social networks allows the assertion that the total number of effective links is far from reaching this amount[14]. Moreover, for the same reason, the calculations can take much longer if they are made on an adjacency matrix.

The objective of the field of *graph algorithms* is to develop techniques in order to sensitively optimize the number of operations carried out (and thus, processing time) during calculations on graphs. Rudimentary in the years following the birth of social network analysis, these techniques are extremely efficient today[15] and allow a thorough understanding of the structure of graphs. This field in turn has been a victim of a change in scale of another order when analyzing huge networks (of millions of nodes), such as that of Webpages. In this case (and even without using an adjacency matrix), simple operations may require significant computing time and it is often necessary to take the special properties of these networks into account. It

---

13 The fact that there is a short chain joining two individuals does not imply that the probability that they will meet someday is high. The practice seems quite the opposite.
14 For example, in a data set consisting of links between a large number of blogs, there are less than 20,000 links for more than 8,000 nodes, while the adjacency matrix has more than 64 billion values. Storing this matrix in computer memory would be like reserving 64 billion empty boxes (as representative of non-existent links).
15 See [KLE 05] for a reference book.

is this aspect that motivates the development of the study of complex networks within the field of graph algorithms.

8.5. Structure of networks and relational patterns

8.5.1. *Clusters*

One of the most frequently studied issues in the analysis of complex networks is the detection of clusters, understood as groups of highly interconnected nodes, giving the keys to a sort of "geography" of a network[16]. The identification of "dense areas" within a wide range of interactions is the first form of interpretation of modes of formation and assembly of individuals. It allows us to point to both the strength and frequency of exchanges between individuals and, under certain conditions, to qualify the relational set by a property they share (a taste, a cultural practice on a blog, etc.) in particular on Web 2.0.

We will now present an overview of the methods developed in order to group the properties sought in these divisions into clusters. We will then discuss the limits of these methods. The main algorithms allowing the identification of clusters can be placed in two opposing approaches from traditional methods of data analysis. The *agglomerative approach* involves aggregating nodes to form increasingly larger clusters, while the *separation approach* cuts the largest cluster, which is the whole network, into increasingly smaller clusters. Most often, in order to know when to stop the processes of aggregation or separation of clusters (to avoid too large or too small clusters), these algorithms use a criterion to assess the quality of the cutting obtained. Widely used, the criterion of *modularity* defined by Newman [NEW 04] is based on the difference between the number of external links and number of internal links to clusters. Both approaches produce hierarchical divisions, in the sense that each cluster is a collection of smaller clusters. What distinguishes the various algorithms is essentially the strategy used, as appropriate, to agglomerate or divide the clusters already calculated:

– Agglomeration by neighborhood similarity, one of the central methods of analysis of social networks [BUR 82, LOR 71, WHI 76], involves placing in the cluster nodes having "almost" the same neighbors into one main cluster. Following this principle, two webpages with links to the same pages are considered equivalent and will therefore be placed in the same cluster.

---

16 We distinguish the detection of clusters from the single problem of drawing graphs, for which there are many methods for placement of vertices and arcs, as well as software for their implementation [BRA 01], or even the visualization of social networks (see for example [FEK 06]).

– Among the most recent techniques based on agglomeration, we can include, for example, the idea of a gravitational model where the nodes exert an attraction depending on their number of links or that of random walks, modeling a hypothetical horde of strollers who would follow all links at random, thus letting themselves be captured with a higher probability in the denser areas of the network [PON 05].

– The separation method proposed by Girvan and Newman [GIR 02], which has been the starting point of a renewed interest in the problem of cluster detection, uses the *centrality* of links (the removal of a link through which a large number of the shortest paths pass is likely to separate the somewhat interrelated parts of the network) as the separation criterion.

The identification of clusters in the network of Webpages greatly occupied computer scientists at the turn of this century, given the importance of search engines [FLA 00]. Today, sociologists are interested in using these techniques to study online communities. One of the most significant contributions is the work conducted by Heer and Boyd on the Friendster community [HEE 05]. From a study of user behavior, the authors have developed, using the agglomerative algorithm of Newman, a tool for the interactive visualization of exchanges among participants which, unlike a directory or search engine, allows people to see their relational neighborhood and the neighborhood of their neighbor on a graph. Research into networks of blogs also use algorithms for the identification of relational clusters [CHI 06]. They have shown, from a single platform of blogs, that Chinese bloggers are engaged in denser and larger community forms than American bloggers [LEN 06].

To finely grasp the overall structure of a network, its boundary lines, its dense areas and their intersections, is one of the major challenges of the study of complex networks, to which cluster detection only partly responds. Current cluster detection techniques only offer cuttings in which a given node belongs to one group, thus denying the very nature of the network, consisting of cross-linking and multiple ownership. Indeed, one of the fundamental research issues of network study today is the possibility of identifying more complex forms by allowing us to take individuals belonging to several overlapping "communities" into account. Similarly, it is often essential to take the symmetry of relations (if blog A has put blog B in its blogroll, has blog B put blog A in its blogroll?) into account to be able to conclude the existence of relational proximity between people on the relational Internet [LEN 06].

The choice of the term "community" to refer to clusters is thus highly problematic and even appears to contradict the concept of a network in social sciences, which was born from exactly the opposite concern. At a time when "community" sites are multiplying on the Internet, we must therefore not be content with such a simplistic reading of automatic clustering. These analysis tools provide a first approximation of collective forms that emerge from the fabric of links of the relational Internet. They allow us to identify relational configurations that are more or less dense and then observe the way users do or do not perceive the existence of these sets of relationships, do or do not give a specific meaning to the collective that gathers them.

8.5.2. *Between communities: the "bridges"*

It is therefore necessary to develop sharper tools to identify more specific forms of organization of networks of links. Thus, from the early anthropological studies focusing on interpersonal relationships, a special place was reserved for mediators, facilitators and brokers[17], real turning points of the network. This theme would then be developed in a quantitative form by Granovetter [GRA 78] and Burt [BUR 92]. Burt made it the heart of his theory of *structural holes*, creating a direct link between these relational configurations and the ability to innovate. The assumption of Burt is that new ideas are generally born on the fringe of constituted groups and that the individuals most likely to bring success are those whose position in the network makes them indispensable and unavoidable, therefore those who are the link between groups of individuals that are not otherwise connected. This mediating role is reflected in the expression of debates of ideas on the Internet, where the structural holes sometimes filled by sites that give themselves the twofold objective of impartiality and information, thus offering links to holders of difficultly reconcilable positions. The localization of these key nodes of the network may be a more useful than identification the "communities", which are often more visible to the Internet user.

The detection of these specific positions can be done through indicators, among which are those developed by Burt to determine how a node is found in the position of a structural hole. Similarly, the *betweenness centrality* measures the extent to which a given node is a privileged passage point between the other nodes of the network. From an algorithmic viewpoint, Burt's indicators have the advantage of being computable on a limited portion of the network (the neighborhood of the node considered), while *betweenness centrality* involves all the nodes of the network. Moreover, to determine the influence of a node in a network, it seems reasonable limit analysis, as in Burt's measures, to the nodes that are not too far away. (In other words, are there noticeable influences between two nodes separated by three or four intermediaries?)

---

17 *Broker* is the term used by Boissevain, who devotes a chapter to it in [BOI 74].

The fact remains that the concept of *centrality* is at the heart of network analysis, to the point where there are several definitions, explained by Freeman [FRE 79], each focusing on a particular aspect (*degree centrality* is in fact the number of neighbors of a node, and *proximity centrality* measures the average distance that separates a node from each of the other nodes). A recent study on the community of bloggers [CHI 06] was based on each of these definitions to measure three elements supposed to be constitutive of the concept of community: belonging, influence and mutual satisfaction of a need. Combined with individual questionnaires, this school-type hypothesis was reinforced by the fact that the questionnaires, on the one hand, and the measures of centrality, on the other hand, identified the same individuals as part of a strong community. While it is still about identifying communities (which are this time not only clusters), the approach calls for another line of study, not only for the nodes with the highest centrality scores for all of the definitions (betweenness, degree, proximity, etc.), but also other patterns such as those that would be good "bridges" while having few contacts.

8.5.3. *Relational patterns*

The search for the "key individual" in a network should not overshadow the other individuals, who are not necessarily central but whose links constitute the backbone of the network. The various ways in which individuals organize their sociability sometimes leaves a recognizable mark on the networks in which they enter. Thus, a network of teenagers' blogs whose authors form a group in "real life" will have a form that totally differs from a "personal diary" blog, whose commentators do not know one another [CAR 06]. A relational point of view using cluster analysis, thus allows us to return our analysis towards individuals. The study of relational forms on the Internet is an alternative way of perceiving individuals faces by identifying the different relational patterns of participants. This field also allows us to combine the sociological and algorithmic approaches of relational pattern identification in complex networks.

The works of the late 1990s on Web structure have led to the identification of relational forms characteristic of various phenomena. They allow us to distinguish different types of pages (pages with multiple links to one another, pages frequently cited, pages citing many pages themselves often cited, etc.). The algorithms that are derived mainly concern the finding of information on the Web, however[18].

Furthermore, the idea of relational forms is similar to that of the *roles* of individuals in a network, which was at the heart of preoccupations of social network analysis in the 1970s. *Structural equivalence*, defined by Lorrain and White [LOR

---

18 For a concise and accessible overview, see [EFE 00].

71], considers two nodes having the same neighbors as equivalent. *Regular equivalence* [SAI 78], generalizing this idea, considers two nodes as equivalent if their neighbors are equivalent. This last concept is motivated by the desire to compare networks whose participants are distinct but some of which have similar positions, or roles. On networks consisting of discussion forums, for example, we can invariably distinguish, by their respective positions in the network of discussions:

– the forum moderator;

– occasional contributors;

– the regulars who intervene in many discussions;

– experts of polemic;

– etc.

It is mainly for methods dividing networks into clusters that these tools have been used. Indeed, structural equivalence does not allow us to compare nodes that would not be found in the same area of a network (for two nodes to be equivalent, they must have the same neighbors). The definition of regular equivalence also raises interpretation problems: to determine whether two nodes A and B are equivalent, we must first determine whether their neighbors are so, but, to do this we must determine whether A and B are equivalent. Algorithms to overcome this circular definition sometimes have a crippling cost in terms of computation time. One possible approach is to arrange the nodes in categories *a priori* fixed, the computing of equivalences being done based on these categories. It is in this knowledge of an *a priori* network that we can determine sociological data from the field study and the particular structure of complex networks.

8.6. Conclusion

The achievement of empirical research involving sociologists and computer scientists in the exploration of significant relational structures is still in its infancy. Indeed, it raises many questions, primarily relating to the ability to retrieve quality data from Websites. The diversity of data formats and site structures make it sometimes difficult to extract information and in most cases information quality deteriorates. On the basis of this, the analysis of social networks on blogs and new applications of Web 2.0 has just begun[19].

---

19 See in particular the scientific community around the *Workshop on the Weblogging Ecosystem*: http://www.blogpulse.com/www2006-workshop/index.html.

It then remains to question the conditions necessary to produce some form of understanding of results obtained by algorithms from graph theory. Depending on their design, these can give significantly different results and suggest representations that sometimes contrast with the organizational and thematic structure of large interaction networks. This variability usefully emphasizes the "constructed" character of algorithmic models, these still enclosing a vision of the network and a particular conception of the ways to create proximity between its nodes. It especially calls to us to develop representation tools that are directly linked to the experiences of users (in the form interactive viewing services, for example) in order to produce a validation or an invalidation of the representations proposed by the individuals themselves. In this sense, the job of sociological interpretation of collective activity on the Internet could benefit from computer scientists working with users to create an original and applied form of hermeneutic loop.

8.7. Bibliography

[ALB 02] ALBERT R., BARABASI A.-L., "Statistical mechanics of complex networks", *Reviews of Modern Physics*, vol. 74, no/ 1, p.. 47, 2002.

[ARA 78] ARABIE P., BOORMAN S., LEVITT P., "Constructing blockmodels: How and why", *J. Math. Psychol.*, vol. 17, pp. 21-63, 1978.

[BER 67] BERGE C., *Théorie des Graphes et ses Applications*, Dunod, Paris, 1967.

[BOI 74] BOISSEVAIN J., *Friends of Friends, Networks, Manipulators and Coalitions*, Basil Blackwell, Oxford, 1974.

[BOU 80] BOURDIEU P., "Le capital social. Notes provisoires", *Proceedings of Research in Social Sciences*, vol. 31, p. 2-3, 1980.

[BOY 03] BOYD D., "Reflections on Friendster, trust and intimacy", Workshop Application for the *Intimate Ubiquitous Computing Workshop, UBICOMP*, 2003.

[BOY 04] BOYD D., "Friendster and publicly articulated social networks", *Conference on Human Factors and Computing Systems (CHI 2004)*, Vienna, 24-29 April 2004.

[BOY 06] BOYD D., HEER J., "Profiles as conversation: Networked identity performance on Friendster", *Proceeding of the Hawaii International Conference on System Sciences (HICSS-39)*, IEEE Computer Society, 4-7 January 2006.

[BRA 01] BRANDES U., RAAB T., WAGNER D., "Exploratory network visualization: Simultaneous display of actor status and connections", *Journal of Social Structure*, vol. 2, p.4, 2001.

[BUR 82] BURT R., *Toward a Structural Theory of Action: Network Models of Social Structure, Perception, and Action*, Academic Press, New York, 1982.

[BUR 92] BURT R., *Structural Holes. The Social Structure of Competition*, Harvard University Press, Cambridge, 1992.

[BUR 01] BURT R., "Structural holes versus network closure as social capital", in N. LIN, K. COOK and R. BURT (eds.), *Social Capital: Theory and Research*, Aldine de Gruyter, New York, 2001.

[CAR 03] CARDON D., GRANJON F., "Eléments pour une approche des pratiques culturelles par les réseaux de sociabilité", in O. DONNAT and P. TOLILA (eds.), *Les public(s) de la Culture*, Presses of Po Sciences, Paris, p. 93-108, 2003.

[CAR 05] CARDON D., SMOREDA Z., BEAUDOUIN V., "Sociabilités et entrelacement des médias", in P. MOATI (ed.), *Nouvelles Technologies et Modes de vie. Aliénation ou Hypermodernité?*, L'Aube, Paris, p. 99-123, 2005.

[CAR 06] CARDON D., DELAUNAY-TETEREL H., "La production de soi comme technique relationnelle : un essai de typologie des blogs par leurs publics", *Réseaux*, no. 138, 2006.

[CHI 06] CHIN A., CHIGNELL M., "A social hypertext model for finding community in blogs", *Hypertext and Hypermedia Conference'06*, Odense, 22-25 August 2006.

[COL 88] COLEMAN J., "Social capital in the creation of human capital", *American Journal of Sociology*, 94, vol. S, 1988.

[DEG 94] DEGENNE A., FORSE M., *Les Réseaux Sociaux*, Armand Colin, Paris, 1994.

[DIE 05] DIESNER J., FRANTZ T., CARLEY K.M., "Communication networks from the Enron email corpus", *Journal of Computational and Mathematical Organization Theory*, vol. 11, 2005.

[EBE 02] EBEL H., MIELSCH L., BORNHOLDT S., "Scale-free topology of email networks", *Physical Review*, vol. E 66, 2002.

[EFE 00] EFE K., RAGHAVAN V., CHU H., BROADWATER A., BOLELLI L., ERTEKIN S., "The shape of the web and its implications for searching the Web", *Proceedings Int. Conf. Advances in Infrastructure for Electronic Business, Science, and Education on the Internet*, Scuola Superiore Guglielmo Reiss Romoli, 2000.

[ERD 59] ERDÖS P., RENYI A., "On Random Graphs I", *Publ. Math. Debrecen*, vol. 6, 1959.

[FEK 06] FEKETE J.-D., HENRY N., "MatrixExplorer: Un système pour l'analyse exploratoire de réseaux sociaux", *Proceedings of IHM2006, International Conference Proceedings Series*, Montréal, Canada, septembre 2006.

[FLA 00] FLAKE G., LAWRENCE S., LEE GILES C., "Efficient Identification of Web Communities", *ACM SIGKDD International Conference on Knowledge Discovery and Data Mining*, Boston, United States, 2000.

[FRE 79] FREEMAN L., "Centrality in social networks: conceptual clarification", *Social Networks*, no. 1, 1978/79.

[GEN 03] GENSOLLEN M., "Biens informationnels et communautés médiatées", *Revue d'économie politique*, no. 113, p.9-40, 2003.

[GIR 02] GIRVAN M., NEWMAN M., "Community structure in social and biological networks", *Proc. Natl. Acad. Sci. USA 99*, 2002.

[GRA 78] GRANOVETTER M., "The strength of weak tie", *American Journal of Sociology*, vol. 78, no. 6, 1978.

[HAS 02] HASSADI H., BEAUDOUIN V., "Comment utilise-t-on les moteurs de recherche sur Internet?", *Réseaux*, vol. 20, no. 116, 2002.

[HEE 05] HEER J., BOYD D., "Vizster: visualizing online social networks", *IEEE Symposium on Information Visualization (InfoVis)*, 2005.

[KLE 05] KLEINBERG J., TARDOS E., *Algorithm Design*, Addison Wesley, Reading, 2005.

[LEN 06] LENTO T., WELSER H., GU L., SMITH M., "The ties that blog: Examining the relationship between social ties and continued participation in the Wallop weblogging system", 3rd *Annual Workshop on the Weblogging Ecosystem: Aggregation, Analysis and Dynamics, WWW06*, Edinburgh, Scotland, 2006.

[LIN 06] LIN Y.-R., SUNDARAM H., CHI Y., TATEMURA J., TSENG B., " Discovery of blog communities based on mutual awareness", 3rd *Annual Workshop on the Weblogging Ecosystem: Aggregation, Analysis and Dynamics, WWW06*, Edinburgh, Scotland, 2006.

[LOR 71] LORRAIN F., WHITE H., "Structural equivalence of individuals in social networks", *Journal of Mathematical Sociology*, vol. 1, 1971.

[MIL 67] MILGRAM S., "The small world problem", *Psychology Today*, vol. 1, May 1967.

[NEW 03] NEWMAN M., "The structure and function of complex networks", *SIAM Review*, 45, vol. 167, 2003.

[NEW 04] NEWMAN M., "Fast algorithm for detecting community structure in networks", *Physical Review E*, vol. 69, no. 6, 2004.

[PAR 1896] PARETO V., *Cours d'Economie Politique*, Droz, Geneva, 1896.

[PAS 05] PASQUIER D., "La culture comme activité sociale", in E. MAIGRET and E. MACE (eds.), *Penser les Médiacultures. Nouvelles Pratiques et Nouvelles Approches de la Représentation du Monde*, Armand Colin, Paris, p. 103-120., 2005.

[PON 05] PONS P., LATAPY M., "Computing communities in large networks using random walks", 20th *International Symposium on Computer and Information Sciences, Lecture Notes in Computer Science*, vol. 3, p. 733, 2005.

[SAI 78] SAILER L., "Structural equivalence: meaning and definition, computation and application", *Social Networks*, vol. 1, 1978.

[SCO 00] SCOTT J., *Social Network Analysis, A Handbook*, Sage, London, 2000.

[WAS 94] WASSERMAN S., FAUST K., *Social Network Analysis: Methods and Applications*, Cambridge University Press, Cambridge, 1994.

[WAT 99] WATTS D., "Networks, dynamics, and the small-world phenomenon", *American Journal of Sociology*, vol. 105, 1999.

[WHI 76] WHITE H., BOORMAN S., BREIGER R., "Social structure for multiple networks I. Blockmodels of roles and positions", *American Journal of Sociology*, vol. 81, 1976.

[ZIP 49] ZIPF G.K., *Human Behaviour and the Principle of Least Effort. An Introduction to Human Ecology*, Addison-Wesley, Cambridge, United States, 1949.

Chapter 9

Analysis of Heterogenous Networks: the *ReseauLu* Project

9.1. Introduction

The use of new computerized methods of information processing has been in response to twin needs in the last 10 years. On one hand, we can take advantage of the constant evolution of computer tools and the simultaneous increase in type and quantity of data available for research in the social sciences, in technological intelligence, in strategic innovation management and in communication management: the number of databases on the Internet keeps growing. On the other hand, during the same period, natural and biomedical sciences have undergone a profound transformation in their working patterns, going from a configuration dominated by small laboratories headed by a scientist with a limited number of collaborators to configurations where large research structures (consortia, networks, etc.) play an increasingly central role [AMI 04, BOU 05, CAM 04, GAU 04, KEA 03, MOG 05, SAL np, VIN 92]. This is also true for research organizations and institutions. Vague or even frankly polysemous concepts such as *network economy* are now part of the official discourse of bodies like the European Union [BAR 01].

These two developments are obviously not independent: thus, the fairly recent beginnings (1989) of the World Wide Web are related to the efforts of researchers from CERN (the European Center for Nuclear Physics), who found a way to counter

---

Chapter written by Alberto CAMBROSIO, Pascal COTTEREAU, Stefan POPOWYCZ, Andrei MOGOUTOV and Tania VICHNEVSKAIA.

the problems arising from the widely dispersed nature of huge collaborative networks that characterize research in this field[1]. Experts interested in the socio-technical dynamics of biomedicine, for example, are no longer content with traditional qualitative analysis methods, such as interviews or ethnographical observations: they must now complement these methods with large number statistics, permitting an understanding of very diverse data without reduction to a few statistical indicators [CAL 01]. These data come mainly from databases of scientific information (publications, patents), Internet sites, news flows, blogs, discussion lists and forums, text corpora, traditional media (TV and radio), enquiries and interviews. Although the general trend is towards the standardization and structuring of these different sources (concentration in databases, use of XML, etc.), which should allow them to be integrated in analytical databases, the development of tools to read and transform these sources remains necessary.

Statistical analysis software offers the possibility of managing data coming from national and international statistical collections using well-defined analytical protocols. Whereas these "standard" data allow the definition, based on a set of indicators, of stable technical-economic configurations, they are of little or no use for the analysis of emerging networks [CAL 02]. Traditional statistical analysis software is insufficient and inappropriate when faced with the heterogeneity of much of the available data. The need thus arises to develop tools that take the following aspects into account:

– heterogeneity of sources: databases, information collected through questionnaires with closed and open questions, interviews, texts and text corpora;

– heterogeneity of structures and levels of structures: relational databases, raw qualitative and coded data, quantitative data;

– heterogeneity of contents: personal data, geographical location, affiliation of people and other entities, different types of links between entities, semantic or lexical data;

– heterogeneity of scale – an individual, an institution, a field, etc.;

– ontological heterogeneity: human and nonhuman entities (objects, substances, etc.).

Taking these different forms of heterogeneity into account forces us to reconsider both the techniques used to analyze networks and the status of the results thus obtained. Indeed, the analysis of homogenous data, for example in fields such as the analysis of social networks or co-citations, is made possible by the use of reliable methods (specifically vector methods) that allow calculation of the distance

---

1 http://www.zeltser.com/web-history/#Origins_WWW, consulted in February 2006, accessed February 5, 2010.

between different elements [WID 03]. Thus, human participants are more or less closely linked depending on the number of people with whom they maintain common friendships, or scientific documents are more or less connected by their citation profile. By their nature, heterogenous data defy this type of metric and thus require the use of an algorithmic method of data analysis whose validity reflects more that of a demonstration than of a mathematical proof [ROS 03]. The simultaneous mapping of objects of different types keeps the advantages offered by methods such as social network analysis – in particular the use of the formalism of analysis and interpretation of network structures, such as centrality, opposition, distinction between groups, the intermediary position of some elements – while not requiring the formulation of *a priori* hypotheses on the nature and role of the different elements.

9.2. The *ReseauLu*[2] project

The methodology proposed by the *ReseauLu* project is based on a well-known approach in computer science, called "object oriented" programming. The "objects" are essentially sets of analytical entities defined by potentially different specificities (people characterized by gender and age; computers by operating system; and molecules by molecular mass, trademarks, therapeutic effect, price, etc.). For a set of entities, a set of links exists that may be direct (a computer belongs to an individual), statistical (two molecules are often prescribed for the same disease), oriented or not, and that may possess characteristics such as frequency, co-occurrence or intensity.

The analytical plan of *ReseauLu* is built from several complementary dimensions of investigation – relational, temporal, textual and statistical – presented in a single interface (see Figure 9.1) with separate transformation and analysis modules:

– *The relational dimension* corresponds to the relationships between analytical entities, whatever the nature of these relationships. It includes the analysis of organizational structure, market structure, actual or perceived relationships in a social network, and the structure of "communities of practice" [WEN 98].

– *The temporal dimension* is present in biographical interviews, family histories, analysis of careers, history of institutions, and in general in databases relating to the timing of events.

– *The textual dimension* refers to the processing of language, the lexical and semantic content of interviews, of speeches, text corpora and open questions of questionnaires. It is also a helpful tool for content analysis.

---

2 *ReseauLu* software, developed by Mogoutov, is marketed by the SARL (Société à Responsabilité Limitée) *Aguidel*, www.aguidel.com.

140   Digital Cognitive Technologies

– *The statistical dimension* in *ReseauLu* introduces a different concept of relationships, which becomes an expression of the relative importance of different forms of data, as well as the expression of the strength of the association.

– *ReseauLu* data processing uses relational database technology. In addition to the processing and generic analysis modules, *ReseauLu* contains two domain-specific modules:

– *Scientometric module*, currently based on the bibliographical databases *Web of Science*, *CAB Abstracts* and *PubMed*, as well as any type of bibliographical data managed by the software *EndNote* and *proCite*;

– *Media analysis module*, based on *Google News*, *Google Blog Search*, *Factiva*, *Yahoo Websearch* etc.

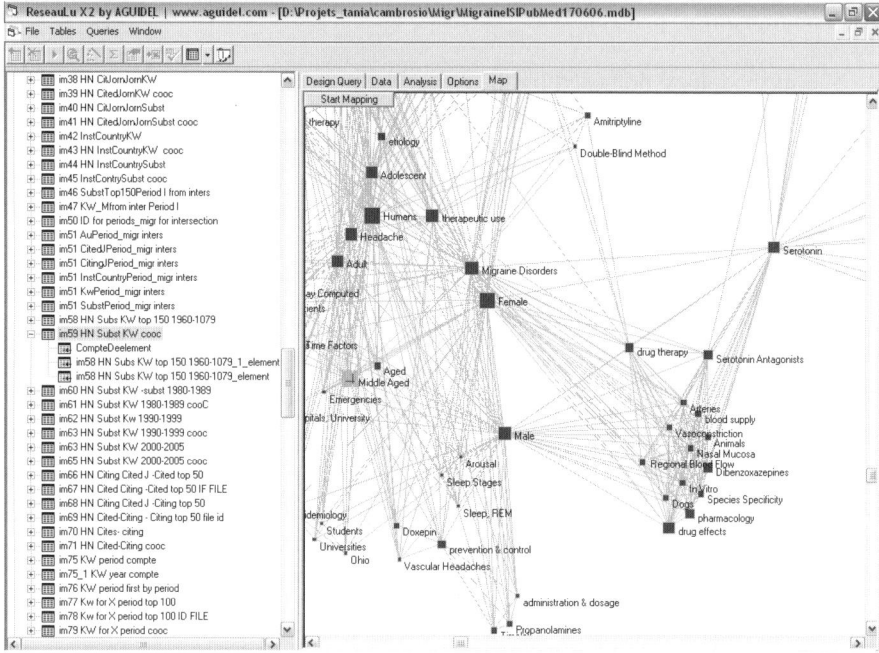

Figure 9.1. *The interfaces of ReseauLu: data management, analytical module, graphical map editors, tools for data import and a set of predefined analytical scripts*

The purpose of the process, in most cases, is to produce a set of maps showing the structure of the studied object taking into account temporal, textual and relational dimensions as well as the scale at which the object is being viewed. Each map is a directed graph or digraph. In this chapter, we will limit our examples to the operations of the scientometric module[3] since this module illustrates the different dimensions of *ReseauLu* in a common thematic framework.

9.2.1. *Scientometric analysis*

The scientometric module of *ResearchLu* helps the researcher in the analysis of the structure and dynamics of a laboratory or a research institute through:

– its publications,

– its positioning in an international context, and

– strategic analysis of its research fields, technological intelligence, research strategy and innovation management[4].

The starting point is the gathering or definition of data sources. There are several possibilities: generic bibliographic catalogs, domain-specific bibliographical catalogs and bibliographic databases of researchers and institutes.

The next step is conceptual, consisting of defining the boundaries of the subject under study, the body of publication of its researchers, its disciplinary or interdisciplinary nature, etc. In this regard, it is often appropriate to combine several sources, allowing a more complete list of references and their content as well as a more robust confirmation of the domain boundaries. Thus, for example, the *Web of Science* contains valuable information on the citations in each article and contact details of the authors, while *PubMed* provides a very sophisticated system of keywords (the hierarchical thesaurus MeSH, for Medical Subject Headings) and a detailed list of the chemical and pharmaceutical substances referred to in the article.

The methodology of scientometric analysis of *ReseauLu* treats each selection of scientific articles as a set of heterogenous data composed of names of authors, institutions, geographical locations and keywords (MeSH, as well as words found in titles and abstracts). By exposing the most relevant co-occurrences of words and names, defined by absolute and relative frequency as well as specificity, we try to reveal the elements of self-organization in scientific and technical fields. The method, known as "co-word analysis" [CAL 86], is based on the hypothesis that when two words appear simultaneously in a set of articles, the subjects they represent are associated and reflect the themes of research defining a domain at a

---

3 For the methodology of relational, biographical and textual analyses, see [MOG 06]; for an example of media analysis, see [DEV 06].
4 We can also use non-bibliographic databases containing, for example, lists of substances and the researchers who produced them, see [CAM 04].

point in time. By adopting this method we can highlight the main areas of interest to researchers and the evolution of these methods over time.

The analysis of associations between human or non-human entities raises fundamental questions about the morphology and dynamics of a scientific field. Should the field allow itself to be limited by the social network of its researchers or, in seeking to understand the activities of laboratories and the evolution of research topics, should we not also make room in the analysis for the techniques used, the substances mobilized and pathologies studied? In contrast to the analysis of social networks, which raised the first hypothesis to the status of an axiom by considering social networks as the sole arena in which an action occurs, the analysis of heterogenous data does not assign any ontological priority to the "social" elements [LAT 05]. In a recent study [BOU 06], the analysis of the dynamics of heterogenous networks of authors and of topics over time has helped illuminate the global dynamics of the field.

Another benefit of the *ReseauLu* methodology is its robustness with respect to missing data [KOS 06]. For example, in a recent study by Cambrosio *et al.* [CAM 06], analyzing the inter-citations between journals specializing in cancer, two types of object were used – homogenous in nature, but heterogenous in status – namely the citing journals and the cited journals. The traditional analysis of inter-citations produces different maps for each of these; in each case, the networks are constructed from an analysis of the similarity profile of citations made or received by each journal. In this type of calculation, modifying the list of journals included in the analysis may have major consequences on the network structure, and it thus becomes difficult to produce maps of an overall domain.

Our approach bypasses this problem; because it brings together on one map both cited and citing journals where the list of cancer journals that we used as a starting point has been expanded and developed by the analysis itself. The robustness of our analysis was moreover confirmed when we included not only journals specific to cancer, but also multidisciplinary journals and those of other disciplines that publish articles on cancer. Even if these new maps revealed journals that were not on the original maps, the relational structure of the global network – and in particular the relationships between journals pertaining to clinical, fundamental and "translational" research (which aims to facilitate exchanges between clinical and laboratory research) – remained unchanged. This relational structure was also confirmed by a heterogenous analysis exploring links between journals and the semantic network characterizing their content[5].

In total, this mapping approach has allowed us to observe the development of the domain thanks to the evolution of inter-citations between publications and transformation of the semantic network generated by them. It has thus enabled us to visualize complex configurations without reducing the information from statistical parameters.

Apart from purely academic use, these approaches have applications in the evaluation of research:

– Analysis of the structure and dynamics of a laboratory or research institute through its publications. This type of analysis normally uses the Web of Science and PubMed databases. After exporting relevant data in text format, ReseauLu extracts data by field and inserts it into predefined tables. To be able to combine several databases, it is necessary to create complex sorting criteria and to standardize the analyzed fields (names of authors, titles of articles, names of journals). From this data, the modules of ReseauLu can obtain results for each database and across databases. Some of these results are simple tables, such as lists of authors classified by number of citations or publications. Other more complex products are presented in different forms, notably: by maps of homogenous networks of collaboration among authors, institutions and countries; or maps of links between topics; or again by analysis of productivity across research groups, laboratories, departments and projects, while all along taking evolution over time into account.

– *Positioning of institutes or research laboratories in an international context.* The data preparation and methodology of analysis are the same, but the process is strengthened by a comparative dimension that situates the output of a laboratory in a national, European or global context, in principle allowing the evaluation of potential partnerships. The results are again in the form of comparative tables of productivity and various other indicators, as well as in the form of maps (collaboration networks, mapping of heterogenous networks).

– *Strategic analysis of research fields.* This type of analysis, compared with the previous two, involves deeper exploration of the semantic (textual) component of data, textual analysis of titles, summaries and keywords. These elements reveal emerging domains, specific domains and weak signals. The results obtained allow us to discern the characteristics of emerging fields at a local, regional, national, European or global level. They help to guide the research of specialists.

Finally, before presenting an example of analysis, here are some technical details on the construction of maps and their interpretation:

– *Mapping of relational data.* Specific algorithms have been implemented to allow the analysis and representation of data structure and the schematization and mapping of analytical entities. These algorithms define the modalities of the categorical variables and different fields of the database from which these variables

---

5. In this study, the terms were extracted from titles and abstracts of articles by using a text-mining software program (*SPPS LexiQuest Mine*).

are extracted. From a matrix of links between different elements being explored, the cartographic module seeks to graphically place elements according to algorithms combining several approaches[6]. The final representation takes into account criteria aiming to:

- represent all the links between elements while minimizing the number of crossovers between these links;

- place the elements as symmetrically and clearly as possible while minimizing the overlap of points;

- place fully connected elements in a space of uniform density;

- cluster strongly connected elements; and

- group together elements that share a similar structural position.

The visualization can be preceded by a selection of the most specific links using the statistical module[7];

– *Principles of reading and interpretation.* The first interpretation technique uses the detection of groups defined by the presence of links of various tightness between their nodes. It is supplemented by the recognition of boundaries between different types of nodes and of symmetrical relations between and within groups. The second technique involves a structural reading; it looks for key elements that occupy a central position in the global structure or within a group, or those that have an intermediary position between different groups.

9.2.2. *Example of the scientometric analysis of a domain of biomedical research concerning migraine*

The last part of this chapter is devoted to a short example from an ongoing study on migraine. The choice of this field is justified, among other things, by our interest

---

[6] Using on the one hand physical simulation of the displacement of geometrical objects subject to random forces and, on the other hand, non-linear planar projection of a multidimensional structure, scaling of the distance between objects, calculation of groups of objects and of the order in which they appear in the structure.

[7] From a 2D table formed between two variables, the software builds a matrix of weighted values that is transformed into a matrix of expected values followed by the construction of a matrix of associations. The matrix of expected values corresponds to the zero hypothesis of statistical independence between the columns and rows of the table. Cell values are defined as a product of marginal values (multiplication of totals corresponding to the rows and columns divided by the general total). Cell values of the association matrix are calculated as the difference between the observed values (Vo) and the expected values (Ve) according to the formula $(Vo - Ve)/\sqrt{Ve}$.

in the problem of biomedical classification [BOW 99, KEA 00]. Indeed, research on migraine was revolutionized in the second half of the 1980s by the production of an international classification of headaches. This process, notable initially for the controversy between researchers in North America and Europe, has given a new impetus to the field, in the context of clinical trials, by enabling comparison of the action of various substances on the same types or subtypes of migraines [POP 04]. It is, therefore, a relatively well-defined field, fairly homogenous in certain ways, but cut across in others by geographical (Europe and North America) and time divisions (before and after the classification).

The *PubMed* (Medline) database previously mentioned has the advantage of a very sophisticated system of keywords, ordered hierarchically, that clearly identify the different fields from a set of terms. However, the entries only give the address of the first author and do not give the list of references cited in each article.

The *Web of Science* (Science Citation Index), which is not limited to the biomedical field, gives the full list of the authors and – its defining feature – the list of references cited in each article. Unfortunately, its system of keywords is very weak compared to that of *PubMed*. Thus, in the present case, a request to establish a database of articles on the field of migraines and headaches during the period 1960-2005 resulted, in the case of *PubMed,* in 32,494 references and, in the case of *Web of Science*, in 23,538 references. Moreover, the distribution over time of the latter is very uneven, as shown by Figure 9.2. We notice a significant jump in the number of articles from 1991, caused by a change in indexing criteria[8]. Thus, the overlap between the two databases is not optimal.

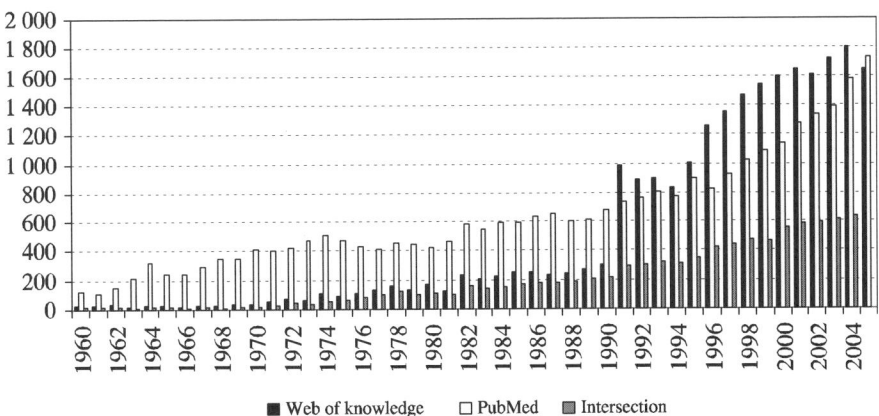

Figure 9.2. *Changes in the number of publications in the field of migraines in the Web of Science and PubMed and their intersection*

We can make the hypothesis that *Web of Science* underestimates the number of articles in the field before 1991 and overestimates it after 1991. For some types of analysis, such as co-authorship networks, the use of *PubMed* data would probably be more appropriate given the higher degree of specificity and sensitivity of the data, but more advanced analysis requires the type of data available only in *Web of Science*. For this example, we have therefore analyzed references from both databases to ensure good robustness of results (high specificity/low sensitivity). A compound key was built from the name of the author, ISBN, year of publication, volume number and page number to overlap the references. The sample thus selected contains 9,590 articles.

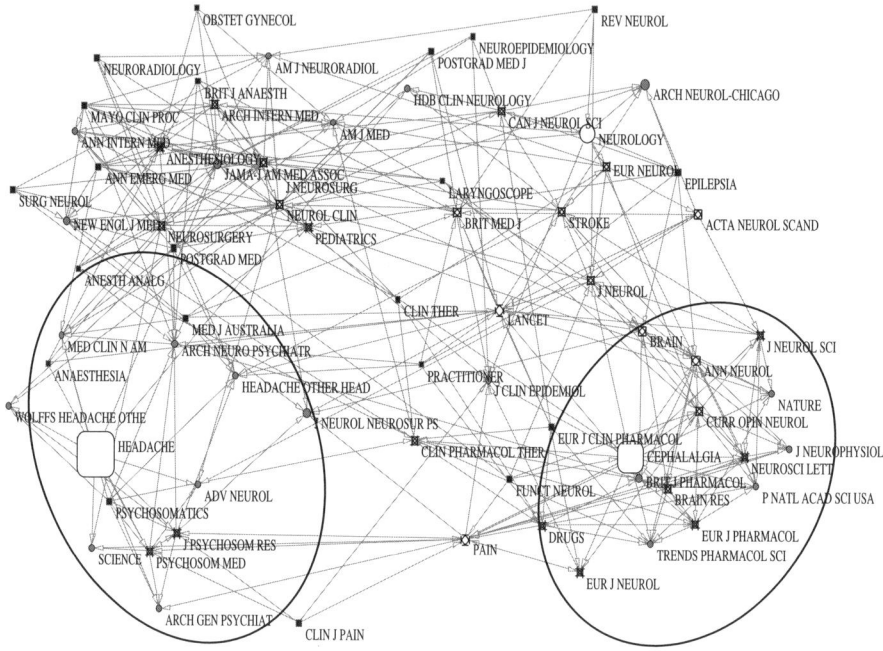

Figure 9.3. *Network of inter-citations between scientific journals as part of the study on migraine (50 most frequent journals among the citing journals and 50 most frequent journals among the cited journals; the network shows 20% of the most specific links)*

To illustrate this, we refer to maps generated from our bibliographical sample. One of them shows the inter-citation network between journals that have published articles in the field based on information from the *Web of Science*. It is a

---

8 This change consisted of taking the terms appearing in the title or abstracts, and also in the titles of articles cited in the list of references of each article into consideration.

homogenous network (all nodes refer to journals), but also heterogenous since it gathers journals that are only cited, journals that cite without being cited, and journals that cite and are cited. The size of the nodes is proportional to the number of inter-citations per journal. The map makes it possible to obtain a global view of the structure of a field as defined by the different publications. On the resulting graph (Figure 9.3), we notice a polarization between the two main journals in the field, *Headache* and *Cephalalgia*, as well as their respective areas of influence. This polarization is all the more interesting knowing that *Cephalalgia* is produced by the European society that prepared the new international classification of headaches, while *Headache* is a journal based in the United States.

By producing maps according to time periods and supplementing them with maps of authors and thematic maps describing the content of journals, we can follow the emergence and evolution of this polarization between areas of geographical influence in greater detail despite the internationalization of scientific practices.

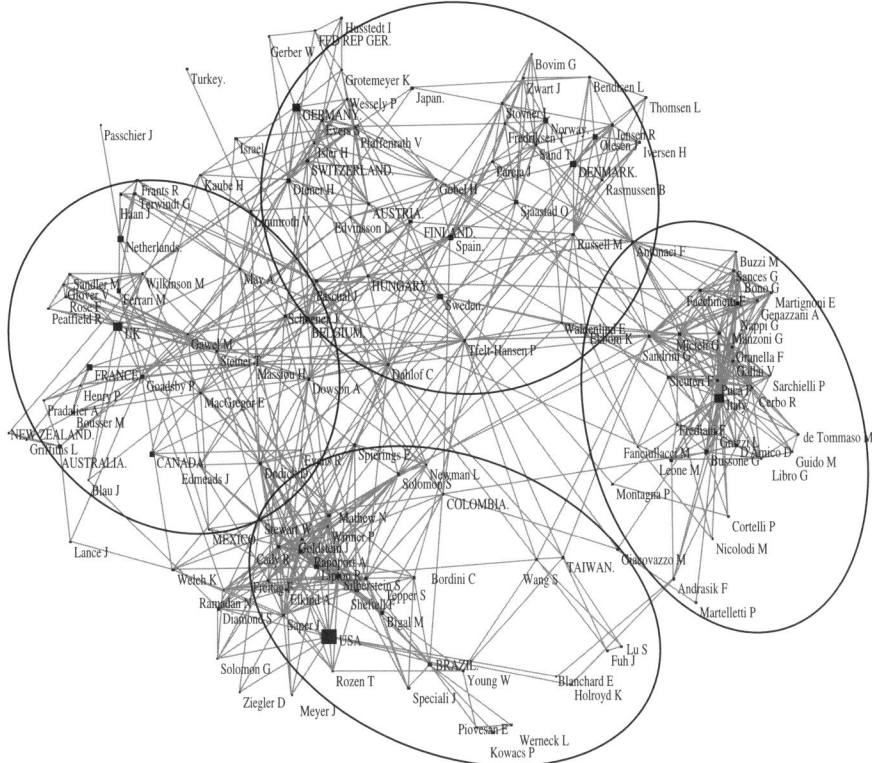

Figure 9.4. *Heterogenous network of the 150 most productive scientists and countries of their professional residence (50% of the most specific links)*

Another form of mapping enables us to examine the geographical dynamics of the field of headaches more precisely by uniting the names of the 150 most productive authors in the field and the countries where they work on the same map. It is thus a heterogenous map (Figure 9.4) that shows the links between authors and between authors and countries, thereby positioning the collaboration networks in a geographical space. This map shows the presence of Anglo-Saxon, Latin-American, Asian and European collaborative networks and, in particular, the Italian research network that has played a very important role.

Figure 9.5. *Collaboration network of researchers in 2000-2005 for the 150 most productive researchers in the field of research on migraines (100% of the links)*

Other series of maps also illustrate the development of the collaborative network of researchers over the 45-year period between 1960 and 2005. We are aware of the existence of different forms of collaboration among scientists [KAT 97], but the term "collaboration" here refers specifically to the co-signing of articles. These maps are thus homogenous. By limiting the demonstration to the last two periods (1990-1999 and 2000-2005), it appears that the first maps of the period show the presence of small groups of researchers, even isolated researchers, who progressively develop links that grow into a relatively well integrated network. Then a very dense group of Italian researchers develops that, through a limited number of individuals (for example Micieli and Nappi) playing the role of a bridge, are linked to other groups. These individuals are linked to the group containing the most productive researcher in the field – Danish researcher Olesen, who is one of the key players in the new classification of migraines. We still distinguish some geographical polarization (Figure 9.5), but the network is now characterized by the presence of a dense core of researchers.

ReseauLu also makes it possible to view heterogenous maps on this topic, showing research topics and the authors who are most specifically associated with them over the period 2000-2005. Information on the content of research can be produced in two ways: either by using the co-occurrence of MeSH keywords available in the *PubMed* database, or by using the co-occurrence of words extracted from titles and abstracts of articles. This second method requires the use of software for natural language text analysis, and hence additional processing that may be quite heavy and complicated. It does, however, have the advantage of providing access to concepts used by the authors of the articles. The first method is much simpler, because the keywords are already included in our database. It nevertheless has a major disadvantage in the sense that the keywords are assigned to articles by professional indexers who use a controlled and uniform vocabulary. Such a procedure tends to blur the lexical specificity of each article. By using, for example, the MeSH keywords, we can distinguish different research topics arising from epidemiological and therapeutic research, along with the names of authors who have contributed to these fields. The headache classification serves as a bridge between these two sets as well as other topics, such as genetics and physiology. Other studies we have conducted based on natural language analysis of articles have allowed us to refine this type of thematic analysis.[9]

Thus, by using a particular field of the *PubMed* database containing the name of chemical substances mentioned in an article[10], we were able to identify the

---

9 This is why we have used this approach in [BOU 06], where we used the software *TextAnalyst* of *Megaputer*, and in [CAM 06], in which we used a more powerful text analyzer, *LexiQuest Mine* of *SPSS*.
10 As mentioned above, in migraine research the development of an international classification of headaches stimulated clinical trials focusing particularly on a family of compounds known as triptans, of which sumatriptan (Imitrex) was the first to be launched in the US market in the early 1990s.

evolution of pharmacological treatment of headaches, showing which substances are most closely associated with each period.

Other variations on this theme are possible, notably maps simultaneously showing links between substances and periods, as well as the co-occurrence of substances within the same article (combined treatments), and, in another example, maps showing associations between substances and periods in a subset of the database containing only articles arising from clinical trials.

9.3. Conclusion

Our objective in this chapter was to present the main features of the analysis of heterogenous networks, concentrating on scientometric-type applications. Compared to more traditional analyses of social networks, heterogenous networks present a dual challenge: methodological and theoretical. At the methodological level, it is a matter of developing tools that take account, in the same space, of concepts that, due to their ontological dissimilarity, do not respond to a common metric. The *ReseauLu* software addresses this challenge by using a set of algorithms resulting in a cartographic representation allowing visual inspection of the relations between different objects. At the theoretical or conceptual level, these techniques permit the lifting of taboos that, in the social sciences, limited our analysis to "social" entities, allowing us to pay proper attention to the tools and objects without which any interaction would be impossible.

9.4. Bibliography

[AMI 04] AMIN A., COHENDET P., *Architectures of Knowledge. Firm, Capabilities, and Communities*, Oxford University Press, Oxford, 2004.

[BAR 01] BARRY A., *Political Machines. Governing a Technological Society*, Athlone Press, London, 2001.

[BOU 05] BOURRET P., "BRCA patients and clinical collectives: New configurations of action in cancer genetics practices", *Social Studies of Science*, vol. 35, p. 41-68, 2005.

[BOU 06] BOURRET P., MOGOUTOV A., JULIAN-REYNIER C., CAMBROSIO A., "A new clinical collective for French cancer genetics A heterogeneous mapping analysis", *Science, Technology and Human Values*, vol. 31, p. 431-464, 2006.

[BOW 99] BOWKER G.C., STAR S.L., *Sorting Things Out. Classification and Its Consequences*, MIT Press, Cambridge, United States, 1999.

[CAL 86] CALLON M., LAW J., RIP A. (eds.), *Mapping the Dynamics of Science and Technology*, Macmillan, Houndmills, 1986.

[CAL 01] CALLON M., "Les méthodes d'analyse des grands nombres peuvent-elles contribuer à l'enrichissement de la sociologie du travail?", A. POUCHET (ed.), *Sociologies du Travail: Quarante Ans Après*, Elsevier, Paris, p. 335-354, 2001.

[CAL 02] CALLON M., "From science as an economic activity to socioeconomics of scientific research. The dynamics of emergent and consolidated techno-economic networks", in P. MIROWSKI and E.M. SENT (eds.), *Science Bought and Sold. Essays in the Economics of Science*, University of Chicago Press, Chicago, p. 277-317, 2002.

[CAM 04] CAMBROSIO A., KEATING P., MOGOUTOV A.. "Mapping collaborative work and innovation in biomedicine: A computer assisted analysis of antibody reagent workshops", *Social Studies of Science*, vol. 34, p. 325-364, 2004.

[CAM 06] CAMBROSIO A., KEATING P., MERCIER S., LEWISON G. ET MOGOUTOV A., "Mapping the emergence and development of translational cancer research", *European Journal of Cancer*, vol. 24, p. 3140-48, 2006.

[CAU 04] GAUDILLIÈRE J.-P., RHEINBERGER H.J. (ed.), *From Molecular Genetics to Genomics. The Mapping Cultures of Twentieth-Century Genetics*, Routledge, London, 2004.

[DEV 06] DEVEREAUX Z., RUECKER S., *Online Issue Mapping of International News and Information Design*, Human it, 2006, http://www.hb.se/bhs/ith/3-8/zdsr.htm.

[KAT 97] KATZ J.S., MARTIN B.R., "What is research collaboration?", *Research Policy*, vol. 26, p 1-18, 1997.

[KEA 00] KEATING P., CAMBROSIO A., "Real compared to what? Diagnosing Leukemias and lymphomas", in M. LOCK, A. YOUNG and A. CAMBROSIO (eds.), *Living and Working with the New Medical Technologies. Intersections of Inquiry*, Cambridge University Press, Cambridge, p. 103-134, 2000.

[KEA 03] KEATING P., CAMBROSIO A., *Biomedical Platforms. Realigning the Normal and the Pathological in Late-Twentieth-Century Medicine*, MIT Press, Cambridge, United States, 2003.

[KOS 06] KOSSINETS G., "Effects of missing data in social networks", *Social Networks*, vol. 28, p. 247-268, 2006.

[LAT 05] LATOUR B., *Reassembling the Social. An Introduction to Actor-Network-Theory*, Oxford University Press, Oxford, 2005.

[LEY 07] LEYDESDORFF L., "Visualization of the citation impact environments of scientific journals: An online mapping exercise", *Journal of the American Society for Information Science and Technology*, vol. 58, no. 1, p. 25-38, 2007.

[MOG 04] MOGOUTOV A., VICHNEVSKAIA T., "Présentation de la technique d'analyze des données d'enquête *Reseau-Lu*", J.-C. CHASTELAND and M. LORIAUX, L. ROUSSEL (eds.), *Démographie 2000. Une Enquête Internationale par Internet Auprès des Démographes*, LLN, Academia-Bruylant, 2004.

[MOG 05] MOGOUTOV A., CAMBROSIO A., KEATING P., "Making collaboration networks visible", B. LATOUR and P. WEIBEL (eds.), *Making Things Public. Atmospheres of Democracy*, MIT Press/ZKM, Cambridge (United States)/Karlsruhe, p. 342-345, 2005.

[MOG 06] MOGOUTOV A., VICHNEVSKAIA T., "ReseauLu, outils d'analyse exploratoire des données", M. DE LOENZIEN and S.D. YANA (eds.), *Les Approches Qualitatives dans les Études de Population. Théorie et Pratique*, AUF, Paris, 2006.

[POP 04] POPOWYCZ S, Classifying migraines: The International Headache Society and the production of the international headache classification system, Masters essay, Department of Sociology, McGill University, Montreal, 2004.

[ROS 03] ROSENTAL C., *La Trame de l'Évidence. Sociologie de la Démonstration en Logique*, PUF, Paris, 2003.

[SAL np] SALONIUS A., "Social organization of work in biomedical research labs in leading universities in Canada: socio-historical dynamics and the influence of research funding", *Social Studies of Science*, forthcoming.

[VIN 92] VINCK D., *Du Laboratoire aux Réseaux. Le Travail Scientifique en Mutation*, Office des Publications Officielles des Communautés Européennes, Luxembourg, 1992.

[WEN 98] WENGER E., *Communities of Practice. Learning, Meaning, and Identity*. Cambridge University Press, Cambridge, 1998.

[WID 03] WIDDOWS D., *Geometry and Meaning*, University of Chicago Press, Chicago, 2003.

PART IV

Computerized Processing of Speeches and Hyperdocuments: What are the Methodological Consequences?

Chapter 10

Hypertext, an Intellectual Technology in the Era of Complexity

10.1. The hypertextual paradigm

The term hypertext must first of all be replaced in its lexical field from a diachronic and synchronic point of view. This word has a history and it is – or was – delimited by competing terms. It was first used in 1965 by Ted Nelson, a visionary American philosopher, who defined hypertext as "a form of non-sequential writing". This link with writing and, consequently, with reading is essential. Hypertext, in its origin, is first a text or a set of texts, more or less fragmented, with non-sequential links among them.

The prefix hyper has no superlative value here. It must be understood in the mathematical sense: hypertext is a text of n dimensions. This property can only be implemented by computerized support and in a device capable of exploiting it. Any prior complex editorial provision, like an encyclopedia for example, can be regarded as a primitive form of hypertext – a proto-hypertext – in some way.

When, in the 1980s, computers were able to display more than alphabetical signs on screen, different media came to compete with the monopoly of text. Graphics, photos, animated pictures and movies gradually emerged in hypertexts. Since then, the term "hypermedia" has seemed more appropriate to reflect this novelty. Among the general public, this term has been supplanted by the term "multimedia". This term, which was first used to describe a variety of media (in the expression "a

---

Chapter written by Jean CLÉMENT.

multimedia advertising campaign", for example), quickly replaced that of "unimedia" that appeared, at one time, to mean the joining of media on one medium, that of the computer.

Today, in common language, hypertext has changed from a noun to an adjective. Thus we speak of a "hypertext link" in a Webpage, more often than we call the hypertext itself by this term. If it is legitimate to continue to talk about hypertext, we must consider that the word covers a wider semantic field than the image given by the Web in its current form. In France, hypertext is often regarded as a category of "hypermedia" but, in the United States it is almost the reverse. Here hypertext has the value of a paradigm. It may designate a piece of writing, a documentary or a new way of organizing knowledge. It evokes a cognitive dimension of the human mind when the latter gives up a hierarchized structuring of the thoughts or of writing for a model more focused on intuitive or analogical thinking, on the complex thought. Ted Nelson wrote: "I build paradigms. I work on complex ideas and make up words for them. It is the only way"[1]. It is this new paradigm that will be discussed here[2].

10.2. Cognitive activity and evolution of textual support

To understand what makes it unique, hypertext should be considered in the development of the media of thinking and memory within Western civilization. This history appears to be oriented by a growing freedom in the way people read documents and the increasingly complex way in which they interpret and analyze information.

The first alphabetical scripts were based on oral language, transcribing the linearity and flow of speech as closely as possible, providing just one possible shaping of thought. This is evidenced by the absence of any visual aid to reading, of any provision that could suggest a disruption of the flow of speech. Until the 12th Century, text is enclosed between its beginning (the *incipit*) and its end (the *explicit*). It occurs in a continuous, linear way, and is focused on the model of oral speech. Punctuation was introduced very gradually. It was not until the 16th Century that it stabilized in its present form. Thus without punctuation to aid in reading the text, long ago it was only possible to read text if the syllables were pronounced aloud, in a continuous muttering.

The first evidence of reading in silence dates from the 4th Century. Saint Augustine indeed told how he was surprised to discover that his master, Saint

---

1 [NEL 06].

2 There is an illustration of the vitality of this paradigm in the proceedings of the 5th Hypertext Hypermedia Products Tools and Methods (H2PTM) conference [SAL 05].

Ambrose, was reading "without moving his lips". The passage from linear reading aloud to tabular visual reading was made possible by the invention and then the generalization of the codex in the 4th Century. The codex defines the page as a space of separated reading, autonomous and discontinuous, and by the emergence of typo-dispositional markers that organize the textual matter in this space. Therefore, the reader could browse through the text more freely and find the information he or she was looking for more easily.

It was mainly the birth of the book[3] in the 12th Century that marked the greatest intellectual turning point in the West. Scholastic needs then generated new forms of publishing and page layout that facilitated access to texts without overloading the memory. This mode of reading reached its maximum effectiveness with the encyclopedia in the 18th Century. Cross-references, references, alphabetical and thematic classification of materials all contributed to favor the freedom of the reader. The author leafs through, forages or browses between the articles according to a network designed by the authors to guide the reader through the diversity of the texts.

The computer, finally, is the most recent stage of this process. By drawing up the references and indexes of the encyclopedia, it offers easier reading but, at the same time it multiplies and disorganizes them. The reader no longer faces a system of hierarchical organization or of information classification, he or she is at the centre of a network that reconfigures depending on its journey. It is part of a complex system of which the reader has only a local and limited view and that he or she contributes change to with each of his or her choices.

10.3. The invention of hypertext

The idea of a hypertext reading and writing system was born well before the term hypertext appeared. It is generally accepted that the precursor is Vannevar Bush. This advisor to President Roosevelt published an article entitled "As we May Think" in 1945 in which he developed the idea of a system in the form of a desktop, called *Memex* (Memory eXtender) capable, among other things, of displaying microfilms on several screens on request, recording and restoring sound. He imagined that documents could be linked between themselves by links posed by the user so as to enable him or her to go through documentation again by activating these links. This is what he called an "associative indexation":

---

3 On the birth of the book in the 12th Century, see [ILL 91].

"It affords an immediate step [...] to associative indexing, the basic idea of which is a provision whereby any item may be caused at will to select immediately and automatically another."[4]

He thus thought of solving the difficulty researchers had in finding documents relevant to their investigations and information within documents in books or libraries. We had to wait until 1965 for Theodor Holm Nelson and Andries van Dam, of Brown University, to develop and implement the first hypertext system (FRESSE). They launched the *Xanadu* project, a hypertextual device for archiving and consultation of documents on the network at this time. In 1968, Douglas Engelbart, the inventor of the mouse and of interactivity, gave a demonstration of his system NLS/AUGMENT that draws up some of Bush's ideas.

In 1987, Apple delivered the software for hypertextual writing (HyperCard) with all its personal computers that contributed in spreading the use of hypertext in the public domain. In the same year the first international conference on hypertext was held.

Finally, in 1990, Tim Berners Lee developed what would become the Web – a publication system of documents on a network using the Internet, which rapidly developed to what we know today. This invention has changed our relation with hypertext by spreading a system that is rather crude, but very simple to use and rapidly replaced earlier systems.

What is most striking in this brief history of hypertext is the concomitant link between the growing need to increase human cognitive abilities to cope with the proliferation of information and the technical development of networks that was made possible by the rapid progress of computer science. In this regard, it is significant that the names of the systems devised by Bush (*Memex*) and Engelbart (*Augment*) refer to an increase in our cognitive and memory abilities. Memory expansion and increase in intellect are at the heart of the development of hypertext.

10.4. Hypertext and databases

From the point of view of its technical architecture, hypertext can be considered as a three-tier or three-level system. Several theoretical models have been proposed [CAM 88, GRØ 94, HAL 94] that we will not discuss in detail here. They can be simplified into the following organization:

– the first layer consists of information organized in a database;

---

4 [BUS 45], reprinted in [NEL 92], p. 51.

– the second, by conceptual hypertext;

– the third, by the user interface.

According to this scheme, it is obvious that the layer of the conceptual hypertext is the most characteristic and justifies the term "intellectual technology". Indeed, it organizes data in a system that links them between themselves, gives them meaning and produces new information that was not contained in the first layer.

Although a traditional database contains information, it is formatted and only accessible through formal requests. Thus, a database needs to define data types (number, string, date, etc.) as well as their size in order to optimize the space available in the machine's memory. This constraint of *a priori* formatting of data reduces their heterogeneity and forces formatting work upstream. Building a database thus requires a local and differentiated definition of the needs of users. This is why databases are based on metadata, that is, information on the data, information involving a point of view on the database. It is on this model that catalogs and bibliographies are built, but also more complex databases structured in thesaurus or ontology with keywords-indexing systems.

A good example is the database of literary history[5]. By selecting a date it allows us to find authors, works, reviews and literary movements indexed to this date or, thanks to an indexation by themes, to suggest connections between literary works. In this type of database, information retrieval is based on a system of queries. The user defines the limits and characteristics of his/her object of research and in response receives a list of items corresponding to his/her criteria. To continue his/her research, he/she then needs to renew the request in a back and forth movement between the questions and answers given by the machine, each answer allowing him/her to refine the query criteria.

Anyone who engages in this work can be compared to a fisherman who, from the edge of the water or of his boat, casts his line or nets then examines what the mesh brings back to him. He never enters the water himself; he stays on the shore or on the boat. In this situation, the user is not part of a system with the machine and remains outside it. Interaction here is limited to the reiteration of requests and their possible refinement. The cognitive activity involved is based primarily on the ability to imagine what the organization of the database allows him/her to search. To simplify, we could say that in a database we only fine what we know will be there. Today, many databases are interfaced with the Web, thus allowing easier and expanded access to information. They are not however real hypertexts.

---

5 http://phalese.univ-paris3.fr/bdhl.

10.5. Automatic hypertextualization

In addition to this formalized model of a database, there is another organization based not just on *a priori* structuring of data, but also on the exploration of their textual content. This exploration can de done in two ways: either by a "full text" search of occurrences of words or strings of characters, or by a search exploiting a semantic analysis of contents.

The first case is well known and can be illustrated by the database *Frantex*[6] and the accompanying software *Stella*. The database contains more than 3,700 books in French. After defining a set according to criteria proposed by the database (author, dates, genre, etc.), the user can search for occurrences of words or forms, find their context of appearance, obtain statistical results on the vocabulary of a subset, determine the lexicon of theme words of a work and others[7].

The second case, which falls within the field of automatic text understanding, poses much more difficult problems. It no longer seeks information from occurrences of forms, but semantics contained in the text. The difficulty can be addressed using three approaches: structural, linguistic and statistical[8]. The structural approach relies on the logical structure of documents. It is therefore only convenient for documents previously prepared for this purpose. The SGML (standard generalized markup language) grammar, conforming to the ISO 8879 standard, allows the definition of the rules of such a markup. The logical entities of documents, such as titles, subtitles, paragraphs and lists, can thus be identified as such and constitute the nodes of a hypertext, while the hierarchical relations that organize them and cross-references that structure them, are expected to provide hypertextual links. The limits of this approach stem from the fact that semantics is entirely dependent on logical units. If it can provide interesting results for certain types of technical documents, whose degree of granularity[9] is important and that are structured according to an *a priori* model, it is ill-suited for more literary or heterogeneous documents. The linguistic approach seeks to create semantic networks between a set of documents.

Where methods of full text search were able to identify only isolated forms, methods based on semantics cannot just interpret forms as a phrase or compound word ("*chemin-de-fer*", for example, the French word for railway, literally "road-of-iron"), but also solve problems of polysemy depending on the context and link

---

6 http://atilf.atilf.fr/frantext.htm.
7 See Chapter 11.
8 There is a good introduction to these techniques in [BALB 96]. See also Chapter 13 of this book.
9 The granularity is the minimum size of an element that can be manipulated by a system.

words to a lexical field. It thus becomes possible to search from a concept and not just from a term, and obtain the set of forms that are part of its semantic network. The statistical approach aims to produce, automatically and on a large number of documents, clusters based on their thematic proximity, regardless their linguistic expression. It is based on complex neural algorithms that allow establishing a semantic mapping in the form of a hypertext, whose nodes are the clusters of text and whose links reflect the thematic proximity.

Automatic hypertextualization appears to offer a solution to handle large volumes of information that would be impossible to process manually. It is therefore adapted particularly to large corpora, in particular on the Web, and it is in this direction that works of researchers are focused[10]. Search engines were the first to benefit from these methods. The XML (eXtensible markup language) standard is now used to index by mark-up the semantic contents of documents. If automatic hypertextualization is a promising avenue of research, however, it remains closely linked to the philosophy of databases. The results it provides are presented as lists or maps that allow us to reach the required information, or discover information that we did not expect to be there, but that are struggling to form a hypertext. They lack the navigational dimension, the possibility to interact intuitively with the system, even if the interfaces offered to the user make use of clickable links.

10.6. The paradigm of complexity

As sophisticated as they are, methods of automatic construction of hypertexts from databases remain trapped in a very basic logical organization and fit into the paradigm of a vision of the arborescent and classificatory knowledge inherited from Cartesians. According to this perspective, the object of knowledge must be separated from the knowing subject and each discipline must be established independently. Furthermore, to understand the world we must simplify it according to two principles. The principle of reduction emphasizes the knowledge of the constituents of a system rather than the system as a whole. The principle of abstraction needs to reduce everything to equations and formulae governing quantified entities. Nature, according to Galileo, is "a book written in mathematical characters", while Descartes dreamt of a "very geometrical Physics". This research on laws of nature, as illustrated by Newton in an exemplary manner, was continued until the 19th Century. It had the merit of allowing great progress in scientific knowledge, but at the cost of a simplification that lead us to reject everything that was not in the model and would be contrary to the order thus projected on the world.

---

10 Read Tim Berners Lee's founding article on the semantic Web [LEE 01].

Complexity only made its reappearance in science in the early 19th Century. It was the works of Sadi Carnot on thermodynamics that first questioned the idea of an orderly world. The second law of thermodynamics, formulated in 1824, introduced irreversibility in physics. This principle of degradation of energy or increasing entropy suggests that the most likely state of any system is disorder. This discovery is fundamental. It establishes that disorder is no longer the negligible residue of our attempt to understand the world, but is irreversibly at the heart of the universe, which is a complex system.

Since this time it has become legitimate to believe in the paradigm of complexity. From this I will take three examples to guide my remarks on hypertext: noise, system and chaos[11]. In a communication situation, noise is first a factor of disturbance and disorder. Lately, the concept of noise has been considered more positively. Its role could be to produce diversity and complexity. Changeux [CHA 83] and then Trabary [TAB 83] were able to study the role of noise in the self-organization of the brain, particularly in the learning process. As early as 1943, Wiener had highlighted the principle of feedback. This "effect" reacts on its "cause": any process should be designed according to a circular pattern. This simple idea proved to be fertile and gave rise to cybernetics. Besides the work of Wiener, a group of researchers led by Von Bertalanffy [VON 93] were thinking about a "general systems theory". They reached this definition:

"A system is a compound of elements in interaction, these interactions being not random by nature".

General systems theory and cybernetics progressively intersected to become what is known today as systemic.

More recently, the concept of chaos, theorized among others by Lorenz [LOR 96][12], has enabled us to highlight how a small variation in initial conditions can lead a system to have very different evolutions. The explanation of chaotic phenomena is found in non-linearity and feedback which, modifying the cause, leads to the production of a new effect that, in turn, provides feedback on its cause, and so on. The study of chaotic systems leads us to reconsider the relation between order and disorder. In living systems, the strongest is that which is able to incorporate chaos. Chaos by itself can generate order. The order that is formed in chaos, is probably proof that systems are not pure disorder but that they carry with them a virtual or

---

11 For a development on these concepts and their relation to hypertext, see [CLE 00].
12 He is the author of the famous "butterfly effect" theory: the flapping of wings of a butterfly in the Amazonian rainforest can trigger a meteorological phenomenon thousands of kilometers away.

potential order that, in some circumstances, may be updated and appear or disappear.

Noise, systems theory and chaos thus appear as a limitation of the fundamental principle of classical science: determinism. At the same time, they expand the scope of rationality and show evidence of a consideration of real, dynamic, open, often unstable and volatile systems. These concepts allow a better understanding of what was previously excluded or ignored by a deterministic and simplistic system of rationality. Noise and chaos form the epistemological foundation of hypertext.

10.7. Writing of the complex

The paradigm of complexity disrupts our relation with knowledge and questions the tool that has always been the expression of it: the classic book. Indeed, because of its layout and structure, the book is almost always the expression of a logical and rational vision of the world. As the databases that are inspired by it, they format knowledge by breaking it into chapters or paragraphs that are strictly ordered. Beyond its architecture, it helps to support speech organized as a simplistic rhetoric that tends to reduce complexity in formulas.

The classic book is the culmination, the achieved form of a simplification process. For some modern thinkers, this form is now sterile. This is how Deleuze and Guattari, in *Mille Plateaus*, spoke of the "root-book":

"The tree is already the image of the world, or the root is the image of the tree world. It is the classic book, as beautiful organic interiority, significant and subjective [...]. The law of the book, is that of reflection, the one-who-becomes-two [...]. Each time we encounter this formula [...], we find ourselves in front of the most classic and most thoughtful thinking, the oldest, the most tired" [DEL 80][13].

This "formula" from which Deleuze and Guattari want to liberate us is that of Descartes, that of the paradigm of simplification. They oppose the figure of the rhizome: an underground stem, bulb or tuber, it obeys a principle of connection and heterogeneity. It is multiplicity, leakage path, adequate to the reality instead of representative of this reality[14].

Criticism of the book as an instrument of deterministic thinking thus leads us to imagine another form of organization of knowledge and the appropriation that would replace the book. To some extent, hypertext meets this expectation and can

---

13 p. 11.
14 See the fine example presented by P. Maranda in Chapter 12.

appear as a more appropriate support to complex thinking than the classic book. This complexity is reflected in two basic characteristics. On the one hand hypertext is an unstructured set of *a priori* elements (nodes) that, being connected to one another, form a system: any action on the elements reconfigures all. On the other hand, each hypertext activation by a user determines a singular path and causes a temporary structuring of the whole. It is in this constructive interaction of a subject with a variable and volatile set of knowledge that hypertext can be regarded as an appropriate response to the challenge of complexity.

10.8. Hypertextual discursivity

The hypothesis suggested here is hypertextual discursivity of thinking and writing as a response to the challenge of complexity. Indeed, intellectual history teaches us to what extent linearization of thinking in the mould of speech constructed according to the constraints of the book has been seen as an obstacle to the freedom of thinking. From Antiquity and Pre-Socratic philosophers to contemporary thinkers, there is no lack of examples that illustrate the rejection of linearized writing to a discontinuous speech. Fragmentary genres such as the aphorism, thought or maxim often seem better able to translate and encourage thought than better built types. Heraclitus, Montaigne, Pascal, moralists of the 17th Century, Novalis, Nietzsche and Wittgenstein have all, to different degrees, practiced fragmentary writing. Some have regretted doing so while others have fully claimed it. All have found a way to be closer to the truth of thought. The fragment here is the place the thought, germ and parcel of truth is produced. There is more: in the space handled by fragments, it is the reader who is invited to reproduce for him- or herself the instrument of thought, and which contributes, more than in linear writing, to the construction of meaning. It is the opposite of the classic argumentative rhetoric for which the order of propositions is crucial. The philosophical speech, for example, often requires strong organization to support the reader [KOL 94].

Hypertext is in line with fragmentary texts. By giving up the linearization of speeches, the author fragments, whatever name we give it, node, lexis [BAR 70] or "texton" [AAR 97], the minimum unit of writing. This fragment must by itself hold a sufficient degree of autonomy to be read in different contexts depending on the journey of each reader. It is in this search for fragmentary autonomy that one of the keys to the success of hypertextual writing is found. However, hypertext cannot only be regarded as a collection of fragments independent of each other that could be read in a random order. Between the fixed order of the book and the deconstruction of the fragmentary collection, hypertext is a semi-constructed feature because the units of speech are organized in network. They form a directed graph whose path contributes to the formation of meaning. The links connecting each of the nodes of hypertext with each other are carriers of a discourse, a metarhetoric that is at the heart of hypertextual writing and is an activity of the author [CLE 04]. The hypertextual theory distinguished "calculated" (by the machine) links from "edited" (by an author) links. It will be understood: it is the latter that concern us here. Automatic hypertextualization is not writing; the links that it produces being only the result of a calculation of the machine.

Speaking of writing regarding links may seem paradoxical. Links, by themselves, are hardly carriers of information. Their anchor on words of text contributes to their semantization (by the metonymic relation between the word and fragment to which it refers). Features such as typing (that allow the reader to anticipate the type of content that he or she will find) or contextual menus can make them more explicit. The links have a substantially elliptical character. The rhetorical figure that best characterizes links is parataxis: the removal of any link word likely to emphasize the logical relation between two sentences of speech. It is only the comparison between two fragments proposed by the author that makes sense and it is up to the reader to discover or apply it.

The role allotted to the reader in hypertext is more important than in the classic book. It is his/her personal journey which, by taking a set of information and propositions, contributes to construct meaning[15]. Not only, as in traditional reading, by a purely intellectual activity but through an interaction with the system that updates the units of reading and raises statements. Two readers never read the same hypertext, not because their readings differ but because what they were given to read is different. This production of meaning is sometimes problematic because the reader can get lost in the hypertext. His/her interaction with the system assumes that every time he/she chooses, depending on his/her route and moods, to click on a particular link the reader does not have an overview of what is proposed. To remedy this disorientation or cognitive overload, features exist to provide a view overlooking the hypertext, to draw up a map or to list the nodes. These tracking attempts in browsing risk losing sight of one of the interesting points of hypertext: the production of a singular statement, the result of a pedestrian path in which the state of the cognitive system of the user interacts with the state of the hypertextual system to determine the route to take. Only this interaction is likely to build a knowledge that the reader did not expect, or was not even looking for, but which he/she needed. This encounter with the unexpected is essential. It represents what is sometimes called serendipity [ERT 03] or the art of finding the right information accidentally.

---

15 On the problem of construction of meaning in hypertext, read [PIO 04].

10.9. Conclusion

Hypertext is not the only form of writing that allows the reader to find relevant information, or the only answer to the demands of complex thought of modern times. However, it is characterized by its ability to integrate some of the concepts of the paradigms of complexity.

The disorientation of which he/she is often accused can be equated with the concept of noise that we know can cause a reorganization of a system to a higher level. The possibility offered by links to branch off at any time recalls a fundamental property of chaotic systems: the difficulty in motivating the choice of path leaves room to hazard and allows unexpected encounters.

The concept of a system, finally, allows us to take into account the mode of operation in which the reader, the feature and the machine interact to produce knowledge.

Thus hypertext is much more than a metaphor of complexity. It is the preferred tool for it. In the era of Web 2.0, the Internet user is no longer content to consult sites by remaining passive in front of his/her screen. With new software tools, the development of broadband and new services, he/she participates in a huge exchange system. The producer of information and documents, belonging to various communities, he/she participates in the creation of social networks in cyberspace. The rapid development of forums, blogs and wikis reflects this "hypertextualizing activity" by which we organize the sense of the world [BALT 03].

10.10. Bibliography

[AAR 97] AARSETH E., *Cybertext, Perspectives on Ergodic Literature*, John Hopkins University Press, Baltimore, 1997.

[BALB 96] BALPE J.-P., LELU A., PAPY F., SALEH I., *Techniques Avancées pour l'Hypertexte*, Hermes, Paris, 1996.

[BALT 03] BALTZ C., "In-formation", in J.-P. BALPE, I. SALEH, D. LEPAGE and F. PAPY (eds.), *Hypertextes, Hypermédias, Créer du sens à l'Ère du Numérique, Proceedings of H2PTM'03*, Hermes, Paris, 2003.

[BAR 70] BARTHES R., *S/Z*, Seuil, Paris, 1970.

[BUS 45] BUSH V., "As we may think ", *The Atlantic Monthly*, July 1945.

[CAM 88] CAMPBELL B., GOODMAN J., "A general purpose hypertext abstract machine", *Communication of the ACM*, ACM Press, New York, 1988.

[CHA 83] CHANGEUX J.-P., *L'Homme Neuronal*, Fayard, Paris, 1983.

[CLE 00] CLEMENT J., "Hypertexte et complexité", *Etudes françaises*, Presses Universitaires de Montréal, Montreal, vol. 36, no. 2, 2000.

[CLE 04] CLEMENT J., "Hypertexte et fiction: une affaire de liens", J.-M. SALAÜN and C. VANDENDORPE (eds.), *Les Défis de la Publication sur le Web: Hyperlectures, Cybertextes et Méta-éditions*, ENSSIB, Villeurbanne, 2004.

[DEL 80] DELEUZE G., GUATTARI F., *Mille Plateaux*, Editions de Minuit, Paris, 1980.

[ERT 03] ERTZSCHEID O., "Syndrome d'Elpenor et sérendipité, deux nouveaux paramètres pour l'analyse de la navigation hypermédia", in J.-P. BALPE, I. SALEH, D. LEPAGE and F. PAPY (eds.), *Hypertextes et Hypermedias, Créer du sens à l'Ére Numérique, Proceedings of H2PTM'O3*, Hermes, Paris, 2003.

[GRØ 94] GRØNBÆK K., TRIGG R., "Design issues for a Dexter-based hypermedia system", *Communication of the ACM*, ACM Press, New York, 1994.

[HAL 94] HALASZ F., SCHWARTZ M., "The Dexter hypertext reference model", *Communications of the ACM*, 1994.

[KOL 94] KOLB D., " Socrates in the labyrinth", in G. LANDOW (ed.), *Hyper/Text/Theory*, John Hopkins University Press, Baltimore, 1994.

[ILL 91] ILLICH I., *Du Lisible au Visible, sur l'Art de Lire de Hugues de Saint-Victor*, Editions du Cerf, Paris, 1991.

[LEE 01] LEE T.B., HENDLER J., LASSILA O., "The semantic Web. A new form of Web content that is meaningful to computers will unleash a revolution of new possibilities", *Scientific American*, May 2001.

[LOR 96] LORENZ E., *The Essence of Chaos*, University of Washington Press, Washington, 1996.

[NEL 92] NELSON T., *Literary Machines 91.1*, Mindfull Press, Sausalito, 1992.

[NEL 06] NELSON T., What I do, available at: http://ted.hyperland.com/whatIdo, 2006 (accessed February 8, 2010.)

[PIO 04] PIOTROWSKI D., *L'Hypertextualité ou la Pratique Formelle du Sens*, Honoré Champion, Paris, 2004.

[SAL 05] SALEH I., CLEMENT J. (ed.), "Créer, jouer, échanger, expériences de réseaux", *Proceedings of H2PTM'O5,* Hermes, Paris, 2005.

[TAB 83] TABARY J.-C., "Auto-organisation à partir du bruit et système nerveux", *L'Auto-organisation. De la Physique au Politique*, Seuil, Paris, 1983.

[VON 93] VON BERTALANFFY L., *Théorie Générale des Systèmes*, Dunod, Paris, 1993.

Chapter 11

A Brief History of Software Resources for Qualitative Analysis

11.1. Introduction

This chapter is intended as a resource for qualitative analysis of corpora of texts in human and social sciences. We will concentrate on the computerized analysis of text corpora[1] by obscuring the tools to analyze documents that are not strictly textual. These can, for example, be objects confided by informants, memories that the observer retains from his/her contact with the field or recordings (audio, photographic or video) of interactions or settings.

Tools to analyze these non-exclusively textual materials (as *Aquad*, *Anvil*, *Atlas*, *Porphyry*[2], *Transana*, *Transcriber* and *Videograph*) are therefore not analyzed here. This methodological work leads to the opening of some black boxes on textual analysis. Such a clarification will bring to mind the circumstances in which different techniques[3] develop and shows that many tools under different names, resume very similar features. For this reason, I prefer to identify families of features rather than families of tools (such typologies are available in [JEN 96, KLE 01, POP 97, WEI 95]).

Chapter written by Christophe LEJEUNE.
1 A corpus is a set of documents. For this chapter, it is a set of texts.
2 See Chapter 1.
3 Here I focus on the methodological dimension of these tools.

In this chapter, I would like to show those scared of this technique that these tools are close to their daily work without software. To the enthusiasts, this chapter points out that none of the proposed techniques is magic, or a Pandora's Box or intelligent, because they are no more than tools. Finally, those of my readers who are already using some of the mentioned tools will perhaps interpret this chapter as a strong overview. Indeed, like any list, this overview covers tools developed in very different worlds. I hope it will encourage dialogs among users from different communities.

11.2. Which tool for which analysis?

As the anthropology of science has brilliantly shown for the other disciplines, scientists transform materials observed into inscriptions. From translation to translation, introductory measurements are moved (mobile) from the phenomenal field to the laboratory [LAT 88]. The inscription (on a support) ensures their stability during this transfer. Given that their ability to represent the phenomenon is not altered, they are said to be immutable. The converging movement from the phenomenal field towards the laboratory is therefore attached to a movement that is moving away from it: the researcher mobilizes the inscriptions in the assembly of argumentative features that are articles [CAL 86]. It is the quality (both mobile and immutable) of these inscriptions that ensures fidelity to the starting observations. By vocation, these new inscriptions start a centrifugal displacement in relation with the laboratory.

In social sciences, the series of transformations into "immutable mobile" starts with the collection (and registration) of the testimony of the informant. This is then taken to the laboratory; scientists conduct other transformations (translations) on it that produce new inscriptions. Among them, the transcription of interviews prototypically occupies the introductory position of translation, making the following transformations possible. Even when no other tool is then used, this operation is indeed necessary to the quotation (without which the empirical basis of sociological arguments is largely weakened). Once the interviews are transcribed, the researcher often opts for the analysis of these intermediary inscriptions.

The following sections review the devices likely to accompany this task. The features presented will be:

– the felt-tip (section 11.2.1);

– text processing (section 11.2.2);

– the operating system – and, in particular the regular expressions – (section 11.2.3);

– the cotext – in particular, the concordances – (section 11.2.4);

– the co-occurrence (section 11.2.5);

– Benzécri analysis of data (section 11.2.6);

– text segments (section 11.2.7); and

– dictionaries (section 11.2.8).

11.2.1. *The felt-tip*

The minimum equipment the qualitative researcher needs in order to analyze his/her textual material is the felt-tip (or, if preferred, the pen or pencil). Without this starting point being the opportunity to discuss the link between science and writing [GOO 77, SER 02], it nevertheless recalls that empirical sources manipulated by the researcher are material. Faced with empirical elements that are transcriptions of interviews, the analyst proceeds to the identification of themes that seem relevant to him/her.

The felt-tip here is the minimum technology required to allow this annotation. Since the present chapter focuses on computer tools, this borderline case stands as non-included endpoint (which means that this chapter does not further study work with felt-tip itself)[4].

11.2.2. *Text processing*

Some researchers propose to take advantage of the familiarity and spread of text processing and use it to analyze the corpus of text [LAP 04, MOR 91]. This specific use of an office tool consists of delimiting segments of text and associating them with labels chosen by the researcher[5]. Two main features are used for this purpose: invisible marks and tables.

Initially designed as an adjuvant to writing, invisible marks allow the user to comment on his/her text (without it appearing in printing); they are thus particularly taken advantage of in the case of collective writing. Used as part of a qualitative

---

4 In section 11.2.7 I show that some software design makes reference to the technology of the felt-tip.
5 Close to work "with felt-tip", this deterritorialization of text processing is consistent with tools widely-spread among people (that I present in section 11.2.7).

analysis, these revision marks can annotate the text, transposing – on screen – the use of the felt-tip. Depending on the strategy of the researcher, these annotations may go from a simple account of sayings to an interpretation.

Second, as shown in Figure 11.1, tables can also be used as a counterpart: annotations then no longer lay in the invisible marks but in a dedicated column (which also resembles paperwork but, this time, comments made in the margin)[6].

The use of text processing to analyze qualitative material is attested and thus possible. However, experience shows that this diversion is not the most appropriate tool[7].

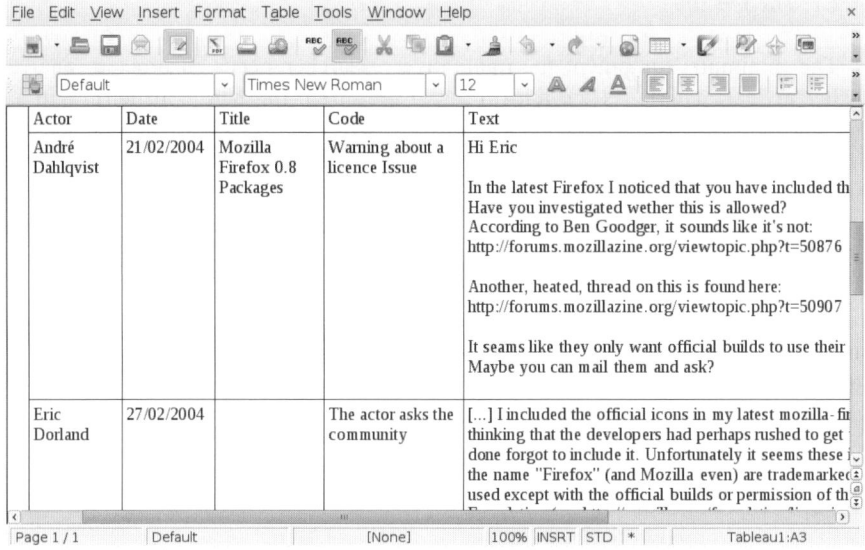

Figure 11.1. *Word processing (Open Office).(The codes of the researcher appear in the fourth column)*

11.2.3. *The operating system*

In practice, researchers who use text processing as an analysis tool take advantage of the availability of other word processor features. The exploration and coding of their textual material involves location of words (or groups of words or parts of words) within the text. As such, these search operations on strings of

---

6 The illustration presented here uses the free word-processing program *Open Office*, http://www.openoffice.org/, accessed February 8, 2010.
7 On large corpora, tabular files become less easy to handle.

characters belong to the know-how of each person rather than to the scientific analysis of qualitative material.

The search for alphanumerical segments is not limited to typographical strings. It includes famous truncations – that allow us, for example, to locate all the conjugated forms of the same verb (or, more generally, all the inflections[8] of the same lexeme [LYO 95]) – as well as operators, such as the logical "or". The range of patterns that can be gathered is thus infinite.

Such localization of generic patterns ("pattern matching") uses what computer scientists call "regular expressions" [FRI 06]. These have been available on all operating systems since the birth of microcomputers. They were born from the scientific study of neurons. In the 1940s, the neurophysiologist Warren McCulloch browsed various disciplines to understand how thought develops in mankind. His meeting with the talented logician Walter Pitts lead to an article [MCC 43] that tried to model the nervous system. Propositional calculation was applied to these neuronal machines (communicating between themselves by electrical pulse).

For McCulloch, this work fits into a constant questioning of the (material) basis of the human mind and the willingness to give to this question a scientific (rather than metaphysical) answer, thanks to experimental psychology helped by biology. These latter disciplines will ignore this work[9].

The works of McCulloch were then resumed in mathematics. Stephen Kleene attached the concept of finite automation to them and included them in his algebra of regular sets[10]. After several developments in mathematics, the regular expressions were gradually introduced into the concerns of computer scientists. The co-developer of *Unix* introduced them in the 1960s both in scientific literature

---

8 Among the morphological transformations, linguists distinguish the inflection of the same lexeme in different word-forms (of which the German versions and the conjugation of verbs in English are examples) of the derivation of a new lexeme (to go from a noun like "territory" to a verb like "territorialize").

9 The aura surrounding the 1943 article – considered a precursor of neural networks and cognitive sciences [AND 92, DUP 09] – calls for some restrictions: with such a force of attraction, it can be quoted without necessarily being the most relevant reference for the foundation of pattern matching. Its history shows nevertheless how various concerns (scientific, philosophical and epistemological) were able to converge in the development of a generic tool.

10 In French, "regular" is sometimes translated as "rational", which led some French researchers [SIL 00] to speak of rational expressions [FRI 06].

[THO 68] and in a series of computer applications (first the editor *qed*, then *ed*, that popularize them). In these editors, the following command should have been entered to execute a connection of patterns:

```
g/Regular Expression/p
```

This function gave its name to the application *grep* (for *global regular expression print*).

The power of regular expressions then increased [FRI 06]. In the late 1980s, the language Perl[11] played a decisive role in their spread. However, Larry Wall, who designed this programming language, is a linguist [SCH 08]. Regular expressions can thus boast about their interdisciplinary origins.

Today, they are part of the analysis of text corpora. This section thus suggests that the researcher can be satisfied with using the operating system to analyze his/her data, without having recourse to other more specific tools than the search functions[12].

11.2.4. *The cotext*

By leaving the use of conventional office tools, the present picture offers a class of more specific tools. Close to previous search features, these tools are used to identify words, expressions or patterns. Their specific contribution lies in the surrounding words of the element sought. Exhibiting the sentence or paragraph in which the target appears allows the researcher to relate each occurrence to a context of apparition (what linguists call "immediate topological environment" or cotext [KER 02, MAI 98, WIL 01]).

This mode of visualization of the passages coming before and after each of the occurrences is called "concordances" [LEB 98] or "index of keywords in context" [WEIT 95]. Concordances[13] are typically presented with vertical alignment of the target (or pole-shape), so that different cotexts can easily be sorted, brought closer to one another and compared [PIN 06] (see Figure 11.2).

---

11 http://www.perl.org/.
12 Jocelyne Le Ber thus deterrotorialized the *grep*, *freq* and *diff* commands of the operating system in order to analyze literary works, in particular *Antigone* by Jean Cocteau [LEB 06].
13 *AntConc* by Laurence Anthony (http://www.antlab.sci.waseda.ac.jp/); *Glossanet* by Cédrick Fairon (http://glossa.fltr.ucl.ac.be/) [FAI 06].

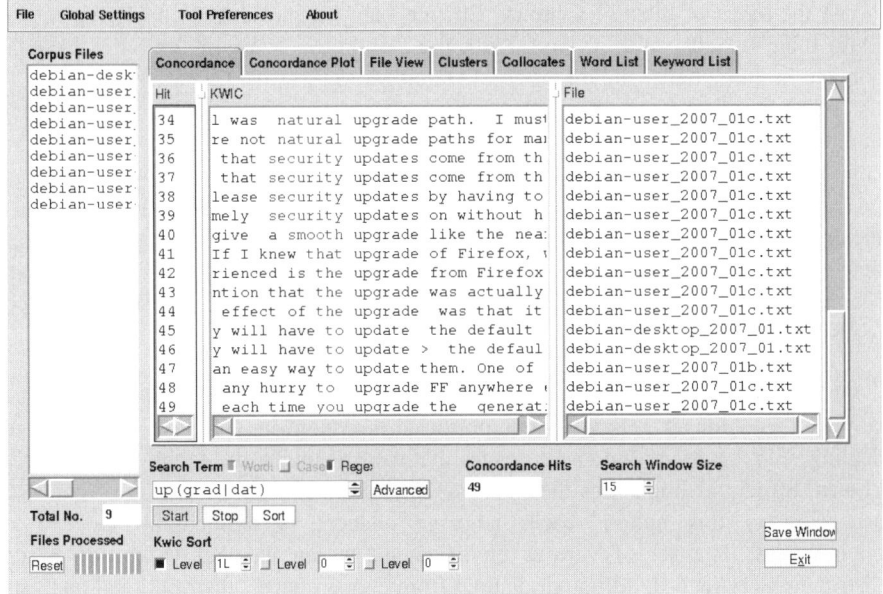

Figure 11.2. *A concordancer (AntConc)*

The history of indexes dates back to the beginning of our era. The first indexes organized in alphabetical order date from the 4th Century AD. The Greeks compiled (non-alphabetical) indexes of geographical names in the 6th or 7th Century but the invention of concordances and indexes of keywords in context fits into the tradition of exegetes of sacred texts. A division of biblical texts into chapters (called "capitulation") was tested in the 4th Century [MEY 89].

The text of the Carolingian *New Testament* contains, in the margin, a capitulation as well as references to similar passages in other Gospels. The proliferation of books and the reduction of volume sizes reduced the room available for such information. In parallel, exegetes produced specific tools to ensure this intertextuality. In the 10th Century, the Masoretes – the Jewish "masters of tradition" – created alphabetical lists of words accompanied by their immediate cotext[14]. They therefore foreshadowed (or invented) the keywords in context [WEIN 04].

---

14 These lists were not indexed, given the absence of numbering of chapters or verses.

At the dawn of the 13th Century, Etienne Langton introduced chaptering in the Latin bible[15]. Between 1238 and 1240, the Dominican friars of the Saint Jacques convent in Paris, under the direction of Hugues of Saint-Cher, made an alphabetical list of concordances involving these new references. These include the chapter number and a letter (A to G), indicating the position of the word in the chapter. The cotextual environment being unavailable in this edition, the system in question prefigures for its part the indexes of keywords *out of* context. In subsequent editions, the framing proposition was added as a third element. It is precisely the debates between Christians and Jews that supported the development, by Rabbi Nathan Mardochee, of a Hebrew concordance in 1523, which was the first to be printed [SEK 95].

In the 1930s, the scientific study of the cotext rose in the United States. The context is no longer religious, but political: the sponsors of these studies were dealing with American Indian languages that were as numerous as they were poorly known. Structural linguistics then developed and introduced the notion of context: according to this notion, each language element is defined by its textual environment. The set of various environments of an element is known as its distribution [HAR 64]. Although distinct from concordances, distributional analysis marks the integration of cotext into scientific tools.

The next innovation in concordances lay in automation. The first automated generations of concordances were proposed by late 1950s[16]. In 1959, Hans Peter Luhn proposed a setup of concordances focusing on the desired form, which he qualified under the acronym *KeyWords In Context* or KWIC [LUH 60, LUH 66]. Some years later (in 1963) the first indexes of keywords designed according to this principle appeared (the target words were from titles of 10 years of publication of the *Association of Computer Machinery* [YOU 63]).

After centuries of existence, the construction of concordances is now automated. The concordancer is thus the tool with the oldest tradition. Its interdisciplinary nature is comparable to regular expressions. It is even currently being extended beyond the study of textual material, with specific applications in genetics and chemistry[17]. The concordances are used to support many inferences of corpus exploration. Their graphical properties make them particularly appropriate to discovering the recurrence of phrases, idioms or expressions (consisting of several simple words).

---

15 The verses were inserted, much later, by Robert Estienne in 1551.
16 In 1951, Roberto Busa, author of the Postscript of this treatise, using tabulating tools made a concordance of four poems of Thomas d'Aquin.
17 In genetics, concordancers help to locate genes sequence alignment (named "synteny").

11.2.5. *The co-occurrence*

Co-occurrence is a variation of the previous cotext-based tools. Although different algorithms exist in this field, the principle is still based on the same logic: identifying the "proximity"[18] of two terms. This proximity is neither semantic, nor syntactic or pragmatic. As with cotextual tools, it refers to a topological dimension. The two terms are even closer than the fact they are separated by a few characters. This proximity is measured along the syntagmatic[19] axis. To go from this measurement to co-occurrence, we aggregate the proximities of each appearance (or occurrence) of the two terms. Contrary to the study of environments that, through the operating concept of distribution, find a justification in American structuralist linguistics, the proximity of co-occurrence is the subject of relatively few linguistic studies in the strict sense. Attested works are those of Maurice Tournier. His design of proximity is based on the psychological discoveries linked to Pavlovian conditioning and epistemologically accompanied the *Lexico* software by André Salem.

Co-occurrence is therefore a feature closely linked to textual analysis tools. Algorithms vary according to how they measure proximity. Some actually take the number of characters into account; others are based on a count of words separating the two terms. Some consider that proximity can be measured throughout the text; others limit the relevance to one chapter, one paragraph, one sentence or one proposition (these different units of context being mostly defined according to typographical criteria). In contrast to concordance, co-occurrence most often does not take the order of words into account; we speak of pairs of co-occurrents or of the network of words associated with a pole. Some software (such as *Tropes*[20] [MAR 98] and *Weblex*[21] [HEI 04]) nevertheless offer features that distinguish association according to the order in which words appear; we therefore speak of *ordered* pairs of co-occurrents.

In contrast to concordance, co-occurrence is not attached to a typical layout of results. It can be presented either as lists or graphs. The graphs are often not oriented [OSG 59, TEI 91]. The use of oriented graphs is nevertheless attested when the order of words is taken into account. Just like concordances, co-occurrences are used to make inferences. In a logic close to automatic categorization, some tools use this measurement to build aggregates of elements that are strongly linked. These

---

18 We also speak of the (lexical) attraction (or repulsion) of two terms [HEI 98].
19 Linguists [SAU 83] distinguish the horizontal (syntagmatic) axis of a sequence of words in a sentence of the vertical (paradigmatic) axis of combinations of words that "go together" (for example, words that have the same sound, signification or distribution).
20 http://www.acetic.fr/, accessed February 8, 2010.
21 http://weblex.ens-lsh.fr/, accessed February 8, 2010.

aggregates (or clusters) offer a synthetic view of the corpus and are subject to interpretation by the analyst. In sociology, such topics maps are mobilized by the anthropology of sciences [LEJ 04], in particular within *Candide* [TEI 95] which is itself based on developments of *Leximappe* [CAL 91, LAW 88, VAN 92][22].

11.2.6. *Data analysis*

In France, the works of the mathematician Jean-Pierre Benzecri gave birth to a series of techniques such as principal component analysis and factor analysis. Their use by Pierre Bourdieu deeply influenced French sociology. For the author of *The Distinction*, the factorial design has become a system of representation of the social space as a force field [BOU 87]. Two (orthogonal) axes cut through this space and trace four quadrants opening the now-famous combinatorial distribution of cultural and economical capitals. Besides the stroke of genius (proceeding from the comparison of the table and theory of social classes), the encounter of the author and data analysis introduces a graphical inscription of what gives a field of positions relative to one another to sociology. This relational design, dear to Bourdieu, is embodied in a two-dimensional space where there are both individuals and variables, as well as social classes and practices [BOU 98].

Bourdieu's analyses deal with figures and, in general, the data analysis belongs to the arsenal of quantitative methods. This tool was, however, originally developed to be used in linguistics. It is therefore not surprising that it has inspired the development of software commonly used as qualitative tools. It is for this reason, and with respect to the related sociological field, that these tools are mentioned here, although these programs are very close to statistics.

In France, one of the sociological[23] tools most able to take advantage of Benzecri's teachings is *Alceste*, developed by Max Reinert[24]. Even closer to automats than the felt-tip, the heart of *Alceste* lies in the descending hierarchical classification of lemmatized forms[25] of full words[26] of the analyzed corpus. This leads to a series of formally constructed classes. The ascending analysis modules (to

---

22 These tools are based on research on information systems [CHA 88, MIC 88].
23 It is in the language sciences, more than sociology, that the most orthodox applications of correspondence factor analyses are found, like the *Data and Text Mining* (DTM) tools developed as a result of *SPAD-T* by Ludovic Lebart.
24 *Alceste* is distributed by *Image*.
25 Lemmatization consists of bringing different inflected or derived forms to a common root. This operation responds to the morphological phenomena mentioned previously.
26 Automated approaches sometimes exclude a list of words from their procedures. These discarded words consist of articles, prepositions, pronouns and conjunctions or, more simply, of the most frequent words of the corpus. They are called "empty words" or "stopwords".

highlight the most typical words for each class) were then mobilized and modules of correspondence factor analysis provide the graphical representations.

In addition to the factorial plans, the results are expressed in the typical form of oriented graphs called dendograms. This tree-like setup provides the researcher with an illustration of nested aggregates [KRI 04]. As in the case of co-occurrences, these may serve as basis for inferences of the analyst. Max Reinert, who designed *Alceste*, maintains that the interest of these classes is essentially exploratory and heuristic. In so doing, he insists on the necessary complementarity of a deep knowledge of the corpus and computer tools (which help to make assumptions rather than to instrument the administration of evidence).

11.2.7. *Segments of text*

Co-occurrence and data analysis demonstrated automatic ways of labeling and segmenting texts corpora. The manual annotation of segments of texts adopts a totally different strategy. Widespread among researchers, tools based on the coding of segments of text bear the name of *Computer-Assisted Qualitative Data Analysis Software* (CAQDAS)[27]. Often claiming a relationship to the grounded theory [COR 07], CAQDAS advocates that the researcher immerse him- or herself in the corpus to be analyzed[28].

Texts belonging to the corpus are read by the researcher. He/she focuses on the passages he/she wishes to: this annotation is done by selection (nowadays, usually with the mouse) and association with a label chosen by the analyst (Figure 11.3). This way of proceeding is very similar, both in ergonomics and philosophy, to the use of the felt-tip. Concerning gesture, the selection recalls the highlighter and labeling, the annotations in the margin. Numerous software programs show a space to the right or left of the text that reproduce the layout of a sheet of paper[29]. In the analysis strategy, these tools keep the coding for the researcher.

This way of proceeding is certainly more refined than the automation described in the previously. However, it requires a demanding work, that is laborious when following an amendment of the analysis framework it involves reverting to the whole of the corpus. According to the testimony of its practitioners, its limits are a tendency to focus on coding at the expense of analysis, interpretation and theorizing.

---

27 The use of word processing, presented in section 11.2.2, is a particular case.
28 *WeftQDA* by Alex Fenton (http://www.pressure.to/qda/); and *TamsAnalyser* by Matthew Weinstein (http://tamsys.sourceforge.net/), accessed February 8, 2010.
29 *NVivo* by Lyn and Tom Richards (http://www.qsrinternational.com/) and *MaxQDA* by Udo Kuckartz (http://www.maxqda.com/), accessed February 8, 2010.

Some researchers consider this limitation to thematic statements as a pledge of scientificity; on this pont they become comparable to radical ethnomethodology that prohibits any interpretation [GAR 02].

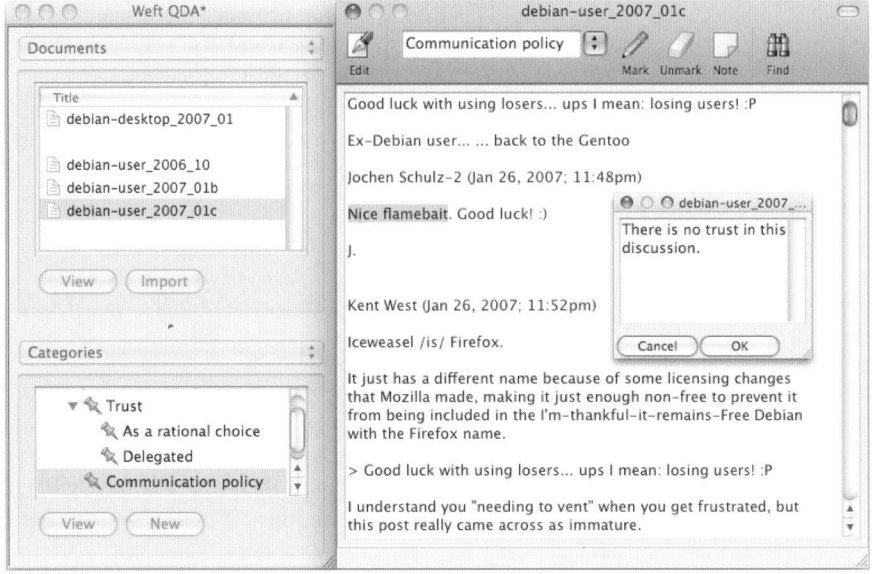

Figure 11.3. *Segments of text highlighted in the text (WeftQDA)*

If the path of the material is similar to reading on paper, these programs offer advantages linked to digitizing the corpus. It thus becomes easy to locate a passage, through both its content and the labeling that was assigned to it[30]. In the same way, limitations inherent to paper no longer apply; the researcher can thus annotate at leisure a passage that particularly inspired him/her, even if this note is long. As I have mentioned, these tools refer to the analysis "by hand" and are thus part of the techniques of content analysis. As a precursor event to these techniques, specialists have identified the controversy that accompanied the publishing of 90 hymns in Sweden in the 1640s [KRI 04]. Although published with the endorsement of the Swedish censor, the "songs of Zion" bothered the Lutheran organization. The content of the collection was the subject of controversy between supporters and critics: for each theme, the frequency and treatment in both the incriminated hymns and classic sources were evaluated. Operated in a contradictory way (by the different protagonists of the controversy), this confrontation is regarded as a forerunner of content analysis.

---

30 Such features make use of the tools presented in section 11.2.3.

Techniques of content analysis appeared in the United States at the end of the 19th Century in the quantitative, diachronic and comparative study of mass communication, in particular the press. One of the first investigations of this type was conducted on a corpus covering editions of the *New York Times* over more than 10 years. The magnitude of the subjects examined was then measured in inches (length of articles) [BER 52]. The author of this study lamented on the tendency of newspapers to give an exaggerated importance to short news items and to sensational articles at the expense of substantive articles on politics, literature and religion.

Political sciences then took over these techniques, in particular to study the propaganda broadcast during the two World Wars of the first half of the 20th Century. In doing so, the tool was sharpened in order to meet the criteria of a scientific discipline. It is from this period that the flourishing studies of political speeches are dated. It is also during this period that the first automatic systems to systematize coding operations appeared.

From the beginning of computing, researchers in human sciences have developed tools of this type [STO 66]. Of course, they were helped in this endeavor by pioneering engineers in computer science (the very ones that I have shown that forged the regular expressions in particular)[31]. CAQDAS is therefore not very young! Moreover, most of the issues related to qualitative analysis had already been formulated by this period: from the selection criteria of relevant passages and sharing them among analysts, to the necessity of interpreting coding and the validity of these interpretations, through the congruence of analysis with the content of texts.

As I have stated in this section, the central virtue of these segments of text is the appropriateness between coding and the fine-tuned meaning that the interpreter is able to extract from an intimate knowledge of the corpus. The investment required for this return is exhaustive, thorough, even (often) repeated reading. Segments of text tools are thus time-consuming. At the opposite end of the spectrum, automated tools – such as co-occurrence or Benzécri statistics – propose to save time on this substantial work. Such automation however sacrifices some finesse, since the subtleties of the participants' expression and the context of each statement are often lost. There is an intermediate way between reading (and annotation) of the whole corpus by the researcher and the transfer of this crucial task to a machine: this solution is dictionaries.

---

31 Developed in the 1990s, the *General Inquirer* is still active today (http://www.wjh.harvard.edu/~inquirer/, accessed February 8, 2010). Its designer, Philip Stone, died on January 31, 2006

11.2.8. *Dictionaries*

Dictionaries allow automation of coding while not necessarily sacrificing the finesse (and indexicality) of the common meaning studied. The use of dictionaries is based on the fact that there is a stable meaning on which we can lean. These vocabularies may include words or phrases depending on their grammatical category, their (almost) synonymy, or their relevance to the theory of the analyst. Sometimes these lists operate a selection of units on which the analysis is carried out (this is the case of lists of "stopwords" or when only one grammatical category – most often that of nouns – is taken into account).

In a predictable way, tools using grammatical categories are mainly developed within the language of sciences[32]. Those that include synonyms, argumentative records or logics of actions tend to be found in the sciences of culture (such as sociology, history or anthropology). These categories then compose an analysis framework. The question of their relevance to the corpus is variable depending on whether these directories are provided as they are in the tool[33] or are built according to the idiomatic reality of the ongoing research[34].

11.3. Conclusion: taking advantage of software

In social sciences, the features reviewed in this chapter are rarely mobilized independently of each other. It is most often by combining them that software can effectively assist the researcher. For instance:

– the Provalis Research Suite (by Normand Peladeau) provides Benzécri's analyses, concordances, dictionaries and segments of text;

– T-Lab (by Franco Lancia) combines co-occurrences and Benzécri's analyses;

– sophisticated segments of text – such as MaxQDA – often propose statistical features (or, at least, allow us to export intermediate results to statistical software);

– CAQDAS such as AtlasTI even includes dictionaries [LEW 07].

Inferences therefore proceed by crossing, comparing and cross-checking.

---

32 For example in linguistic engineering, *Unitex* by Sébastien Paumier (http://www-igm.univ-mlv.fr/~unitex/) or *Nooj* by Max Silberstein (http://www.nooj4nlp.net/); in linguistics of the corpus, *Xaira* by Lou Burnard (http://www.xaira.org) and, in statistical analysis of textual data, *Hyperbase* categorized version by Etienne Brunet or *Weblex* by Serge Heiden. Websites accessed February 8, 2010.

33 This former option is consistent with the methods of content analysis [STO 66].

34 This latter case is relevant for interpretative and grounded approaches; for this reason, I propose to speak of registers [LEJ 08].

At the end of this range of possibilities, I hope I have answered the readers' questions or have at least dispelled the shadow hovering around these mysterious software programs in sociological analysis[35]. Whatever the strategy chosen, I hope I have shown that there is no technique that ensures the scientificity or originality of research. Whether or not it is computerized, the method requires discipline and, in all cases, the quality of interpretation is always the responsibility of the scientist[36].

In qualitative sociology, the asset of analysis software lies ultimately in the facilitation offered for exchange and discussion among researchers[37].

Rather than presenting the tool, like a shield against criticism, it is important to open the black boxes and share experiences.[38]

11.4. Bibliography

[AND 92] ANDLER D. (ED.), *Introduction aux Sciences Cognitives*, Gallimard, Paris, 1992.

[BER 52] BERELSON B., *Content Analysis in Communication Research*, The Free Press, Glencoe, 1952.

[BOU 87] BOURDIEU P., *Distinction: A Social Critique of the Judgement of Taste*, Harvard University Press, Cambridge, 1987.

[BOU 98] BOURDIEU P., *Practical Reason: On the Theory of Action*, Stanford University Press, Stanford, 1998.

[CAL 86] CALLON M., "Some elements of a sociology of translation: Domestication of the scallops and the fishermen of St Brieuc Bay", in J. LAW (ed.), *Power, Action and Belief: A New Sociology of Knowledge*, p. 196-223, 1986.

[CAL 91] CALLON M., COURTIAL J.-P., TURNER W.A., BAUIN S., "From translations to problematic networks: An introduction to co-word analysis", *Information on Social Sciences*, vol. 22, no. 2, p. 191-235, 1991.

---

35 This chapter is an invitation to use the available tools: not only domestic software (presented in sections 11.2.2 and 11.2.3), but also free software, whose (spreading and modification) logic is congruent with the scientific mind.

36 Even when we use computers, the construction of an interpretation that is beyond the thematic analysis remains a real challenge. It may only be identified by avoiding the excess theory that forgets (or overwrites) the ground, and naïve empiricism, which neglects interpretative work.

37 See also [DEM 06].

38 I thank Aurélien Benel, Alex Fenton and Raphael Leplae for their advice. I dedicate this chapter to the memory of my professor of methodology and director of my doctorate thesis at the Institute of Human and Social Sciences at the University of Liege, René Doutrelepont, who died on the April 1, 2005.

[CHA 88] CHARTRON G., Analyse des corpus de données textuelles, sondage de flux d'informations, PhD Thesis, University of Paris VII, Paris, 1988.

[COR 07] Corbin, J., Strauss, A., *Basics of Qualitative Research: Techniques and Procedures for Developing Grounded Theory*, Sage, Thousand Oaks, 2007.

[DEM 06] DEMAZIERE D., BROSSAUD C., TRABAL P., VAN METER K.M., *Analyses Textuelles en Sociologie*, PUR, Rennes, 2006.

[DUP 09] DUPUY J.-P., *On the Origins of Cognitive Science: The Mechanization of the Mind*, The MIT Press, 2009.

[FAI 06] FAIRON C., SINGLER J., "I'm like, "hey, it works!": Using Glossanet to find attestations of the quotative (be) like in English- language newspapers", in A. RENOUF and A. KEHOE (eds.), *The Changing Face of Corpus Linguistics*, Rodopi, Amsterdam/New York, p. 325-336, 2006.

[FRI 06] FRIEDL J., *Mastering Regular Expressions*, O'Reilly, Sebastopol, 2006.

[GAR 02] GARFINKEL H., *Ethnomethodology's Program: Working Out Durkheim's Aphorism*, Rowman and Littlefield, Boston, 2002.

[GOO 77] GOODY J., *The Domestication of the Savage Mind*, Cambridge University Press, Cambridge, 1977.

[HAR 64] HARRIS Z.S., "Distributional structure", in J.A. FODOR and J.J. KATZ (eds.), *The Structure of Language. Readings in the Philosophy of Language*, Prentice-Hall, New Jersey, p. 33-49, 1964.

[HEI 98] HEIDEN S., LAFON P., "Cooccurrences. La CFDT de 1973 à 1992", *Des mots en Liberté, Mélanges Maurice Tournier*, vol. 1, p. 65–83, 1998.

[HEI 04] HEIDEN S., "Interface hypertextuelle à un espace de cooccurrences: implémentation dans Weblex", PURNELLE G., FAIRON C., DISTER A., (eds), *Le Pouvoir des Mots. Actes des Journées Internationales d'Analyse Statistique des Données Textuelles*, p. 577–588, Louvain, 2004.

[JEN 96] JENNY J., "Analyse de contenu et de discours dans la recherche sociologique française: pratiques micro-informatiques actuelles et potentielles", *Current Sociology*, vol. 44, no. 3, p. 279–290, 1996.

[KER 02] KERBRAT-ORECCHIONI C., "Contexte", CHARAUDEAU P., MAINGUENEAU D., (eds.), *Dictionnaire d'Analyse du Discours*, p. 134–136, Seuil, Paris, 2002.

[KLE 01] KLEIN H., "Overview of text analysis software", *Bulletin of Sociological Methodology*, vol. 70, p. 53-66, 2001.

[KRI 04] KRIPPENDORFF K., *Content Analysis. An Introduction to Its Methodology*, Sage, Thousand Oaks, 2004.

[LAP 04] LA PELLE N., "Simplifying qualitative data analysis using general purpose software tools", *Field Methods*, vol. 16, no. 1, p. 85-108, 2004.

[LAT 88] LATOUR B., *Science in Action: How to Follow Scientists and Engineers through Society*, Harvard University Press, Cambridge, 1988.

[LAW 88] LAW J., BAUIN S., COURTIAL J.-P., WHITTAKER J., "Policy and the mapping of scientific change: a co-word analysis of research into environmental acidification", *Scientometrics*, vol. 14, p. 251-264, 1988.

[LEBA 98] LEBART L., SALEM A., BERRY, L., *Exploring Textual Data*. Kluwer, Dordrecht, 1998.

[LEBE 06] LE BER J., "L'adaptation comme contraction. Une analyse informatique d'Antigone", Duteil-Mougel C., Foulquié B. (eds), *Corpus en Lettres et Sciences Sociales. Des Documents Numériques à l'Interprétation. Actes du Colloque International d'Albi "Langages et Signification"*, p. 257-268, 2006.

[LEJ 04] LEJEUNE C., "Représentations des réseaux de mots associés", PURNELLE, G., FAIRON, C., DISTER, A. (eds.), *Le Poids des mots. Actes des Journées Internationales d'Analyse Statistique des Données Textuelles (JADT)*, p. 726-736, 2004.

[LEJ 08] LEJEUNE C., "Au fil de l'interprétation. L'apport des registres aux logiciels d'analyse qualitative", *Swiss Journal of Sociology*, vol. 34, no. 3, p. 593-603, 2008.

[LEW 07] LEWINS, A., SILVER, C., *Using Software in Qualitative Research. A Step-by-Step Guide,* Sage, London, 2007.

[LUH 60] LUHN H.P., "Keyword-in-Context Index for Technical Literature", *American Documentation*, vol. 11, n° 4, p. 288-295, 1960.

[LUH 66] LUHN H.P., "Keyword-in-Context Index for technical literature (KWIC Index)", in D.G. Hays (ed.), *Readings in Automatic Language Processing*, Elsevier, New York, p. 159-167, 1966.

[LYO 95] LYONS J., *Linguistic Semantics: An Introduction*, Cambridge University Press, Paris, 1995.

[MAI 98] MAINGUENEAU D., *Analyser les Textes de Communication*, Dunod, Paris, 1998.

[MAR 98] MARCHAND P., *L'Analyse du Discours Assistée par Ordinateur*, Armand Colin, Paris, 1998.

[MCC 43] MCCULLOCH W., PITTS W., "A logical calculus of the ideas immanent in nervous activity", *Bulletin of Math. Biophysics*, vol. 5, pp. 115-133, 1943.

[MEY 89] MEYNET R., *L'Analyse rhétorique. Une nouvelle Méthode pour Comprendre la Bible. Textes Fondateurs et Exposé Systématique*, Cerf, Paris, 1989.

[MIC 88] MICHELET B., L'analyse des associations, PhD Thesis, University Paris 7, Paris, 1988.

[MOR 91] MORSE J., "Analysing unstructured interactive interviews using the Macintosh computer", *Qualitative Health Research*, vol. 1, no. 1, p. 117-122, 1991.

[OSG 59] OSGOOD C., "The representational model and relevant research methods", in I. DE SOLA POOL (ed.), *Trends in Content Analysis*, University of Illinois Press, Urbana, p. 33-88, 1959.

[PIN 06] PINCEMIN B., ISSAC F., CHANOVE M., MATHIEU-COLAS M., "Concordanciers: Thème et variations", in J.-M. VIPREY, A. LELU, C. CONDE and M. SILBERZTEIN (eds.), *Actes des Journées Internationales d'Analyse Statistique des Données Textuelles*, Besançon, Presses Universitaires de Franche-Comté, vol. 2, p. 773-784, 2006.

[POP 97] POPPING R., "Computer programs for the analysis of texts and transcripts", in C.W. ROBERTS (ed.), *Text Analysis for the Social Sciences. Methods for Drawing Statistical Inferences From Texts and Transcripts*, Lawrence Erlbaum, New Jersey, p. 209-221, 1997.

[SAU 83] DE SAUSSURE F., *Course in General Linguistics*, Open Court, London, 1983.

[SCH 08] SCHWARTZ R., PHOENIX T., FOY D., *Learning Perl*, O'Reilly, Sebastopol, 2008.

[SEK 95] SEKHRAOUI M., Concordances: Histoire, méthodes et pratique. PhD thesis, University of la Sorbonne nouvelle Paris 3 and École normale supérieure de Fontenay St-Cloud, Paris, 1995.

[SER 02] SERRES M., *Origins of Geometry*, Clinamen Press, Manchester, 2002.

[SIL 00] SILBERZTEIN M., INTEX, User Manual, 2000, available at http://intex.univ-fcomte.fr/, accessed February 8, 2010.

[STO 66] STONE P.J., DUNPHY D.C., SMITH M.S., OGILVIE D.M., *The General Inquirer: A Computer Approach to Content Analysis*, MIT Press, Cambridge, United States, 1966.

[TEI 95] TEIL G., LATOUR B, "The Hume machine: can association networks do more than formal rules?", *Stanford Humanities Review*, vol. 4, no. 2, pp. 47-65, 1995.

[THO 68] THOMPSON K., "Programming techniques: Regular expression search algorithm", *Communications of the Association of Computer Machinery*, vol. 11, no. 6, p. 419-422, ACM Press, New York, 1968.

[VAN 92] VAN METER K.M., TURNER W.A., "A cognitive map of sociological AIDS research", *Current Sociology*, vol. 40, no. 3, p. 123-134, 1992.

[WEIN 04] WEINBERG B.H., "Predecessors of Scientific Indexing Structures in the domain of religion", *Second Conference on the History and Heritage of Scientific and Technical Information Systems*, p. 126-134, 2004.

[WEIT 95] WEITZMAN E., MILES M., *A Software Source Book. Computer Programs for Qualitative Data Analysis*, Sage, London/Thousand Oaks/New Delhi, 1995.

[WIL 01] WILMET M., "L'architectonique du 'conditionnel'", *Recherches linguistiques*, vol. 25, p. 21-44, 2001.

[YOU 63] YOUDEN W.W., "Index of the Journal of the Association of Computer Machinery", vol. 1-10 (1954-1963), *Journal of the Association of Computer Machinery*, vol. 10, no. 4, p. 583-646, 1963.

Chapter 12

Sea Peoples, Island Folk: Hypertext and Societies without Writing[1]

12.1. Introduction

We built the http://www.oceanie.org site in collaboration with Oceanians to share our experience of their cultures with them. We created a hypertext method of representing their socio-cultural universe, providing them with a considerable amount of ethnographical data on their continent. The approach that we have defined avoids abstract categories and focuses on the concrete, offering a direct hold on fundamental components of their worldview. We have modeled the representation of these data by drawing from neurosciences.

The many testimonials received from Oceanians and Internet users of diverse backgrounds lead us to believe that we have been able to meet the challenge: that of shaping a better ethnographic tool to explore the human mind that is more in line with the nature of its operations than traditional writing[2]. The creation of our site, a prototype of the project ECHO (Cultural Hypermedia Encyclopedia of Oceania), has mobilized specialized Oceanist colleagues, Oceanians, international professional institutions and associations [MAR 98].

Chapter written by Pierre MARANDA.
1 http://www.oceanie.org, accessed February 8, 2010.
2 Our prototype won a gold MIM at the International Market of Multimedia in 2001, "Training and education" category; *Yahoo* proclaimed it "Site of the Year 2001", in the "Arts and culture" category, and it was a finalist at the Stockholm Challenge 2002.

Its development continues with a team of multidisciplinary researchers consisting of anthropologists, geographers, historians, linguists, archaeologists, semioticians, political scientists and computer scientists, in synergy with Oceanians and the various national institutions of their nations[3].

12.2. A concrete and non-linear approach

The originality of ECHO is, among other innovations, its concrete approach to integration of a mode of non-linear writing and close to orality. The source corpora of the site come from writings and drawings by the first explorers of the Pacific, and more specifically from first-hand ethnographic data – recent sound recordings, photos, videos and field notes. To structure them we opted for tangible titles rather than using ethnocentric concepts embedded in Western thought, such as religion, vernacular architecture and others. Thus, instead of "traditional religion" we use "ancestors", a most significant term for Oceanians. Instead of the label "vernacular architecture", we use "home", a term equally full of meaning – needless to say – not only in Oceania. Instead of "urbanization", we use "*wantok*", which means the community of city dwellers of common origins that communicate in Pijin, the Melanesian *lingua franca*. For the Christian religions, we will have "Jesus", for Islam, "Mohammed"; characters who call out directly to Oceanians. For the global economy, we will use "*dolla*" (dollar).

Obviously, the user who wants to can find the broad categories familiar to Westerners on the site, but each time users are connected to field data. For example when clicking on the node "priest" on the *Ancestors* graph (see Figure 12.1), the Internet user will see several underlined words on the screen – hypertext references – that deploy what Westerners call "religion". Each entry, of great semantic richness, raises resonances reflecting the ways of thinking and living of Oceanians. When we present our approach to them in symposiums or workshops, they tell us that it makes sense to them and that they are inspired by it to recover identities that have become confusing. Being something that they want, they navigate from one development to another in a lexicon that they say is "gut-wrenching". The "cross-mapping" [KUP 04, LEROU 00] that results, ignoring linearity, offers a direct and interactive grasp on existential data. The Internet users – Oceanians, Oceania specialists and general public – can thus build their own knowledge in relational architectures of which they are the prime drivers.

Operating in hypertext and hypermedia "modes of resonance" allows the user to move non-linearly, as advocated by the eminent anthropologist and prehistorian André Leroi-Gourhan, who promoted a return to orality. According to this scientist's

---

3 See "crédits" on www.oceanie.org.

[LER 65] speculations on what we now call information and communication technologies (ICT), the "conventional ICT", based on writing enslaves the hand and harnesses the mind. Writing flattens thought processes by reducing them to linearity, which contrives the naturally expansive doings of the mind. In this regard, Leroi-Gourhan anticipated the advent of computerized systems for speech recognition, now operational as a kind of "return to orality". He has brilliantly demonstrated how the tool "moulds" the gesture that, in turn, moulds the thought. Along those lines he envisioned computer science as a way to create a new approach to the human phenomenon, combining biology and ethnology. He thus in a way anticipated the concept of the "cyborg"[4]. He argued that writing, that has shaped literate civilizations would become obsolete as only a minority of humans are conversant with it. How many billions of men and women remain illiterate nowadays despite all sorts of literacy programs?

However whether we use a pen and two fingers or 10 fingers on a keyboard does the enslavement not remain? Does another type of "cyborgism" prevail with the handling of the mouse? And will thought processes work in a resonance mode as exists among people without writing? A "text" – for example the site www.oceanie.org – then becomes an "echo chamber" where "the scents, colors and sounds respond to one another" (Baudelaire, *Correspondences*).

The new hypertext and hypermedia ICT, like orality, better equips us than "the phonetic commitment of the hand" [LEROI 65][5]. According to Gardin [GAR 94][6]:

---

[4] Cyborg (*cyborg*: *cyb*ernetic + *org*anism): a term invented by the author of *2001, A Space Odyssey*, in his science-fiction novel *Rendez-vous with Rama* [CLA 72]. A cyborg articulates a living organism with a mechanical prosthesis. The two form an integrated whole, such as a hacker or a simple Internet user with his/her computer. Hybrid, the cyborg can be found among people without writing since times unfathomable. We can indeed observe such "hybrids" today in societies like those of the Papua New Guinea, where men wearing masks merge with them to make them alive, as documented by Marylin Strathern who, inspired by Donna Haraway [HAR 85, MAR 93, STR 91], uses the cyborg concept in that context. Let us mention *Biomuse* systems in this regard. These were developed mainly in Australia and allow the transplantation of artificial parts to the human body. There also exists at the University of Ulm, in Germany, the bio-electronic connection of silicon to neurones of leeches that produces a transistor controlled by a neuron.

[5] p. 262. About the enslaving of the mind by writing. See also the book by the anthropologist Jack Goody, *Domesticating the Savage Mind* [GOO 77] and the very interesting contribution of linguist Alessandro Zinna in *The Invention of the Hypertext* [ZIN 02], which deals with hypertextual writing and contains several classical references on the impact of the transition from orality to writing.

[6] p. 30.

"Far from easing research on knowledge representations in our particular fields, they [the new technologies] on the contrary make it compulsory that we give precise, practical, operational answers to questions that we had been accustomed to address in the rather less sharp terms of philosophy".

12.3. Type of hypermedia modeling and implementation

The approach we devised to implement www.oceanie.org uses a non-linear modeling by "attractors" and "attraction basins". These operational concepts are found both in chaos theory and in the neurosciences; we adapted them from the latter and developed them with complements derived from resonance theories as we will explain below.

12.3.1. *A model derived from neurosciences*

Moving away from the traditional form of encyclopedias, we have designed ECHO by adapting neural network concepts linked to those of category ART (Category Adaptive Resonance Theory [MAR 10, WEE 97]). To us this tack seemed more appropriate than others to represent sociosemantic universes with the least possible distortion. We could, of course, with that strategy incur the charge of reverting to a kind of biological reductionism similar to that of Emile Durkheim and his colleagues of *L'Année Sociologique* in the early 20th Century. In their case, however, the metaphor led them to read societies as living bodies while neurosciences scrutinize the functioning of the human mind. Far from biological reductionism we line up with epistemology and the theory of morphogenesis [PET 92, PET 94], which integrates and models the cognitive and physical components of meaning beyond language and speech.

Developments of the law of Hebb [HEB 49] on cellular assemblies opened new research horizons on which artificial intelligence still draws and which motivated our approach to the construction of our notions of "attractors" and "attraction basins".

"A cellular assembly consists of a group of cells connected in a reverberation circuit, that is, a complex and interconnected loop. When an external trigger excites the cells of the loop, they are mutually excited and the whole circuit goes into reverberation" [BOW 00].

For the neuropsychologist Donal Oldings Hebb, the strength of connections (synapses) between neurons can vary diachronically. The synergistic activity of an emitting neuron (pre-synaptic) and of a receiving neuron (post-synaptic) produces a strengthening of the synaptic connection that is designated by the term *Hebbian learning*. Furthermore, the inertia of one of the two neurons causes a weakening of the connection. Hebbian learning also applies to large groups of neurons in the whole of which loops are formed. If neuron *A* excites neuron *B* which, in turn, excites a neuron *C* which returns to *A*, then the synapses that connect these neurons become stronger and increase the probability of reiterations of this loop. This positive feedback reinforces the self-stabilization of neuronal networks [JAG 99, KAM 90, PET 94].

In terms of representations, this "looping" (or "cycles" in terms of graph theory) consolidates the associations of ideas, behaviors and strategies tested in the context of pragmatics [MAR 94b], such as the "empirical deductions" that Levi-Strauss notices in the structuring of cultural universes [LEV 71]. The same holds regarding policies of matrimonial alliances: systems of generalized exchange expand the economical, social and political "circuits of reverberation" that the restricted exchange does not allow for [LEV 49]. Thus these "cycles" generate semiospheres [FON 03, LOT 90, LOT 98][7] of variable amplitudes in all societies. Take, for example, the case of the "priesthood" semiosphere: it includes women in the Anglican church while it does not in the Catholic church; the amplitude of the former's sacerdotal semiosphere is broader than the latter's[8].

The approach that we defined can model the connectivity and recursivity that generate stability in distributive networks in such semiospheres. In other words, such models developed in neurosciences and in the theory of resonance describe the consolidation of memory and cognitive structures[9], some relatively inert, others relatively flexible and innovative. Furthermore, models of the same type can account for not only neuronal dynamics but also, extrapolated, for the dynamics of societies: of their representations of the world and of themselves. They can also map out

---

7 The semotician Jacques Fontanille summarizes Lotman's [LOT 90, LOT 98] concept of semiosphere as follows: "The semiosphere is the domain in which the subjects of a culture experience meaning. The semiotic experience in the semiosphere precedes, according to Lotman, the production of speech, because it is one of its conditions. A semiosphere is primarily the domain that allows a culture to define and position itself so as to be able to interact with other cultures" [FON 03], p. 296. See also the socio-cosmic approach of the *Erasme* group [COP 98].

8 For developments concerning the semiospheres of menstruation, of extra-long penises, of widowhood, etc., see [MAR 81] and [MAR 94c], Chapter 1.

9 *Memory organization packages* and *memory organization processes*: organizational processes of the memory in synergy with *imagination structuring processes*. The scripts that imagination builds – prospective, retrospective, compensatory, etc. – build on cell assemblies to generate *memory association packages* [MAR 86, MAR 89, MAR 90] in which *imaginary structuring processes* find their groundings.

vectors of disturbance stemming from discriminatory pragmatics (for example, in regard to women).

12.3.2. *Implementation: "attractors" and "attraction basins"*

Positioning ourselves beyond linearity we address social facts not as coherent and well-articulated structures, but rather as semiospheres or constellations of representations and actions. Relations that reverberate onto one another structure the universe of meaning, which gives the impression of understanding each other to those sharing it when they communicate[10]. We operationalize the concept of semiosphere by using those of attractors and attraction basins (see below, Figure 12.1). The latter spread and radiate around the former, "words full of meaning" that "are gut-wrenching" as we are told by the Oceanians to whom www.oceanie.org provides a partial overview of their socio-cosmic universe. Each society and each culture has a repertoire of words full of meaning or "carrier categories" [LET 92] that configure them and which they configure through feedback, like women, men, gods, work, sex, etc. The semiotician François Rastier has spotted about 350 of them in industrial cultures [RAS 92][11]. We believe that it will require more than a thousand such concrete terms to represent the universe of the people of Oceania.

The attraction basins, structured by those carrier categories, overlap to varying degrees in a given population. These overlaps create communities of interpretation and of behavioral orientation, some consonant, others dissonant, the latter leading to tensions, conflicts and other social dysfunctions. Our definition of attractors and their basins is congruent with the theory of "mental places" and "lexical worlds" of the computer scientist Max Reinert, author of the text analysis system *Alceste*[12]. Reinert wrote:

"[…] we make the assumption that the speaker, during his speech, invests various successive mental places and that these places, by imposing their objects, impose at the same time their type of vocabulary. Accordingly, the statistical study of the distribution of this vocabulary should be able to trace these 'mental rooms' that the speaker has successively inhabited, noticeable trace in terms of 'lexical

---

10 The feeling of mutual understanding between speakers requires their semiospheres to overlap, whether they are consonant (for example, political correctness) or dissonant. Hence the recourse to the theory of resonance and that of circuits of reverberation which, depending on the degree of sharing of sub-semiospheres, generate friendly communities, more or less closed groups or ghettoes – consonances – and also antagonisms – dissonances – [MAR 94a, MAR 05, MAR 10, WEE 97].
11 See also [RAS 91].
12 See Chapter 11.

worlds' [...].[They are associated] with collective references absorbing codes of a culture, ideologies of a group, etc."

Reinert then brings these mental places closer to "the notion of 'social representations' or what Grize calls 'cultural preconstructs'" [REI 92]. We shall also emphasize how this theory fits with that of Lévi-Strauss' "pre-stressed symbols" [LEV 62][13] and, up to a certain point, to the concept of "keyword" in documentary language and in thesauri used for discourse analyses [GAR 91], mainly when we aim to transform a database – i.e. information – into knowledge.

We are therefore looking at the ways Oceanians have traditionally built these "mental places" that they inhabit and in which they think out tangible and intangible universes. Yet, such places – such "cultural preconstructs" – form "clusters of meanings", components of semiospheres. The same goes for cultures: ideas about ancestors or houses differ greatly in terms of their polysemy according to the semiosphere within which they develop. Thus, no European worships his ancestors like Oceanians, Asians or Africans do.

For us, a cluster of meaning takes shape around a central seed, which we refer to by the term "attractor". As we see below in Figure 12.1, an "ancestors" attractor occupies the center of a cluster of meanings. The concepts that revolve around it emanate from it at the same time that they bounce back to it (reflecting barriers in terms of Markov chains [MAR 00], "circuits of reverberation" in neurosciences and in resonance theory [MAR 05, MAR 10]). The basin of this attractor thus unfolds as "nodes" that emanate from it and which, in an inverse motion, consolidate it through their convergence on its polysemy. This modeling yields a representation of intersections of vectorial planes within a "culture" according to dynamics that are in no way linear. Indeed, the repercussions of components impacting each other at various amplitudes define the scope of the basin that the attractor generates[14]. The configurations of such universes of meanings vary spatiotemporally. On the one hand, in our model they differ synchronically from one region of Oceania to another; on the other hand, they also vary diachronically over the history of Oceania.

---

13 See Chapter 8.
14 We thus reach the philosopher Simondon's concept of transduction [SIM 64]. p. 30: "We mean by transduction a physical, biological, mental, social operation, by which an activity propagates gradually within a domain, by grounding this propagation on a structuring of the domain from place to place: each region of such structures is used in the following region as a principle of constitution". We shall note the congruity of this theorisation with that of morphogenesis (above, section 12.3.1); see also the works of François Rastier on semiotics and cognition [RAS 92]. As for computer systems moving in some way along the same lines as we do, see *Atlas.ti* (Berlin), *Thinkmap* of *Plumbdesign* (New York) and *Verbatim* [LEROU 00], which show a similar concern to represent knowledge multidimensionally.

Let us introduce the concept of "appetence" here. This is a concept of oriented graph theory proposed by Frank Harary to map out the dynamics of the attraction of a point by another. Associations between semantically high-load terms repeated from generation to generation – as, indeed, between neurons according to *Hebbian learning* – consolidate their directed linkages, i.e. appetencies (for example mussels and French-fries for Belgians or veil and women for some Muslims, etc.). Some associations have so much semantic weight – so much potency – that they eventually acquire a stereotype force with robust identity functions that "are gut-wrenching". Such appetencies generate "isosemies", that is, ways of giving the same connotations to both terms and behaviors. They polarize either consent and approval or rejections that can lead to ostracism, even conflicts[15].

Representations and cognitive mappings find an instrument that minimizes distortions in this semantic resonance approach (*Category ART* [LEB 94, MAR 94a, MAR 05, MAR 10, WEE 97]). This is because, unlike classical logic, it works in loops and cycles in accordance with the exercise of our memory, our imagination, our mind (the memory organization processes and imaginary structuring processes mentioned in footnote 9). The connections of the nodes of a basin to the attractor, as well as their distance, require the same probabilistic calculations of semantic proximity as those that Reinert's system operates and that structure clusters of meaning[16]. We can thus offer Oceanians a hypermedia "mental map" in the form of some attractors and their basins. Stimulated or at least intrigued by our graphs, they use them as sources of inspiration for writings in which they reactivate their cultural roots. It is worth noting in this connection that www.oceanie.org is used in the Intranet of high-schools in French Polynesia[17].

---

15 Some of these appetencies remain unidirectional ("unilateral" in terms of oriented graphs). For example, for cultures where we can find the metaphor "this woman is an angel", "woman" is in appetence of "angelic character". But this metaphor is not reversible to "this angel is a woman". It is therefore a unilateral appetence of "woman" to "angel". A bidirectional appetence ("bidirectional" in terms of oriented graphs), however, is reciprocal, as in the reversible metaphor "youth is the morning of life" compared to "the day is still young".

16 I designed the computer system *DiscAn – Discourse Analyser –* to implement these probabilistic calculations by combining the theory of oriented graphs with first-degree Markov chains [MAR 93, MAR 94c]. For Bayesian models of probabilistic coherence and probabilistic inference rules that regulate the levels of confidence and surprise, doubt or disconcertment see [MAR 79, MAR 93, MAR 05, TAL 01].

17 Taken aback, the reader may point out that we have reduced the thought to its associative capacity. We are working at the underlying level of speech, however, which is deployed effectively on associative bases that "pre-stress" it [LEV 62]. In this, our approach works upstream of the linearity that structures argumentative, deliberative and other forms of discourse that literate thought imposes on its subjects.

Some nodes in a basin are only connected to an attractor by relays when others feed it and are fed by it through direct inputs. For instance, in Figure 12.1 there is a direct link connecting "ancestors" and "mana" while "parures" (ornaments) on the third circle transit through "first fruits" or through "dances" on the second circle, to make it to the first circle where the immediate associates of "ancestors" are found. The length of these "paths" (a term of graph theory) – direct or indirect – manifests the degree of semantic proximity between different nodes, including their connectivity. Oceanians find food for thought and discussion in these representations of connections and the bouncing of nodes on each other.

12.4. The construction of attractors and their basins

We entrust the definition of an attractor and the construction of its basin to a "workshop leader", a recognized specialist of the subject. He proposes a first sketch of the graph for Oceanians and colleagues. Receiving contributions from various sources in connection with his draft, he ensures their harmonization and then submits an enriched version to his collaborators. Together with the members of our virtual college on the Internet and in conjunction with the Steering Committee of ECHO, the workshop leader convenes work sessions (generally in the context of annual congresses like those of the *Association for Social Anthropology for Oceania* or *European Society for Oceania*). Contributors then comment on the proposition of the graph based on the societies to which they belong or that they know well. Synthesizing this monographic knowledge, the workshop leader builds a new, revised version of the attraction basin of the attractor for which he is responsible. The result is a "holistic" vision that is relatively Pan-Oceanic that Oceanians and colleagues discuss or review according to variable contexts due to historical circumstances. This last operation remains to be done in the course of further site developments.

12.4.1. *An example of an attractor and of its basin: ancestors*

On the home page of www.oceanie.org, clicking on the "Tour" path gives access to various subjects:

– physical geography of Oceania;

– its settlement by the Oceanians;

– its colonization;

– contemporary threats to the environment and to local economies (deforestation, over-fishing, etc.); and

– other topics.

196   Digital Cognitive Technologies

Additionally on the right, at the top of the screen, a rectangle gives access to a research engine – for example, "Bougainville" will retrieve information on the journeys of this explorer in Oceania, maps and excerpts from his writings (engravings and texts). The user can compare his journeys, texts and engravings to those of other explorers of the Pacific: Mendana, Cook, etc. The search engine works in hypertext, whatever topic you explore (yam, fishing, statues, initiation, etc.) and it provides a list of weighted hypertexts depending on the degree of relevance to the topic under investigation. In www.oceanie.org our means have only allowed us to process three attractors so far: Ancestors, House and Wantok.

Let us now take up *Ancestors* and its attraction basin as an example, configured under the aegis of the workshop leader and universally-recognized specialist Alan Howard from the University of Hawaii.

Figure 12.1 displays our representation of the *Ancestors* semiosphere. We have restricted the graph to three concentric circles to ensure easy readability. The first circle places the nodes at a semantic distance 1 from the attractor, the second at a distance 2, and the third at a distance 3. These attributions of positions come from empirical data collected in the field and reviewed with the Oceanians. Some of them find that that graph suits them well whereas others may discuss it, taking this as a point from which to partially redesign the graph for a better match with their notion of ancestors. As regards diachrony, it happens that the graph will have to be modified to fit the historical contexts the society has lived through.

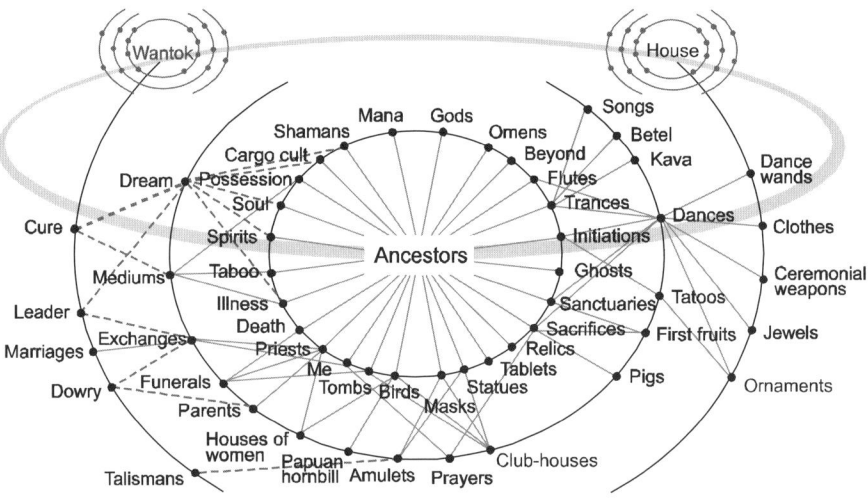

Figure 12.1. *The Ancestors graph, foreman Alan Howard*

By a simple click on a node, we obtain an image and can activate its hypertext. Photographs, video clips, sound recordings and texts show up as does the possibility of virtually manipulating some of the artifacts in the database in three dimensions. In addition, we can contribute to enrich the site by submitting reviews, comments, suggestions or other ways to see Oceania.

12.5. Conclusion

In a virtual space, such as that of www.oceanie.org – which remains nevertheless real – acceptable statements come from concordances between terms that we combine. Within a semiosphere, the "sentences" admissible according to a "cultural grammar" depend on the meaning of "words" (analogous to neurons) that they connect as sentences (analogous to synapses) that are syntactically correct. Events occur, however, that are often disturbing and new, relevant "words" are needed to handle them by forming adaptive new and acceptable "sentences", as Emile Durkheim pointed out, though formulating it differently[18]. The members of a society try to achieve this by processing upsetting events in order to reduce the unusual to the plausible, the culturally dissonant to cultural consonance. For this purpose they use clusters of meaning in which a tradition provides various resources of which the event tests the interpretative power. They will quickly eliminate some inadequate clusters and eventually rely on appropriate sub-semiospheres. They will apply the operational effectiveness of appetencies in the sub-semiospheres that they hope will be adequate to meet the challenges they have to cope with.

Those to whom a semiosphere and its cultural grammar provide the means to express themselves have to promote it or at least assume the appetencies that it upholds. Indeed, a sort of cognitive and pragmatic "jurisprudence" of acceptance, indifference, tolerance or rejection, is consolidated as a society grows and corroborates its traditions. Dynamics perpetuate themselves by sustaining what the historian Jocelyn Létourneau [LET 92] calls "horizons of expectations" stemming from the "carrier categories" of a culture. Knowing what to expect, we can live in a relatively reassuring and comfortable inertia that works as an ideological gyroscope. On the other hand what the same author calls "shameful categories" opens on "rejected horizons". Here we must emphasize that it is not a question of reducing such dynamics to simplistic binary expectations but is a matter of tendencies, the probability coefficient of which depends on a culture's appetencies – some neutral, others relatively flimsy, others more appealing, etc.

---

18 For instance, in his lectures at the high school in Sens in 1884, available on the web at: http://assets.cambridge.org/97805216/30665/sample/9780521630665ws.pdf.

To summarize, some heavy symbols, i.e. attractors and their basins, become so bound together in clusters of meaning and their associations generate such undeniable phrases – ruts of thought and feelings – that they are enshrined in dogmas, laws or other forms of social imperatives. These attractors and their basins operate as *approximation* systems. Some are relatively open, generous, fervent and friendly; others are relatively closed, hesitant, tentative and stern. The latter sometimes consolidates more or less unwavering prejudice where collective delusions, racism and hatred find their bases. The non-linear hypermedia approach that we have proposed to Oceanians attempts to map out the "hypertexts" that they have configured in semiospheres that have enabled them to maintain their identities, perpetuate them through centuries, and which they have to try and keep operational in current strenuous contexts.

In addition, Internet users from other continents use our work and override their own confines of thoughts. Evolving non-linearly from dissonance to consonance, can we hope to become better equipped to explore the founding dynamics of human relations, enlightened faith – and trust – in others.

12.6. Bibliography

[BOW 00] BOWLES R., An Introduction to Cellular Assemblies, 2000. Cell assemblies. A new type of neural network. available at http://richardbowles.tripod.com, accessed February 8, 2010.

[CLA 72] CLARKE A.C., *Rendez-vous with Rama*, Harcourt Brace, New York, 1972.

[COP 98] COPPET D., ITEANU A., *Cosmos and Society in Oceania*, Berg, Oxford, 1998.

[FON 03] FONTANILLE J., *La Sémiosphère. Sémiotique du Discours*, Presses Universitaires de Limoges, Limoges, 2003.

[GAR 91] GARDIN J.-C., *Le Calcul et la raison. Essai sur la Formalisation du Discours Savant*, EHESS, Paris, 1991.

[GAR 94] GARDIN J.-C., "Informatique et progrès dans les sciences de l'homme", *Revue Informatique et Statistique dans les Sciences humaines*, no. 30, p. 11-35, 1994.

[GOO 77] GOODY J., *The Domestication of the Savage Mind*, Cambridge University Press, Cambridge, 1977.

[HAR 85] HARAWAY D., "A manifesto for cyborgs: Science, technology, and socialist feminism in the 1980s", *Socialist Review*, vol. 80, p. 65-107, 1985.

[HEB 49] HEBB D.O., *Organization of Behavior*, John Wiley and Sons, New York, 1949.

[JAG 99] JAGOTA A., "Hopfield neural networks and self-stabilization", *Chicago Journal of Theoretical Computer Science*, Article 6, 1999. (Available at http://cjtcs.cs.uchicago.edu/articles/1999/6/ contents.html, accessed February 8, 2010.)

[KAM 90] KAMP Y., HASLER M., *Réseaux de neurones récursifs pour mémoires associatives*, Presses Polytechniques et Universitaires Remandes, Lausanne, 1990.

[KUP 04] KUPIAINEN J., "Internet, translocalization and cultural brokerage in the Pacific", in J. Kupiainen *et al.*, *Cultural Identity in Transition. Contemporary Conditions, Practices and Politics of a Global Phenomenon*, Atlantic Publishers, New Delhi, p. 344-362, 2004.

[LEB 94] LE BLANC C., "From cosmology to ontology through resonance: A Chinese interpretation of reality", in G. BIBEAU and E. CORINO (eds.), *Beyond Textuality. Ascetism and Violence in Anthropological Interpretation*, Mouton de Gruyter, Berlin, p. 57-78, 1994.

[LER 64] LEROI-GOURHAN A., *Actiona Le Geste et la Parole, I – Technique et Langage*, Albin Michel, Paris, 1964/1981.

[LER 65] LEROI-GOURHAN A., *Le Geste et la Parole. II – La Mémoire et les Rythmes*, Albin Michel, Paris, 1965.

[LER 00] LE ROUX D., VIDAL J., "VERBATIM: Qualitative data archiving and secondary analysis in a French company", *Forum: Qualitative Social Research*, vol. 1, no. 3, 2000. (Available at: http://www.qualitative-research.net/fqs-texte/3-00/3-00rouxvidal-e.pdf, accessed February 8, 2010)

[LET 92] LETOURNEAU J., "La Mise en intrigue. Configuration historico-linguistique d'une grève célébrée: Asbestos", *Recherches Sémiotiques/Semiotic Inquiry*, vol. 12, p. 53-71, 1992.

[LEV 49] LEVI-STRAUSS C., *Les Structures Elémentaires de la Parenté*, PUF, Paris, 1949.

[LEV 62] LEVI-STRAUSS C., *La Pensée Sauvage*, Plon, Paris, 1962.

[LEV 71] LEVI-STRAUSS C., "The deduction of the crane", in P. MARANDA and E. KÖNGAS MARANDA (eds.), *Structural Analysis of Oral Tradition*, University of Pennsylvania Press, Philadelphie, p. 3-21, 1971.

[LOT 90] LOTMAN Y., *Universe of the Mind: A Semiotic Theory of Culture*, Indiana University Press, Bloomington, 1990.

[LOT 98] LOTMAN Y., *La Sémiosphère*, Limoges Univeristy Press, Limoges, 1998.

[MAR 79] MARANDA P., "Myth as a cognitive map: A sketch of the Okanagan myth automaton", *UNESCO, Conference on Content Analysis*, Pisa, 1975, reprinted in W. BURGHARDT and K. HOLKER (eds.), *Textprocessing/Textverarbeitung*, Mouton de Gruyter, Hambourg/Berlin, p. 253-272, 1979.

[MAR 81] MARANDA P., "Semiotik und anthropologie", *Zeitschrift für Semiotik*, no. 3, p. 227-249, 1981.

[MAR 86] MARANDA P., "Imaginaire artificiel: Esquisse d'une approche", *Recherches Sémiotiques/Semiotic Inquiry*, vol. 5, p. 376-382, 1986.

[MAR 89] MARANDA P., "Imagination: A necessary input to artificial intelligence", *Semiotica*, vol. 77, p. 225-238, 1989.

[MAR 90] MARANDA P., "Vers une sémiotique de l'intelligence artificielle", *Degrés*, no. 18, p. a1-a15, 1990.

[MAR 93] MARANDA P., " Mother culture is watching us: Probabilistic structuralism", in E. NARDOCCHIO (ed.), *Reader Response to Literature: The Empirical Dimension*, Mouton de Gruyter, Berlin, p. 173-190, 1993.

[MAR 94a] MARANDA P., "Beyond postmodernism: Resonant anthropology", in G. BIBEAU and E. CORIN (eds.), *Beyond Textuality. Ascetism and Violence in Anthropological Interpretation*, Mouton de Gruyter, Berlin, p. 329-344, 1994.

[MAR 94b] MARANDA P., "Imagination Structuring Processes", in L.J. Slikkerveer and B. VAN HEUSDEN (eds.), *The Expert Sign: Semiotics of Culture. Towards an Interface of Ethno- and Cosmosystems*, DSWO Press, Leiden, p. 169-186, 1994.

[MAR 94c] MARANDA P., NZE-NGUEMA F.P., "*L'Unité dans la Diversité Culturelle: Une Geste Bantu*, Press of the University Laval-Agence of Cultural and Technological Cooperation, Quebec/Paris, 1994.

[MAR 98] MARANDA P., JORDAN P.-L., JOURDAN C., *Cultural Hypermedia Encyclopedia of Oceania*, Laboratory of Semiographic Research in anthropology – Centre for Research and Documentation on Oceania, Quebec-Marseille, 3rd revised and expanded edition, 1998.

[MAR 00] MARANDA P., "The bridge metaphor: Polarity or triangulation? Humanities + semiotics *versus* sciences, or humanities + semiotics + sciences?", in P. Perron *et al.* (eds.), *Semiotics as a Bridge between the Humanities and the Sciences*, Legas, New York/Ottawa/Toronto, p. 129-161, 2000.

[MAR 05] MARANDA P., "Ethnographie, hypertexte et structuralisme probabiliste (avec commentaires par André Petitat et réponse de P. Maranda)", in D. MERCURE (ed.), *L'Analyse du Social: les Modes d'Explication*, Laval Univeristy Press, Quebec, p. 183-220, 2005.

[MAR 10] MARANDA P., "Speak, that I be! Echo chambers and rhetoric", in I. Strecker and C. Meyer (eds.), *Rhetoric Culture. Theory and Exemplars*, Studies in Rhetoric Culture I, Berghan Books, Oxford/New York, 2010, in press.

[PET 92] PETITOT J., *Physique du Sens*, Editions du CNRS, Paris, 1992.

[PET 94] PETITOT J., "Physique du sens et morphodynamique", *Recherches Sémiotiques/Semiotic Inquiry*, vol. 14, p. 387-408, 1994.

[RAS 91] RASTIER F., *Sémantique et Recherches Cognitives*, PUF, Paris, 1991.

[RAS 92] RASTIER F., *La Sémantique Unifiée*, Computer Laboratory for Mechanics and Engineering Sciences (LIMSI), Orsay, 1992.

[REI 92] REINERT M., "La méthodologie "ALCESTE" et l'analyse d'un corpus de 304 récits de cauchemars d'enfants", in R. CIPRIANI and S. BOLASCO (eds.), *Ricerca Qualitativa e Computer. Teorie, Metodi e Applicazioni*, FrancoAngeli, Milan, p. 203-223, 1992.

[SIM 64] SIMONDON G., *L'Individu et sa Genèse Physico-biologique (l'Individuation à la Lumière des Notions de Forme et d'Information)*, PUF, Paris, 1964.

[STR 91] STRATHERN M., *Partial Connections*, Savage, Rowman and Littlefield, Savage, 1991.

[TAL 01] TALBOTT W., "Bayesian epistemology", *Stanford Encyclopedia of Philosophy*, Metaphysics Research Lab, CSLI, Stanford University, 2001. (Available at: http://plato.stanford.edu/entries/epistemology-bayesian/, accessed February 8, 2010).

[WEE 97] WEENINK D., "Category ART – A variation on adaptive resonance theory neural networks", *Proceedings of the Institute of Phonetic Sciences*, no. 21, p. 117-129, 1997.

[ZIN 02] ZINNA A., *L'Invention de l'Hypertexte*, Centro Internazionale di Semiotica e Linguitica, Working papers and pre-publications, Urbino, series F, no. 318, 2002.

PART V

How do ICT Support Pluralism of Interpretations?

Chapter 13

Semantic Web and Ontologies

13.1. Introduction

Since its inception on the Internet, the Web has seen many achievements based on principles, methods and tools, the efficiency of which and the number of potentially available resources[1] that no one would have predicted. The technologies developed are used daily for personal and professional uses, on the open Internet, on corporate networks and for various communities.

Many businesses were partially changed by the development of these technologies. Since the mid-1990s, the World Wide Web Consortium (W3C) has taken the initiative of developing programming languages and formalisms necessary for the communication protocols and markup in languages such as HTML for presentation, XML for certain types of content description and many others. The role of the W3C, which combines many organizations, is to produce recommendations for the Web that often become *de facto* standards after extensive open and complex processes of discussions. Despite the power of the tools made and availability of multiple sources of information, however, we cannot help but notice that the access and exploitation of these resources remain fundamentally the responsibility of human users and are heavy, time-consuming processes, even if some surface processing (statistics, etc.) can help these users, who are the only ones who can interpret their results.

Chapter written by Philippe LAUBLET.
[1] We use the term resource to refer to any file accessible by a URL Web address that contains text, images, sound, audiovisuals or a combination of these media.

"To overcome these limitations, the propositions of semantic Web[2] (SW), thanks to the W3C, first refer to the vision of the Web of tomorrow as a vast area of exchange of resources between human beings and computers, interacting among themselves first, but also with their human users, enabling a qualitatively superior exploitation of large volumes of information and varied services. To achieve this, the Web must first have common communication protocols and programming languages to represent consensual knowledge which must allow expressing through a computer the descriptive and operational knowledge associated with these resources which would allow their use by software agents" [LAU 02].

The vocabulary used to express this knowledge must be specified and shared. Reasoning should then be done automatically (by computer) using this knowledge. Computers must, therefore, be equipped with a semantics that allow some interpretations, from the implicit to the explicit, from the informal to the formal. This is why the works of SW largely focus on ontological research (in the sense of the use of this word in knowledge engineering, as explained in section 13.3). In the context of SW, these ontologies are first proposed to equip us with annotations and metadata on Web resources, particularly textual documents and fragments of these that are then enriched by a semantically-structured descriptive layer. We will take examples in section 13.4. However, different kinds of heterogeneities inevitably arise: differences in formats, syntax, storage modes and vocabulary. Thus, in a complementary way, the automatic integration of knowledge from heterogenous sources is crucial to ensure what is called semantic interoperability between the different sources, for example documentaries. In this context, in addition to technical aspects (communication protocols, languages, etc.), ontologies are essential for achieving mediations that allow exploitation of this set of heterogenous resources in a coherent manner. For example, to organize a trip a user may visit different tourist sites, each having it own data format. The interface offered to him/her uses consists of different techniques by which he/she can query data and get a consistent response.

Priority applications are the collection of documents and search for specific information, the exploitation of digital documents and access to information portals, but also, and maybe increasingly, the use and combination of many types of services for different users of Web resources. Of course, the prospects opened by SW are not simply technological, but also social and economical. They can change and enrich many individual and collective activities, particularly the cooperative ones (see Part VI of this book), by facilitating production, access and exploitation of digital resources in their diversity.

---

2 This use of the word semantic certainly merits discussion like the one developed in [USC 03]. Its meaning is often ambiguous, between its intuitive use and formal semantics in the sense of a logic and more founded approach.

However, the search for solutions can lead to different points of view on the future of SW, laying more or less emphasis on the automation or, conversely, on the user. In this chapter, we will attempt to explain the SW project, which is first a vast program of enrichment of Web resources by a layer of semantic representation of content, even if this will not suffice to describe the magnitude and potential impacts of such a perspective. We will discuss the real scope of SW and of different issues raised by such an approach. We will particularly query the significance of the ontological attempts that oscillate between standardization of descriptions and multiple interpretations. In the following section, however, we will begin to place SW with respect to the current Web.

13.2. Semantic Web as an extension of current Web

It is important to note, as written in its first "proselytes", that:

"The semantic Web is not a separate Web, but an extension of the current Web in which information has a well defined meaning allowing computers and people to work together"[3].

This view allows us to think of and anticipate the coexistence between the Web as it exists today and the current and future achievements of SW. This coexistence is one of the conditions of the acceptance of SW. The latter will only become more widely spread if it relies on the same characteristics that made the Web successful. Among these, we can cite its universal mission and the diversity of its uses, the homogeneity of techniques used that are also transferable in business, the hypertext structure that remains to be extended, to mention only a few.

SW will also have to integrate the important developments that have already occurred. We cite:

– the diversification and structuring of resource offers and diversification of economical models as examples, even if they are still in research and development;

– the improvement of tools for design and construction of sites; and

– the emergence of new technical offers, such as the XML language as a generalized exchange format with its galaxy of associated tools.

More recently, there has been an evolution of research tools with the increased power of generalist engines, but also the emergence of specialized tools per domains or resource type. The most notable aspect of Web developments is perhaps the

---

3 [BER 01]

success of new social practices that rely on a new generation of tools like RSS[4] feed aggregations, already widely used in business, blogs and wikis[5]. We often speak of social tools for the Web[6].

Yet, despite these important developments in Web, there are still many limitations that justify the proposed contributions by the SW. Descriptions of information content are still little used. For example, search engines using keywords leave a lot of work to the user. More generally, finding the right resources is difficult and the help from these engines is not enough. Analyzing the content of pages, extracting the relevant information and combining the different results is time-consuming and sometimes tedious. In addition, the labels of HTML links are only interpretable by users and not wtih semantics usable by machines. We could say, as a paradox, that the information and services on the Web today are not very usable by machines, and perhaps decreasingly so without the aid of machines, given the ever-increasing number of resources.

In summary, we can distinguish three generations in the genealogy of the Web, which is still short. The first consists of mainly handwritten HTML pages, interpreted and used primarily by human users. The second, currently the majority, adds HTML to many pages automatically created, from databases for example. The third, which has not yet reached maturity, will see many more automatically-created pages but they will be interpreted and used both by software and human users, in the service of these users, of course.

SW will undoubtedly be an important element. We could say that the current Web stores its resources, while SW aims to allow software agents to perform different activities. Among the tasks the most studied, we have already noted in a non-exhaustive way:

– information searches based on concepts and their organization as opposed to or in addition to that based on simple keywords;

---

4 RSS technology allows a Web user to subscribe to feeds of information corresponding to his or her interests.

5 A *wiki* is a website where external users can easily add to the content, edit existing content and manage the hyperlinks between pages. A blog is a website where an author often makes relatively short texts available on certain subjects. Existing tools make it easy to write on the Web, including for those who are not computer scientists, with opportunities for comments and adding links.

6 The term *Web 2.0*, fashionable at the time of writing of this chapter, intends to bring together a set of tools and practices under a common name that tend to make the Internet user more active within different social networks to which he or she chooses to participate, thanks to new more flexible technologies and broader interactivity capabilities.

– browsing using semantically-typed links in addition to that based on links without labels, the labels of links indicating the type of relations that link the source and the target, thus allowing to better select the links followed according to the goals of the user;

– search for detailed answers to specific questions;

– customization of pages available to a user according to his/her interests.

These tasks could be considered as part of the activities of knowledge management in enterprises. Many recent projects have focused on the intersection of this problem and SW. Other families of applications are beginning to benefit from the contributions of the propositions and tools of SW. We could cite the integration of enterprise applications, electronic commerce and, more generally, Web services. We will not develop a catalog of these applications in this chapter, whose scope is more focused on knowledge models useful in the context of SW and on computerized interpretation mechanisms to which they can contribute.

In the next section, we present the notion of ontology as it is used by researchers in knowledge engineering and how it was taken over by those who are developing the tools and making achievements in SW.

13.3. Use of ontologies

In the 1990s, the borrowed term[7] of ontology became increasingly widely used in the international community of *knowledge acquisition* (now *engineering*) for the construction of knowledge-based systems. Put simply, ontologies describe *the conceptualization of a domain* in the sense of a formal structure of a part of reality as perceived by an agent. Gruber [GRU 93] defines ontology as the specification of a conceptualization. Ontology is thus an artifact designed to fix a commitment to a certain conceptualization as a set of concepts and relations accompanied by a formal characterization of the meaning ascribed to them. Ontologies as artifacts are therefore intended to be reused in different applications and to facilitate communication and exchange of information between these applications. Thus, once built and accepted, ontology must reflect an explicit consensus allowing a certain

---

7 Computer scientists were often accused of using a term from philosophy in their own way. We will distinguish here ontology as philosophical discipline, or more accurately part of metaphysics that applies to the being independently of his or her particular determinations, from ontologies as theoretical or computational artifacts. Building on ontology, it is also deciding on a certain way of being and existance of objects in the real world that the system models with a set of concepts and relations [CHA 05]. A certain affiliation thus exists between these disciplines.

level of sharing and communication of knowledge between the different agents, whether they are human beings or software.

This consensus can be more or less wide, depending on the people who want to share the conceptual model described in the ontology. They can be subject to an agreement in different communities of practice [WEN 99] that consist of people sharing common interests, cooperating and exchanging their knowledge to create a collective value useful to everyone. Furthermore, the computerized formalization of these ontologies allows the implementation of automatic reazonings in the most complex productions that provide richer interpretations of provided or acquired knowledge.

In the context of SW, although their role is not limited to this, ontologies are first used to deduce the vocabulary used by metadata, their structuring and operation. They serve, more generally, as "pivotal representation" for the integration of sources of heterogenous data, to describe Web services and, in general, wherever it is necessary to support software modules by semantic representations requiring a certain consensus. We can give an initial simplified definition:

"An ontology is a model of a given domain according to a certain view of the world. This ontology is designed as the specification of a set of concepts – for example entities, attributes, process –, their definitions, interrelations and different properties and associated constraints".

From this model, we construct, for a given application, a base of knowledge consisting of the determination and organization of a set of elements, sometimes referred to as individuals, corresponding to the concepts as well as relations between these elements and different logical assertions on these elements. These are often called instances of concepts. An example would be the University of Paris 4 as an instance of the concept of a university. An instance can be identified if necessary by several terms or aliases, for example Paris 4, Paris-4, Paris-Sorbonne, University of Paris-Sorbonne, etc.

Let us give some examples in a non-technical way and without being exhaustive. The constituents of an ontology based in a simplified manner on ontologies built for the creation of portals of several European universities will include:

– a set of concepts structured by hierarchical relations, for example *university, course, person, teacher, researcher, student, city, country*. *Teacher* is a sub-concept of *person*. We see here an example of the main principle of structure that takes different names, depending on the approaches, for example generalization/specialization or even subsumption of a sub-concept by its over-concept. We constantly use this relation as the basis of reasoning mechanisms: *a student is a person therefore ...*;

– relations between two concepts, the first concept being called the domain of the relation, the second, called the range. For example, the relation *teach* with *teacher* as a field and *course* as range, or even *member of* between *person* and *university*;

– attributes of concepts, where each value is of simple type like a number, a string, date, etc.; for *person*, his/her date of birth, e-mail, for *student*, the ones before plus his/her registration number;

– various constraints and axioms that restrict the interpretation of elements: only registered students can follow courses or even a course must have at least five registered students.

Moreover, representation languages for the most expressive ontologies, often from the family of description logics[8], allow the construction of new concepts from previous elements such as *teacher_researcher*, by combination, or even *university in a large city with economics courses* by restriction of the concept *university* using a constraint. The advantage of such constructions are the automatic reasoning mechanisms included in these languages that can be exploited, allowing a better interpretation of the knowledge introduced. Knowledge of various types can be added, following the expressiveness of languages, such as formal definitions of concepts, properties of relations, and rules specific to the domain considered. The roles of ontologies in knowledge engineering are numerous and we cannot cover their diversity in detail here. In the following section, we will focus on their specific role in relation to metadata associated with Web resources as proposed by SW. We will then come back to the notion of ontology[9].

13.4. Metadata and annotations

The establishment of SW goes through the association of descriptions interpretable by software with Web resources. Such information complementing the resources are called metadata or annotations[10], depending on the context. They can

---

8 The version proposed by the W3C, recognized as a *de facto* standard for the SW, is called OWL for Ontology Web Language. Languages are at the heart of the SW approach. By replicating the achievements of the Web for which one of the reasons for its success was standards for languages, the address system and communication protocols, HTML, URL, HTTP, etc. For SW the main languages that are syntactically based on XML are RDF, RDF(S) and OWL. Readers interested in reading more on this topic can consult Chapter 2 of [CHA 05].

9 A general article on ontologies written by Fabien Gandon is available on INRIA's website: http://interstices.info/display.jsp?id=c_17672 (accessed February 8, 2010).

10 The use of these two words varies greatly, and sometimes indiscriminately, depending on the authors.

serve in different roles for both the research of these resources and exploitation for different tasks. These concepts were not invented by the designers of SW and have been widely used in documentary engineering for a long time. They are found at the heart of the SW approach, treated with great diversity according to their role in relation to applications, user objectives and choice of realization[11]. Furthermore, their relational structuring allows a large expansion in their uses.

The first and most easily characterized metadata of a first family can be termed documentaries. They describe the conditions of production of resources: who? when? where? and how? They are obviously widely used in many information search tasks. In the context of SW, the set of documentary metadata most widely reproduced and designed as the standard is the *Dublin* Core[12]. Its main descriptors are: *title, creator, subject, description, publisher, contributor, date, type, format, identifier, source, language, relation, coverage, rights*. Each of the descriptors is defined by a set of 10 attributes from the standard description ISO/IEC 11179[13] for data elements. Of course, the *Dublin Core*, and extensions that exist for different types of resources, can be used independently of SW, but it fits in many achievements of SW. Metadata, corresponding to *Dublin Core* are then expressed in a resource description framework (RDF) and are often integrated with other sets of metadata. The example of CanCore[14] is good from this point of view. It is a standard Canadian source of metadata for research and location of online educational resources (or parts of these resources) called "educational objects". They include:

– general metadata;

– attributes of the educational object (title, language, subject, description, etc.);

– information on the lifecycle of the object;

– the circumstances of its development (name, date of publication, information of publication and of version, etc.);

– meta-metadata on the file of metadata (contributors, language, date, validation);

– technical and educational metadata (technical format, size, location, etc.);

– educational (type of resources, context, age level, etc.); or

– on rights.

---

11 We will not detail here all the types of annotations and metadata proposed by the researchers. Yannick Prié and Serge Garlatti [PRI 05] offer a very useful synthesis.
12 http://dublincore.org (accessed February 8, 2010).
13 http://metadata-standards.org/11179/ (accessed February 8, 2010).
14 http://www.cancore.ca (accessed February 8, 2010).

The second major family of metadata concerns the content of resources, whether these are texts, images or audiovisual resources. They refer to the entire resource or a portion of it as a textual fragment. They tend to indicate what it is about, what it refers to and, more generally, they annotate these resources with concepts, instances or relations. In the approaches that we want to highlight, concepts and relations of the ontology of the domain (see below) are the basis of these annotations. If we simply take the example of a search engine, the latter could better satisfy the needs of its users by using the different relations of ontology and in particular that of subsumption.

In general, a distinguishing factor is that of the choice of specific language of the SW in which these metadata/annotations are expressed. The choice of language corresponds to the desired level of structuring. This can range from simple keywords to conceptual structures of a very different nature and formal complexity. The most frequently used structuring is offered by the RDF language that allows us to associate a set of relational triples to a resource or part of a resource. Of course, the use of metadata and annotations, produced for example in the context of collaborative activities of a work or relational community, can be done outside the framework of SW. However, the use of the same standard representation languages of SW allows the construction of a system and exploitation by the same tools. Similarly, the fact that ontologies support these representations offers a more advanced exploitation.

In this context of the use of ontologies to build this descriptive layer of resources, researches sometimes requiring the involvement of multidisciplinary teams (computer science/human and social sciences) should be conducted on some fundamental themes in the domain. We will mention a few points [PRI 05]:

– Should we annotate with an existing ontology or build an adapted ontology during the annotation?

– What about the notion of indexing the document by the author him- or herself?

– Can we trust a user to correctly describe his/her own documents if we compare him/her, for example, with the function of a documentalist or an archivist?

– How can we take into account the problems of rights of annotations (spreading of metadata, responsibility, privacy)?

A metadata can be published as part of a task, which assumes its author can control its use. However, unexpected uses can arise. There is also the problem of "validity" of these metadata in relation to the evolution of documents, and in terms of time (lifecycle, quality and validation of metadata/annotations).

13.5. Diversity of ontologies debated

Ontologies are at the heart of interpretation processes at work in the realizations of SW. Those carried out so far have been very diverse in nature and it would be difficult to give an exhaustive description of the approaches adopted. Nevertheless, several classifications of ontologies have been proposed in knowledge engineering [GUA 98, VAN 97]. By taking the object of conceptualization as the criterion, we often distinguish:

– The ontologies of a domain are undoubtedly the most numerous ones that can be very specific and correspond to a given occupation in a particular society. They can also cover a wide application domain, such as an ontology of tourism [ROU 02], an ontology to describe audiovisuals [ISA 04], or the conceptual reference model (CRM) for the exchange of information on cultural objects and museums [DOE 03][15].

– Ontologies of problem-solving methods where concepts used in the reasoning appear as the concepts of sign or syndrome within the medical reasoning [CHA 05] or the plans, the steps in the framework of reasoning for planning, or even the hypothesis concept for diagnostic methods. We also speak, with some nuances, of ontologies of tasks.

– Ontologies of application, double specialization of domain ontology and of a problem solving ontology. In these, the concepts of the domain will play the role assigned by the concepts of problem solving.

– Representation ontologies "which identify and organize the primitives of the logic theory to represent other ontologies, for example the *frame ontology of Ontolingua* [GRU 93] or ontology of properties by Guarino and Welty [GUA 00]" [CHA 05].

– Generic ontologies that intend to model more general concepts. A first approach seeks to identify hierarchies of the most abstract concepts allowing a first hierarchical categorization to which we will try to hang the concepts of domain ontology. We will find concepts such as that of animated entities or events. More formal approaches and, probably more interesting without being contradictory, address given cognitive phenomena to obtain ontologies of time, space, events/states or, even, constituent parts of a whole (often called mereologic). It is worth mentioning the work of the Laboratory for Applied Ontology[16] Trento with DOLCE (descriptive ontology for linguistic and cognitive engineering*)*.

---

15 This classification of domain ontologies should be refined. The difficulty is that it is a continuum according to several criteria.
16 http://www.loa-cnr.it/ (accessed February 8, 2010).

It is interesting to recall at this stage the difference between ontologies and lexicons or simple terminologies[17], whose coverage can be very broad, but that only offer descriptions in a natural language that cannot be interpreted by software and with only a very simple relationship between terms. In the case of ontologies, the coverage is often narrower, but with a high level of detail and strict formal relations between concepts that allow software to construct an interpretation from them. With the exception of some generic ontologies that are essentially theoretical, most of the ontologies we are talking about are built to be the semantic support of computer applications created to satisfy different users. This reality is essential in relation to two debates:

– The first concerns built ontologies and their relative scope. It seems worthwhile to note that ontologies that can actually be used by computer applications involve choices, *ontological commitments*, which are justified by the application to be developed and perspective of objects in the concept extensions. For example, the hierarchy of sub-concepts of events in the CRM is specific to museum objects with, for example, concepts of acquisition or transfer of custody. These considerations question the purpose of the ontologies initially displayed and are always followed by those who want to offer general and almost universal models of the world. In these approaches, ontologies should be independent of any task and practice. They thus reach a form of universality that offers a large potential for reuse. Of course, this debate is not irrelevant to long-term reflections in many disciplines on how to categorize, classify, identify and index. This does not mean that ontologies with a certain level of genericity cannot have a use, particularly to initiate domain ontologies. The concepts introduced in domain ontology are introduced as specializations of concepts of generic ontology, which thus offer a structuring framework and a support in knowing what we seek. These reflections question in particular the significance of ontological attempts that oscillate between normalization of descriptions and multiple interpretations.

– The second debate focuses on the necessary level of formalization. It can be relatively weak if the ontology is designed either to be essentially accessed by the human user, or to be used mainly to provide a vocabulary to annotate documents, for example. However, this level of formalization must be higher if different tasks are delegated to software agents. In more complex applications, ontology can also be the basis of automated inferences. For this it should be formally defined in the sense that the language in which it is described must have a formal semantics that describe the authorized manipulations. In the context of SW, this language is OWL (ontology Web language). It respects the expressiveness of *description logics*, the most recent of computer formalisms of knowledge representation from artificial intelligence.

---

17 There are many intermediary realizations between simple terminologies and more complex ontologies. This is not the purpose of this contribution.

These fragments of mathematical logic allow us to associate the expressiveness of graph formalisms with the correction of logical formalisms.

There is thus tension between the two roles of ontologies, both as a subject of consensus for humans who interpret it in the context of their activity and a formal object allowing its use by a computer[18]. In fact, they should allow content usable by the machine to connect significance for humans. This double constraint, which we find wherever ontologies are used, requires that we develop construction methodologies, an issue on which the *knowledge engineering* community is working [CHA 05]. These methodologies must take into account these two aspects by integrating the interests and roles of different people involved in the construction processes.

It should also be noted that, in the context of SW, the issues of evolution and combined use of several ontologies from various sources (fusion, alignment, etc.) is particularly important, as is research on the processes of distributed uses of ontologies. Indeed, as we have already highlighted, the diversity of distributed sources of information and their heterogeneity are one of the main difficulties encountered by Web users today. This heterogeneity can arise from the format or structure of sources, mode of access or semantic heterogeneity between the conceptual models or the associated ontologies. Taking these problems into account is one of the keys to the establishment of a set of applications in the context of SW, particularly for the more complex or ambitious applications. These must be carried out in real and computerized, intrinsically heterogenous frameworks. SW should then largely take advantage of research already carried out in information integration, particularly the creations of so-called mediation systems [ROU 02]. Indeed these are software programs capable of allowing users to access several heterogenous applications in a unified manner, by hiding this heterogeneity.

13.6. Conclusion

The different researchers and applications that are emerging seem to show that the way is open for two complementary views of the SW. The first focuses more on the creation of software tools using representations with formal semantics and powerful inferential mechanisms. They are often expensive to construct and maintain knowledge on. The second focuses on semi-formal representations and is based more on the user for their operational exploitation. For its supporters, it may, in the short term, be more flexible to create and better correspond to the cognitive

---

18 It is from this point of view that ontological researches are part of artificial intelligence, whose languages handle formal representations that are understandable by human beings and usable by software at the same time.

functions of users. The first, however, will allow for better management of different tasks by software agents and should create more trust and better Web security. The debate is open for both views of the SW and for those who express intermediate propositions.

13.7. Bibliography[19]

[BAG 05] BAGET J.-F., CANAUD E., EUZENAT J, SAÏD-HACID M., "Les langages du Web sémantique", in J. CHARLET, P. LAUBLET and C. REYNAUD (eds.), *Web Sémantique. "Interaction, information, intelligence"*, Cépaduès, Toulouse, 2005.

[BER 01] BERNERS-LEE T., HENDLER J., LASILLA O., "The semantic Web", *Scientific American*, May 2001.

[CHA 05] CHARLET J., LAUBLET P., REYNAUD C. (eds.), *Web Sémantique. "Interaction, Information, Intelligence"*, Cépaduès, Toulouse, available at: http://www.revue-i3.org, 2005.

[DOE 03] DOERR M., HUNTER J., LAGOZE K., "Toward a core ontology for information integration", *Journal of Digital Information*, vol. 4, no. 1, 2003.

[GRU 93] GRUBER T.R., "A translation approach to portable ontology specifications", *Knowledge Acquisition*, vol. 5, p. 199-220, 1993.

[GUA 98] GUARINO N. (ed.), *Formal Ontology in Information System*, IOS Press, Amsterdam, 1998.

[GUA 00] GUARINO N., WELTY C., "Ontologies and knowledge bases", *12th International Conference on Knowledge Engineering and Knowledge Management*, Juan-les-Pins, 2000.

[ISA 04] ISAAC A., TRONCY R., "Designing an audio-visual description core ontology", *14th International Conference on Knowledge Engineering and Knowledge Management*, Northamptonshire, 2004.

[LAU 02] LAUBLET P., REYNAUD C., CHARLET J., "Sur quelques aspects du Web sémantique", *Assises du GDR I3*, Cépaduès, Nancy, 2002.

[PRI 05] PRIE Y, GARLATTI S, "Méta-données et annotations dans le Web sémantique", in J. CHARLET, P. LAUBLET and C. REYNAUD (eds.), *Web sémantique. "Interaction, information, intelligence"*, Cépaduès, Toulouse, 2005.

[ROU 02] ROUSSET M.-C., BIDAULT A., FROIDEVAUX C., GAGLIARDI H., GOUASDE F., REYNAUD C., SAFAR B., "Constructeur de médiateurs pour intégrer des sources d'information multiples et hétérogènes : le projet PICSEL", *Revue I3*, vol. 2, no. 1, p. 5-59, 2002.

---

19 Interested readers can find a complete bibliography in the different chapters of [CHA 05].

[USC 03] USCHOLD M., "Where are the semantics in the semantic Web", *AI Magazine*, vol. 24, no. 3, 2003.

[VAN 97] VAN HEIJST G., SCHREIBER A., WIELINGA B., "Using explicit ontologies in KBS development", *International Journal of Human Computer Studies*, vol. 46, p. 183-292, 1997.

[WEN 99] WENGER E. (Ed.), *Communities of Practice: Learning, Meaning and Identity. Learning in doing: social, cognitive and computational Perspectives*, Cambridge University Press, Cambridge, 1999.

Chapter 14

Interrelations between Types of Analysis and Types of Interpretation

14.1. Introduction

As a first step [VAN 99], we will develop a *general outline* of research in human and social sciences (HSS) to try to grasp its complexity without reducing its great diversity. In a second step, we will take into account the important role of information and communication technologies (ICT) [VAN 03a] in the use of this general outline and in the study of interrelations between types of analysis and types of interpretation in HSS [VAN 03b]. To end, we will mention certain issues that future researchers could address [VAN 03c].

The general outline that we are talking about could also be called an "analysis path". We can limit it in a first instance to the field of statistical analysis in social sciences. The four rather "classic" stages of this general outline of social science research are:

– the choice of variables used to characterize the data provided by or on each individual in the population studied;

– the choice of individuals who constitute the population studied;

– the coding and/or re-coding of information contained in the initial data;

– the analysis methods that are used to process or transform these initial data to provide results.

Chapter written by Karl M. VAN METER.

To be complete, it should be recognized that this general outline already has an initial stage (stage 0): the transformation of the information contained in an internal dialog or an abstract mental representation (a "social representation" in the sense of Serge Moscovici [ABR 92, BUS 01, MOS 84, MOS 00, MOS 03]) into recorded and manageable, if not formalized, information. Similarly, this outline should also include a terminal stage (stage 5): the transformation by the researcher of the results – provided by analysis methods – into final results, publicly presented in a speech or scientific text. This is the case, for example, with the written or oral explanation of a chart generated by a computer program. This explicit mention of "speech" and "text" in this last stage is constructed as the counterpart of the inverse transformation of stage 0, where the information passes from a state of "speech" or "text" to "recorded and manageable" format. In the first case, it is a transformation of a speech into formalized quantities while in the second case, it is the inverse transformation: the interpretation and transformation of quantitative or formalized results into a speech of qualitative nature which describes or presents these results"[1].

14.2. ICT and choice within the general outline of HSS research

ICT can intervene at all stages of this outline. The evolution and development of these technologies are of great influence on the progress of research as it is characterized by the general outline of six stages (from 0 to 5). Their inclusion is not new: "Given the increasing availability of computer programs for data analysis methods in social sciences, and in particular programs allowing the use of several analysis methods"[2], we can easily vary the choices made in stage 4 and compare the results thus obtained with those from other methods. This amounts to studying first hand the "interrelations between types of analysis and types of interpretation"[3].

Let us now consider more closely the choices involved in these interrelations thanks to the general HSS research outline and its relations with ICT developments:

– In general, stage 1 (choice of variables) is far from being established as a formalized procedure and thus manageable by ICT. It rather tends to show the incremental character of the construction of this type of knowledge and, equally, the influence of the social and institutional environment on the practice of research. Indeed, we often notice that "schools of thought" (see below) or research teams employ the same types of variables. Nevertheless, the choice of variables remains perhaps the most original and creative work of the researcher, and thus the furthest away from formalization or the construction of a system supported by a computer program or any other form of ICT.

---

1 See [VAN 99].
2 *Ibid.*
3 *Ibid.*

– Compared to stage 1, stage 2 (choice of individuals) has become an almost exact science (sample theory) and is established as a specialized discipline with its own literature and its own software [SHA 02]. It can easily cope with the complex problems of sampling [LEY 93, VAN 90].

– With stage 3, (coding/re-coding of data), software systems associated with what is called CAQDAS (computer-assisted qualitative data analysis systems), the best known of which is NVivo[4] (formerly NUD*IST), have their origins in the computerization of old manual procedures consisting of tagging pieces of text in order to arrange and classify them in various categories designed by the researcher. This comes to "cut and paste" for texts or whole corpora. The development of ICT has helped to develop the possibilities of re-coding and manipulation of texts, and thus to systematizing this type of operation. However, although ICT have played a crucial role in the evolution of stage 3 of the general outline of HSS research, their contribution is not entirely positive, as shown below. Indeed, the multiple possibilities and coding/re-coding facility can trivialize this operation and seem to give it an insignificant impact on the results and interpretations, which is not the case at all[5].

– Stage 4 (choice of analysis method) may be the stage of the general outline the most influenced by the development of ICT. This not only concerns computer development of the various analysis methods already existing[6], but also the computerization of methods previously carried out "by hand" (like the KEB – Key Enlargement Batch – mentioned below) or even analysis methods that only exist as a concept and do not have operational implementation (like some *Monte Carlo* simulation methods, used mainly in nuclear physics[7]). Apart from the development of methods, ICT has enabled the storage[8], cataloging and wider, faster and,

---

4 http://www.qsrinternational.com (accessed February 9, 2010).
5 See our article [VAN 83a].
6 See: the American Statistical Association, http://www.amstat.org/; Survey Research Methods Section, http://www.amstat.org/sections/srms/; Royal Statistical Society, http://www.rss.org.uk; StatsNet, http://www.statsnetbase.com; SOSIG statistics page, http://www.sosig.ac.uk/statistics/; Virtual Library: Statistics, http://www.stat.ufl.edu; Michael Friendly's Statistics and Statistical Graphics Resources, http://www.math.yorku.ca/SCS/Stat Resource.html; and also: http://www.stat.washington.Edu/raftery/Research/Soc/soc_software.html (websites accessed February 9, 2010).
7 Monte Carlo simulation has been used successfully to simulate chain reactions in nuclear reactors. In the environment of computer simulation of social phenomena, it is said that the success of Monte Carlo models in nuclear physics has inspired their use in HSS, leading to ICT progress in these fields. See also [GIL 05].
8 Data sources: Statistical Resources on the Web, http://www.lib.umich.edu/ govdocs/ stats.html; Statistics, http://www.statistics.com/; Social Change, http://gsociology.icaap.org /data.htm; Social Policy Virtual Library, http://vlib.org/SocialSciences (websites accessed February 9, 2010).

sometimes, free[9] distribution of available methods. In addition, important software Websites have been developed[10], such as Sage Publications[11], the International Statistical Institute[12] or even ProGamma[13].

14.3. Questions relating to initial data and presentation of results

The two questions of substance, mentioned earlier, clearly apply to any empirical approach in HSS that is not confined to the statistical analysis in social sciences:

– A. What are the properties and structure of the information contained in the initial data?;

– B. What is the relation between this initial information and the results announced or presented by the researcher following analysis of these data?

Let us therefore consider the relationship between the four "classic" stages of the general outline more closely with regard to these two types of problems, which are of crucial importance.

---

9 This is the case for the software *Trideux* developed by Cibois [CIB 03].

10 For example, for processing missing data: University of Texas Statistical Services, http://ssc.utexas.edu/; AMELIA (software program for estimating missing data), http://gking.harvard.edu/stats.shtml. For graphics software, see: *BTS's Guide to Good Statistical Practice*, http://www.bts.gov/publications/guide_to_good_statistical_practice_in_the_transportation_field/pdf/entire.pdf; US Energy Information Administration, http://www.eia.doe.gov/cneaf/electricity/2008forms/consolidate_923.html#E923; *Gallery of Data Visualization*, http://www.math.yorku.ca/SCS/Gallery/. For learning and teaching, see: HyperStatistics Online, http://davidmlane.com/hyperstat/index.html; http://www.ats.ucla.edu; http://www.statsoft.com/textbook/stathome.html (websites accessed February 9, 2010).

11 At Sage Publications, before the astonishing growth of specific websites, HSS software were grouped on a base (*Scolari*) that included the software *Atlas.ti*, http://www.atlasti.com (now with Scientific Software Development), Educational Consulting Inc, http://www.skware.com, *C-I-SAID*, http://www.code-a-text.co.uk, *Decision Explorer*, http://www.banxia.com, *Diction*, http://www.dictionsoftware.com, *Ethnograph*, http://www.qualisresearch.com (with Qualis Research Associates), *HyperResearch*, http://www.researchware.com, *MAXqda*, http://www.maxqda.com (with *VERBI Software*), *Methodologist's Toolchest*, http://www.ideaworks.com (with Idea Works Inc), N6, *NVivo*, and *NVivo Merge*, http://www.qsrinternational.com (with QSR International), and *SphinxSurvey*, http://www.sphinxdevelopment.co.uk (with Sphinx Development UK). Websites accessed February 9, 2010.

12 See [SHA 02], which includes sections on links of a general nature, links to statistics learning materials, free software and data sources.

13 As with *Scolari* at Sage, the ambitious Dutch project ProGAMMA attempted to group all the HSS software on a single URL, accepting the rapid development of ICT and the ability to create your own URL.

For stages 1 and 2 relative to question A (information contained in the initial data), it is a direct determination: the information coming into the general outline is directly determined by the choice of variables and individuals. Any other determinates, such as the influence of the investigator, structure of the questionnaire, errors that can occur in data entry, the wrong choice of respondent and other such problems, are considered accidental.

For question B (the relation between initial information and results), we had the opportunity to examine the impact on the results of the introduction of fictitious variables of arbitrary values (created by the computer and added to the set of "true" data) using manual testing. We have also been able to analyze the introduction of fictitious individuals (with arbitrary answers for the set of "true" variables) separately [VAN 84, VAN 03b]. It should be noted that ICT can now provide accessible and practical ways to create and introduce individuals or/and fictitious variables in order to test the stability of the results of an analysis method and to compare interpretations. Although certain simulation methods can perform almost similar tasks, to our knowledge, the development of ICT has not yet provided a computer program that allows data reformatting by removing the fictitious individuals or fictitious variables and, with them, the closest real individuals or variables, which statistically comes to removing the data closest to values without significance. In the 1980s, we developed and implemented such a method "by hand", the KEB, which allows this type of data "cleaning" on a formal basis and not at the discretion of the researcher [VAN 83b]. In general, the introduction of fictitious variables changes the general structure of results (and thus their interpretation) very little, except at the borders of classes or partitions (whether it is a classification analysis or a factor analysis). It will, however, affect the statistical indicators (lower statistical significance, statistical variance rises, etc.). Similarly, the introduction of fictitious individuals changes the general structure of results very little but often obscures the boundaries between classes and partitions. It has a similar influence on statistical indicators such as the fictitious variables [VAN 84]. In the context of factor or classification analysis, data "cleaning" using the KEB method is very similar to an "amplification" of the final graph that accentuates the structure of the results. ICT often offer "zoom" as a basic option. Zoom does just that, however, without any other modification or addition of information. In this case, the "cleaning" is often done by grouping and recoding some of the data[14].

Let us look at stage 3 (coding and recoding information) and its relationships with questions A and B. These relationships can be abstractly characterized as a bijection or, respectively, as a function mapping of a discrete finite set onto another. This means that the data can only be modified by grouping. There may therefore be

---

14 See [VAN 03b] for graphs resulting from applying the KEB method.

information loss, but, in principle, no deformation. Compared to A and B, we need to vary the coding (thus recoding) to judge their impact on the results[15].

In the previous section on CAQDAS, we mentioned the influence of ICT in this field. For example, in research on crime cases mentioned above, we have shown that a recoding usually made in HSS (recoding by main modes for each variable) largely transforms the general data structure, while leaving the local structure (fine or basic structure) little changed. It should be noted that this type of recoding, usually called the "cleaning of a data file" is very common in social sciences. The development of ICT has not necessarily solved the problem because the "recoding" options are the basis of almost all current systems and form the heart of the CAQDAS. However, few systems seem to be affected by the consequences of such recording. They let the researcher compare the results to detect any distortion[16].

14.4. Choice of analysis methods and general outline

In stage 4 (analysis method), we can see that for question B (relation between the information at the start and the results at the end), each analysis method uses algorithms, often integrating metrics or indices of similarity unique to it, to generate results. We have stressed that, except for some rare cases, the choice of method determines the choice of algorithm and thus the results generated, without the possibility of comparing what other algorithms would do to the same data[17]. ICT have contributed greatly to this specific field of data analysis in HSS[18]. Almost all systems of analysis software (*Pajek*[19], *Spad*[20], *SAS*[21] and SPSS[22]) offer choices of algorithms at most stages of an analysis to:

– measure similarities between individuals;

– decompose matrices;

---

15 See [SCH 03].
16 [VAN 83a] shows how far such a deformation can go.
17 At that time, there were very few other ways to study the relation between stage 4 and question B than changing the method, or using several analysis methods, from where the term of "multimethod analysis" comes.
18 In 1999, we reported that the choice of similarity algorithm between individuals in a data set largely determines the structure of the classes at the output of a classification analysis (a clustering). We used the example of two different algorithms proposed in the SAS software system which, at the time, was one of the few to offer an algorithmic choice.
19 http://vlado.fmf.uni-lj.si/pub/networks/pajek/ (accessed February 9, 2010).
20 http://www.spadsoft.com/ (accessed February 9, 2010).
21 http://www.sas.com/ (accessed February 9, 2010).
22 http://www.spss.com (accessed February 9, 2010).

– determine the number of classes;

– test the significance; and

– many for other operations.

If we cannot vary the algorithms, we can often use another analysis method and compare the results. This is the case, for example, with the analysis of more than 1,000 abstracts presented at the first congress of the French Association of Sociology; French researchers applied four different methods to this same original data set[23].

In general, in stage 4, the strengths and weaknesses of an analysis method are only considerations in relation to other methods, considerations themselves based on the comparison of results of these different methods. We thus see the fundamental role of multimethod analysis, and also the fundamental contribution of ICT development in this field.

14.5. Methodological choices, schools of thought and interlanguage

Within the context of the general outline of research in HSS proposed at the start of this article, the researcher follows a path defined by his/her choices related to stages 0, 1, 2, 3, 4 and 5. Even with the explosion of opportunities offered by ICT, his/her choices are often neither arbitrary, nor objective or fixed in advance by rules. They are generated through the education and training of the researcher. These are largely, but not exclusively, socially determined and can thus constitute coherent groups of similar paths. Thus we can speak of schools of thought, invisible colleges and paradigms. Indeed, the complexity of HSS research has been historically divided into "schools of thought"[24], most often associated in each case with a specific major sociologist and his/her work:

– "invisible colleges" (well described by Crane [CRA 72] and based on analyses of "co-authorship" and "co-citation")[25]; and

---

23 The first version of the results of various analyses was presented in the *Bulletin de Méthodologie Sociologique* in 2005 [DEM 05; TRA 05; DES 05; BOU 05] and are the basis of a book [DEM 06]. The detailed confrontation and comparison of the results of these four methods have helped to reinforce some of the results and put others into perspective.

24 The term "schools of thought" only seems to have a metaphorical use and not a scientific definition.

25 Following the works of Crane and their significant influence, the analysis of "co-authorship" and of "co-citation" in scientific literature has become the standard model of the structure – and sometimes evaluation – of scientific research itself.

– "ordinary" science as opposed to "revolutionary" science (described by Kuhn [KUH 62] as the common use of scientific practices or paradigms[26]).

We note that these different ways to organize the complexity of HSS make *people* (scientists and their social circles), research *objects* (themes, their vocabularies and associated keywords) and *methods* (scientific procedures applied by *people* to *objects* to obtain results) play different roles.

The first concept of "schools of thought" and its more precise formalization as "invisible colleges" is much more focused on people than the "ordinary" and "revolutionary" sciences of Kuhn, which are more oriented towards objects, methods, and theories. Theory is moreover a fundamental component of each invisible college, paradigm or language community. It influences, even determines, many choices in the general outline of HSS research, including some interactions with ICT development in HSS. However, our analysis focuses on interactions between ICT and the general outline, which are probably less known and less discussed by researchers in HSS.

For his part, Berthelot [BER 03] mentioned the term "language communities" in relation to HSS. It is true that HSS researchers who are used to making similar choices in their paths through the general outline will be able to "talk to each other" and even exchange data and compare results and interpretations. However, if their paths are different and other choices were made, it is unlikely that this dialog would take place without recourse to a type of "translation" or to a "translator" between these methodological or language communities. The development of ICT seems to have only recently addressed this problem of communication between different methodological or language communities[27]. Yet, the modular structure of software in modules[28] and interoperability between modules and software are developments often promoted in ICT. Unfortunately, the strong development of ICT in HSS research has a negative side as ICT propose increasing choices in the context of the general outline. As a result, they potentially offer more distinct paths, making it

---

26 Paradigms are not directly observable since they are part of the social norms of scientific research and considered to correspond to sub-disciplines identified by Crane via invisible colleges.

27 See Chapter 15.

28 The software programs carry out operations that consist of several sub-routines such as addition, multiplication, variance calculation and construction of cross-tabulations. Each sub-routine is done by a computer module, and it is the whole set of these modules that makes the software.

increasingly difficult to find "translations", "translators" or an "interlanguage" that can enable different communities to communicate[29].

The concept of interlanguage developed by Berthelot may not be completely lost, however, since a "common denominator" exists thanks to the presence of the computer and its software, and the current requirement consisting of converting the scientific approach into a computerized procedure built from modules, as we mentioned above. Since these procedures and associated software are systematically modular, and modules have often been "borrowed" from other software, a good knowledge of operations at this level can open a dialog between communities associated with different types of methodologies and different types of software. It is then possible to go beyond the simple level of computer discussion related to modules and grasp and understand the integrality of the analysis process and methodological and theoretical framework described by the general outline of HSS research.

14.6. Conclusion

Thanks to ICT, the categorization and standardization of notions and concepts, such as unemployment, education and socio-professional position used by HSS, have made much progress, particularly in Europe where various national systems of categorization enjoy a historical basis. Similarly, we note that comparative analyses, made possible and more accessible by ICT, are increasingly common[30]. However, these advances, due largely to technologies, only seem to have an indirect influence on the eventual lowering of barriers that define the boundaries between:

– HSS scientific communities,

– schools of thought,

– invisible colleges,

---

29 The concern for sharing interpretations is sometimes incorporated into the software design. It is the case for example with the software *Prospéro* [CHA 03]. Thus problems of effective collaboration arise between different communities, links between a software platform and various sociological theories, evaluation of interpretations and of interpretative closure [REB 04]. Indeed, some programs are strongly committed to a theory, others are more "ecumenical", others more "syncretic", combining various possibilities for the investigation and implicitly calling for an "interpretative turn" (*ibid.*, p. 124). The link between theory and methodology in HSS is sometimes problematic, even without the use of ICT. See for example [REB 05].
30 See also Chapter 15.

– scientific paradigms, and

– language communities,

because these division lines are social and cognitive. The theoretical aspect of these distinctions is important here and often unexplained and debated. These distinctions are clearly not only dependent on technical and scientific fields and influenced by ICT development. This situation is analogous to the example of a word-by-word translation freely provided by ICT, more often without taking cultural bases of the two language communities involved into account. ICT, but more importantly their users, may one day perform this type of translation more often and more systematically.

14.7. Bibliography

[ABR 92] ABRIC J.-C., MOSCOVICI S., *Psychologie Sociale*, PUF, Paris, 1992.

[BER 03] BERTHELOT J.-M., "Language and interlanguage: reflections on modalities of interrelationships between analysis and interpretation in sociology", in K.M. VAN METER (ed.), *Interrelation between Type of Analysis and Type of Interpretation*, Peter Lang Verlag, Bern, p. 207-226, 2003.

[BOU 05] BOUDESSEUL G., "De quoi parlent les sociologiques réunis en congrès? Eléments de complémentarité entre une analyse lexicale ouverte et le cumul de varables fermés", *Bulletin de Méthodologie Sociologique*, no. 85, pp. 68-84, January 2005.

[BUS 01] BUSCHINI F., KALAMPALIKIS N. (ed.), *Penser la Vie, le Social, la Nature. Mélanges en l'Honneur de Serge Moscovici*, Editions de la Maison des Sciences de l'Homme, Paris, 2001.

[CHA 03] CHATEAURAYNAUD F., *Prospéro – Une Technologie Littéraire pour les Sciences Humaines*, Editions du CNRS, Paris, 2003.

[CIB 03] CIBOIS P., "Interaction between a research environment and a statistical technique: The Tri-Deux method", in K.M. VAN METER (ed.), *Interrelation between Type of Analysis and Type of Interpretation*, Peter Lang Verlag, Bern, p. 125-143, 2003.

[CRA 72] CRANE D., *Invisible College: Diffusion of Knowledge in Scientific Communities*, University of Chicago Press, Chicago, 1972.

[DEM 05] DEMAZIERE D., "Des logiciels d'analyse textuelle au service de l'imagination sociologique", *Bulletin de Méthodologie Sociologique*, no. 85, pp. 136, January 2005.

[DEM 06] DEMAZIÈRE D., BROSSAUD C., TRABAL P., VAN METER K.M., *Analyses Textuelles en Sociologie: Logiciels, Méthodes, Usages*, PUR, Rennes, 2006.

[DES 05] DE SAINT LEGER M., VAN METER K.M., "Cartographie du premier congrès de l'AFS avec la méthode des mots associés", *Bulletin de Méthodologie Sociologique*, no. 85, pp. 44-67, January 2005.

[GIL 05] GILBERT N., TROITZSCH K.G., *Simulation for the Social Scientist*, Open University Press, London, 2nd edition, 2005.

[KUH 62] KUHN T., *The Structure of Scientific Revolutions*, University of Chicago Press, Chicago, 1962.

[LEY 93] LEYLAND A., BARNARD M., MCKEGANEY N., "The use of capture-recapture methodology to estimate and describe covert populations: An application to female street-working prostitution in Glasgow", *Bulletin de Méthodologie Sociologique*, no. 38, p. 52-73, March, 1993.

[MOS 84] MOSCOVICI S., FARR R., *Social Representations*, CUP-MSH, Paris, 1984.

[MOS 00] MOSCOVICI S., *Social Representations: Explorations in Social Psychology*, Polity Press, London, 2000.

[MOS 03] MOSCOVICI S., "Serge Moscovici. Le père des représentations sociales. Seize contributions pour mieux comprendre", *Journal des Psychologues*, Special Issue, October 2003.

[REB 04] REBER B., "Interprétations sociologiques partagées, technologies littéraires et évaluation", *Langage et Société*, no. 109, p. 111-126, 2004.

[REB 05] REBER B., "Technologies et débat démocratique en Europe. De la participation à l'évaluation pluraliste", *French Review of Political Science*, vol. 55, no. 5-6, p. 811-833, 2005.

[SCH 03] SCHILTZ M.-A., "Influence of the choice of statistical analysis on basic operations in survey research: Coding and selection of variables", in K.M. VAN METER (ed.), *Interrelation between Type of Analysis and Type of Interpretation*, Peter Lang Verlag, Bern, p. 43-73, 2003.

[SHA 02] SHACKMAN G., "Free statistical tools on the Web", *International Statistical Institute Newsletter*, vol. 26, no. 1 (76), 2002.

[TRA 05] TRABAL P., "Le logiciel prospéro à l'épreuve d'un corpus de résumés sociologique", *Bulletin de Méthodologie Sociologique*, no. 85, pp. 10-43, January 2005.

[VAN 83a] VAN METER K.M., "Sociologie de la criminalité d'affaires – Deux méthodes différentes, deux représentations sociales distinctes", *Bulletin de Méthodologie Sociologique*, no. 1, p. 3-18, 1983.

[VAN 83b] VAN METER K.M., "Le KEB, amplification d'images ou nettoyage du bruit de fond dans les enquêtes en sciences sociales – le cas de l'analyse factorielle des correspondances", *MSH Information*, vol. 45, p. 62-72, 1983.

[VAN 84] VAN METER K.M., "L'introduction de variables arbitraires dans les méthodes d'analyse en sciences sociales", *Annals du Quatrième Congrès Reconnaissance de Formes et Intelligence Articielle*, AFECT-INRIA, Paris, p. 465-479, 1984.

[VAN 90] VAN METER K.M., "Methodological and design issues: Techniques for assessing the representatives of snowball samples", in E.Y. LAMBERT and W.W. WIEBEL (eds.), *The Collection and Interpretation of Data from Hidden Populations*, Washington, Department of Health and Human Services, Research Monograph Series, no. 98, p. 31-43, 1990.

[VAN 99] VAN METER K.M., "Typologie de base, ensembles flous et analyse multi-méthode en sciences sociales", in N. RAMOGNIGNO and G. NOULE (eds.), *Sociologie et Normativité Scientifique*, Presses Universitaires du Mirail, Toulouse, p. 233-265, 1999.

[VAN 03a] VAN METER K.M. (ed.), *Interrelation between Type of Analysis and Type of Interpretation*, Peter Lang Verlag, Bern, 2003.

[VAN 03b] VAN METER, K.M., "Multimethod analysis and stability of interpretations", in K.M. VAN METER (ed.), *Interrelation between Type of Analysis and Type of Interpretation*, Peter Lang Verlag, Bern, p. 91-124, 2003.

[VAN 03c] VAN METER K.M., "Conclusions and future research: A structure for world sociology", in K.M. VAN METER (ed.), *Interrelation between Type of Analysis and Type of Interpretation*, Peter Lang Verlag, Bern, p. 227-246, 2003.

Chapter 15

Pluralism and Plurality of Interpretations

15.1. Introduction

In this chapter we will first distinguish pluralism and the plurality of interpretations. Thereafter, we will focus on the plurality of interpretations that unfold in the chain of operations of discursive materials using a computer. We will show, through two examples, how computer resources allow the validation of interpretations at work in different research operations. We will first examine the categorization process from two research experiments[1] using *Sato* software. We will then focus on the validation of interpreting results using the triangulation methods proposed by various pieces of software: *Sato*, *Alceste*, *Lexico3* and DTM[2].

15.2. Diversity of interpretations

The diversity of interpretations can be represented using two axes. The first, horizontal, refers to the possible viewpoints in the observation and analysis of the

---

Chapter written by François DAOUST and Jules DUCHASTEL.
[1] The first experiment focuses on the exploratory analysis of a corpus of interviews on the use of tobacco and the influence of anti-tobacco messages [DAO 06, GEL 04]. The second is from an exhaustive study of the representation of the political community in federal and provincial Prime Ministers' speeches in the context of the constitutional conferences in Canada from 1941 to 1992 [BOU 96].
[2] See the software reference in the bibliography.

same object, which we will designate as pluralism of interpretations. The second, vertical, designates the plurality of hermeneutical choices that are necessary at all stages of empirical research conducted from the same point of view. The presence of pluralism of interpretations arises from the division of empirical objects according to the various disciplines, as well as being over-determined by epistemological, methodological and hermeneutical points of view [DUC 99a]. Pluralism of interpretations is the product of the complexity of possible choices the researcher must make in various contexts[3].

Plurality of interpretations refers to the interpretative dimension of the overall knowledge operations produced in a research approach [DUC 99b]. We distinguish the overall interpretation for results and interpretations, which are produced at each stage of research by each methodological decision. Let us illustrate this point by showing the path covered in a process of text analysis by a computer. Analyzing a text using computer resources implies successive states of text transformation [MEU 90][4]. The transition between two text states involves transformation operations. The *initial* text is the speech itself, mobilizing cognitive, linguistic and cultural resources with its unique conditions of production and reception. Speech text is a complex object and its analysis involves a series of reductions. The first of these reductions occurs in the transition from speech to *manuscript* text. This transcript of the speech involves choices about the information to be selected. Should we only retain the words used or should we keep track of its expressive forms? Should we retain enunciation marks and the paralinguistic elements that accompany it? These choices not only determine the future of interpretation, but are already a given interpretation of a first text.

The second transition is between *manuscript* text and *electronic* text. Operations of:

– capture (manual or automated);

– normalization (spelling, syntactic, semantic, lemmatized);

– processing (of peritextual or textual information); and

– management (databases, partitions)

are all stages where choices are made on the basis of what appears important or essential to retain. Information loss is again the result of an interpretation of what is

---

3 See also Chapter 14.
4 We take the idea from Jean-Guy Meunier according to which the successive treatments of the initial text would produce so many new versions of the text: from the *speech* text to *manuscript* text, then to *electronic* text, to *representation* text, to *results* sub-texts, finally to the final *interpretative* text.

valid and a restriction of possible interpretations in the subsequent phases of research. The edited text becomes the material to which the subsequent operations of research will be applied. Based on such a matrix, a set of successive transformations of this text will produce enhanced versions of the same text. The *representation* text is thus composed by a succession of enhanced versions based on the description of its units using one or several systems of categories (morphosyntactic, semantic, rhetoric, etc.) linked to descriptive or explanatory theories. Assigning a category to a word or a textual segment based on such theories involves an interpretative choice that will also have an impact on subsequent interpretations.

From the edited text or various versions of the *representation* text, it will be possible to produce as many sub-texts combining partitions of the text and various descriptions. These sub-texts can come from both the explorations of data described and readings guided by *a priori* assumptions. In both cases, the hypotheses represent possible interpretations of the text. Finally, an overall *interpretation* arises from the reconstruction of the text based on the results of all previous sub-texts. At that stage, the object to be known is reconstructed within a new text, different from the initial speech, but with a restored complexity. This final text is that of the social sciences expert. Here lies a possible gap between the data and the meaning bestowed upon it. While the local interpretations can be controlled to the extent that we are aware of producing them through each methodological choice, the overall interpretations require new cognitive, linguistic, cultural aspects, and conditions of production and reception unique to a new speech.

15.3. Interpretations and experimental set-ups

What distinguishes computer-assisted textual analysis from the simple interpretative comment is the construction of experimental set-ups that aim to construct facts that support interpretation. According to Habert, the experimental set-up is an "assembly of instruments, tools and resources designed to produce 'facts' whose reproducibility and status (interpretation) are the subject of controversy" [HAB 05]. In his definition of experimental set-up, Habert indicates that an instrument is "a successful experimental set-up". The instrument is therefore a stabilized set-up whose directions for use and interpretations of output results are subject to a certain consensus. When we carry out a new analysis to answer new research questions or aiming at a speech whose function remains to be explained, we must develop an original experimental set-up adapted to new ideas or, at least, to assumptions that are deployed in specific discursive functions.

From a technical point of view, the experimental set-up is created by means of transparent and replicable calculation procedures, and by procedures assisted by a categorization whose record must be explicit. This means that the criteria of this categorization are clearly expressed and that it is possible to return to the corpus to identify the tags that are the physical hallmark of codification. Thus, the controversy of interpretation can rely on the close discussion of the establishment procedures on which a fact relies. The use of computerized procedures is also intended to allow the coexistence of several experimental set-ups built on the same corpus. These set-ups, which form different viewpoints and theoretical perspectives, can provide support these various viewpoints and multiplicity of interpretative paths corresponding to the intrinsic plural nature of reading.

15.4. Exploratory analysis and iterative construction of grids of categories

To illustrate the process of iterative construction of a grid of categories based on an exploratory analysis of a corpus, we will use the case of a research on a corpus of interviews on the use of tobacco and the influence of anti-tobacco messages [GEL 04]. The interviews were conducted in 2000 among 48 young French people divided into nine groups.

The starting point of the approach is to compare, using simple statistical indices, lexicons associated with sub-texts cut according to the stratification variables established at the start:

– sex;

– smoker/non-smoker;

– before/after the anti-tobacco message.

The establishment of the corpus is thus in an interpretative context that seeks to address a certain number of research questions related to the influence of anti-tobacco messages on a targeted public. Thus each interview is conducted in two phases. It begins with a first exchange where the interviewer asks various questions. Then, the interviewer introduces an anti-tobacco pamphlet and the discussion and continues to discuss of this dissuasive message.

To compare the lexicons according to the profiles of the interviewees and the stages of the interview, researchers have used an algorithm of lexical distance based on the distance from Chi-square. The measure evaluates the difference in the use of a given vocabulary between two subsets of the corpus. The lexical forms are sorted in descending order of contribution to the measurement of distance, which makes it possible to identify, in order of importance, the specifics of each sub-text. The researchers have also used an algorithm of participation that computes the standardized averages of a set of lexical forms, generally corresponding to a lexical category, for each sub-text formed during analysis. This exploratory approach carried out using *Sato* software [DAO 05] is based on an interactive to-and-fro movement between what is shown by the lexical analysis and the context in which words highlighted by algorithms of distance and participation are used.

Thus, the application of the algorithm of distance on the lexical frequencies associated with statements before and after the introduction of the anti-smoking brochure revealed that the words describing physical appearance and health in general are those that most characterize the vocabulary before the introduction of the brochure. We also find words referring to pleasure and addiction. In contrast, after introduction of the brochure, we find the following words at the top of the list: *testimony, concrete, solution, numbers, death*.

The transition of the lexicon of non-described forms to categorized lexicon, resulting from an explicit experimental set-up permits a level of interpretation with a more general scope. For example, Table 15.1 allows us to view the over- and under-representation of a category of the grid in various sub-texts grouping statements of interviewees according to their sociological profile. Here, the *participation* analyzer computes the relative frequency of the category *death*. A and B show before and after the introduction of the pamphlet. We also have the *sm* and *ns* for smoker and non-smoker, as well as *m* and *f* for man and woman. It is noted that the theme of death is more apparent after the introduction of the pamphlet and that this positive salience is characteristic of non-smokers of both sexes, and of women, whether smokers or not.

Based on lexical analysis of these raw data, interpretative models were developed to include lexical anchor points in semantic[5] and enunciative system categories able to translate, in the speech itself, the attitude of young people in relation to smoking and the influence of dissuasive advertising. Categorization was intended to bridge the research problem and textual data by comparing the statements before and after introduction of the brochure according to the sociological profile of the subjects. Following the categorization of characteristic words, researchers conducted a systematic review of words sorted in alphabetical order to identify relevant inflectional variants. For example, the *distance* analyzer showed a significant presence of the pronoun "I" after the introduction of the brochure, which suggested that the brochure could have caused the subjects to feel more personally involved. To validate this hypothesis, it was necessary to add all the pronouns referring to the first person to the category. It was assumed that this would be a sign of appropriation of the statement by the enunciator. To confirm their

---

5 The grid translates, in 28 categories, the lexical anchors of interpretation (the *soc-* prefix refers to a set of categories referring to social relationships identified by young people): appearance, arrest, denial, concrete, danger, addiction, soc-I, illness, death, pleasure, advertising, tobacco, nicotine, drug, prohibition, smoker, soc-friend, soc-family, soc-people, freedom, desire, consciousness, will, soc-young, costs, start, health, education, prevention.

intuitions, researchers repeated distance and participation analyses by applying them on the categories of the grid and their frequency. For the category *soc-I*, the frequency difference observed for the form *I* disappeared. The interpretation of the *I* phenomenon was thus elsewhere.

Property	Coverage	Lexemes	Occurrences	Z Score
Total Freqency	78,703/78,703 (100%)	9/3,985 (0.23%)	80/7,8703 (0.10%)	0.00
A	23,544/78,703 (29.91%)	4/2,087 (0.19%)	19/235,440 (0.8%)	-1.01
B	28,074/7870,335 (67%)	6/2,351 (0.26%)	47/28,074 (0.17%)	3.46
Asm	13,758/78,703 (17.48%)	4/1,580 (0.25%)	13/13,758 (0.09%)	-0.26
Bsm	15,923/78,703 (20.23%)	6/17,490.(34 %)	24/15,923 (0.15%)	1.94
Ans	9,786/7,870,312 (43%)	2/1,240 (0.16%)	6/9786 (0.06%)	-1.25
Bns	11,898/78,703 (15.12%)	3/1,425 (0.21%)	23/11,898 (0.19%)	3.14
Am	14,468/78,703 (18.38%)	4/1,634 (0.24%)	8/14,468 (0.06%)	-1.75
Bm	16,010/78,703 (20.34%)	4/1,797 (0.22%)	21/16,010 (0.13%)	1.17
Af	9,076/78,703 (11.53%)	2/1,153 (0.17%)	11/9,076 (0.12%)	0.58
Bf	11,811/78,703 (15.01%)	5/1,379 (0.36%)	26/1,181 (0.22%)	4.04

Table 15.1. *Participation analyzer (subject = death).Key: sm – smoker, ns – non-smoker, m – male, f – female)*

The construction of a grid of categories is based on a corpus analysis protocol that is both transparent and respectful of the specifics of the enunciation context. It is an iterative approach that combines inductive perspective, often associated with qualitative methods, use of simple tools of lexical statistics, and attention to textual pragmatics. This method has the advantage of producing qualified data that reflect the interpretative approach of the analyst by providing important anchor points for the whole vertical interpretative chain. Once established, the interpretative value of the categorization must still be validated beyond the witness corpus by extensive and methodical application of the grid on a wider corpus, as illustrated in the following section.

15.5. Categorization process from a grid

We will give the example of a categorization process that was conducted as part of research on the Canadian constitutional speech [BOU 96]. The categorization was performed from a grid of socio-semantic categories developed as part of previous research projects on corpora of political speeches in Quebec and Canada and refined at the exploratory stage of research. Two questions arose for the researchers: what impact does categorizing have and how can the stability and reliability of such an operation be ensured?

Assigning a category to a unit of speech is, first, to distinguish this unit from other units whose meaning varies and, second, to combine this unit with other units in a class of objects on the basis of a community of meaning. The researcher needs to produce a local interpretation of the meaning to be given to this unit. The traditional criteria of sound methodology require that categorization is valid, reliable and stable. The validity measures the degree of adequacy between a theoretical problem and the description of objects of observation. The reliability and stability intend to ensure that this measure of appropriateness is constant in time and space. This implies that the same choices on the same object will be made by the same coder at various times, and by different coders at the same time.

It is useful to decompose the act of categorizing along three axes to better grasp the complexity and, subsequently, to design computer strategies allowing the increased stability of interpretative choices. When the researcher is faced with a unit of speech to categorize, he or she has three criteria that can be represented along three axes: syntagmatic, paradigmatic, problematic (see Figure 15.1). Each coding unit is first on a syntagmatic axis (the sentence, the paragraph or any other contextual unit decided by the researcher). The ascription of meaning to a coding unit will take this context into account. Indeed, socio-semantic categorization must more often refer to context because of the possible variations in meaning of each unit. The second axis is paradigmatic, referring to the possible meanings of a unit. The encoder chooses the meaning corresponding to its use in context. In sociological research, it is not sufficient to know all the meanings that can be taken by a same unit; this word must be related to a grid of categories resulting from a research problematic. The encoder will therefore choose on the basis of a possible meaning (paradigmatic axis) in a given context (syntagmatic axis) a category corresponding to a grid (problematic axis) to apply it to a unit of speech (word or segment).

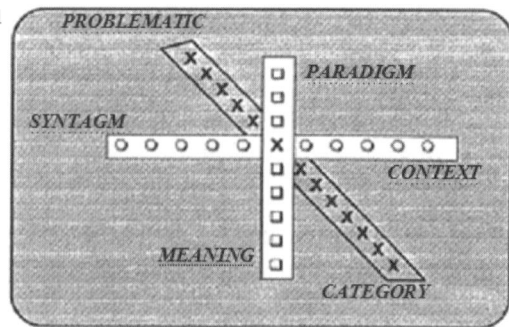

Figure 15.1. *Criteria for the identification of the meaning of words to categorize*

In our constitutional corpus, categorization was performed in a particular way. Candidates for categorization were limited to nouns and adjectives identified following the application of a morphosyntactic dictionary (LDB – Lexical Data Base – in *Sato*) and cleared of any ambiguities in context. A second dictionary of words *a priori* judged irrelevant in relation to the retained socio-political theory was also shown in the lexicon. Finally, a third dictionary of words whose meaning was judged invariable allowed us to reduce the number of candidates for categorization by about 20%. The residual nouns and adjectives were categorized in context. To ensure reliability and stability, we relied on functions of *Sato*. Each different word that was a candidate for categorization (in this case nouns and adjectives not having been eliminated in previous phases) was categorized in the same operation. We obtain, for example, a concordance of the word Quebec, which allows us to examine all the contexts in which the word appears. This method considerably increases the stability of decisions because of the synchronicity of the procedure. The word Quebec can have different meanings depending on context. It may be:

– the province of Quebec as a territory;

– the Government of Quebec (in the form: "Quebec requests...");

– the city of Quebec; or

– the people of Quebec (sovereign Quebec).

In each case, the category will vary. Two additional tools allow us to access the paradigmatic and problematic information. Faced with the task of applying a category to an occurrence of Quebec, the encoder may be interested to know what categories are available. Where such a grid of categories was applied to another corpus or in the case where decisions have already been taken on occurrences of the word Quebec in the corpus under study, it is possible to instantly obtain the list of categories already applied to the word Quebec (see Table 15.2).

Pluralism and Plurality of Interpretations 239

The categories in Tables 15.2 and 15.3 are: Socio: Socio-semantic categories; Fed: Federal Government; Que: Quebec Government; PHQ: Other Provinces' Governments; Aut: Aboriginal People; P-: Probability. The socio-semantic categories included are us7b8: city; et2b2: government; us7b3: province; us7b2: Quebec as a space; et2b2,u*:error; us2c4: nation.

Total freq	Socio	FED	QUE	PHQ	AUT	P-FED	P-QUE	P-PHQ	P-AUT	
13	nil	0	8	5	0	0	0.02	0.00	0	Quebec
36	us7b8	15	7	14	0	0.02	0.02	0.01	0	Quebec
127	et2b2	7	85	31	4	0.01	0.22	0.01	0.03	Quebec
2	us7b3	0	1	1	0	0	0.00	0.00	0	Quebec
335	us7b2	18	165	137	15	0.03	0.43	0.05	0.10	Quebec
1	et2b2,u*	0	1	0	0	0	0.00	0	0	Quebec
7	us2c4	0	7	0	0	0	0.02	0	0	Quebec

Table 15.2. *Lexicon of categories of the word Quebec*

A second type of information is also available. It is the list of words that have received a given category. Thus, the encoder can verify the universe of meaning that is covered by one or other category (see Table 15.3). Here, we ask the software to give us the words that have received the category "territorial space" (us7b2).

Total freq	Socio	FED	QUE	PHQ	AUT	P-FED	P-QUE	P-PHQ	P-AUT	
2	us7b2	0	0	2	0	0	0	0.00	0	East
10	us7b2	1	3	6	0	0.00	0.01	0.00	0	Province
335	us7b2	18	165	137	15	0.03	0.43	0.05	0.10	Quebec
7	us7b2	0	4	3	0	0	0.01	0.00	0	Quebecer (male)
4	us7b2	0	4	0	0	0	0.01	0	0	Quebecer (female)
2	us7b2	0	0	2	0	0	0	0.00	0	territorial

Table 15.3. *Lexicon of words having received the category "us7b2" (territorial space)*

The exercise of categorization is a paradigmatic example of the hermeneutic dimension related to any research operation. The use of *Sato*'s features has made it possible to explain procedures involving interpretative choices and has allowed us to increase the reliability and stability of the overall process.

15.6. Validation by triangulation of methods

To paraphrase the term "triangulation of data" used in qualitative analysis, we could use the expression "triangulation of methods" to identify the need to verify the extent to which treatment procedures affect the stability of constructed facts on which interpretation is based. To illustrate this process, we will again take the example of the corpus of interviews on tobacco use [DAO 06]. In this experiment, researchers used methods and software based on the French tradition of data analysis[7]: *Alceste*, *DTM*, and *Lexico*. In their initial approach with *Sato*, researchers "contrasted" the overall lexicon by dividing the corpus according to variables external to the text and that are related to some social characteristics of the speakers. The *Alceste* method works the opposite way. The software builds a classification of statements, whose statistic approximation corresponds to segments of text of similar length. Thus, *Alceste* tries to bring out the structure of the speech by screening profiles of repetition in simple statements. It is then possible to juxtapose external variables on the classes of statements and their characteristic vocabulary. Applied to the corpus of interviews, *Alceste* produces two classes. The first class is strongly characterized by statements expressed after exposure to the anti-tobacco message. We also find, but more weakly, a significant presence in the of statements of non-smokers. The second class is strongly characterized by statements preceding the presentation of the anti-tobacco message. We also find, but more weakly, a significant presence in the statements of smokers.

It would be possible to interpret the classes created by *Alceste* by trying to characterize the specific vocabulary of classes. For example, the first class emphasizes themes such as awareness (*see, shock, impact, image, testimony*), death and illness (*cancer, lung, death*), and media (*commercials, TV, spot, road safety*). Statements after the anti-tobacco message address more serious issues and represent a response in relation to advertising campaigns. What particularly draws attention here, however, is that *Alceste* confirms that the variable before/after the presentation of the anti-tobacco message represents the first element of the corpus structure, thus confirming the *a priori* assumption used with *Sato*.

To construct their grid, researchers used the algorithm of distance to contrast the vocabulary. They wanted to verify the convergence of this index with the statistical

---

7 See Chapter 11.

test used by *Lexico* [SAL 03]. It is the calculation of specifics that uses the model of hypergeometric law. The researchers noticed that there was a very large overlap between the lexical forms that contribute most to the distance and the specificities calculated by *Lexico*.

The *DTM* (Data and Text Mining) [LEB 05] software is a tool dedicated to the exploratory analysis of multi-varied digital and textual data. The typical example of data to be analyzed by this software is the compilation of polls including answers to both closed and open questions. For the analysis of interviews, the researchers applied this model of coupling open and closed questions. They regarded the corpus as a set of 87 individuals. The open question is unique and its answer is the overall statement of the participant, with before and after the anti-tobacco message being considered two distinct questionnaires. The closed questions reflect the sociological profile of the individual (sex and smoking) and the conditions of enunciation (before and after the introduction of the brochure). *DTM* conducts a factorial analysis of correspondence crossing these 87 individuals and the 903 lexical forms whose frequency is greater than four. Then, it traces the sociological variables on the space built by the factorial analysis of correspondences (FAC). The graph shows that the distribution in the lexical space of sociological characteristics of young people takes the oppositions identified by the algorithm of distance. We thus obtain the same divisions, whether we go with *Sato*, from sociological variables to contrast the lexicon, or with *Alceste*, from statements to construct lexical worlds and put in the sociological traits, or, finally with *DTM*, from text divided into individuals in order to contrast sociological traits. Three different methods confirm the stability of the interpretative model.

The advantage of multi-varied analyses is that they take into account all the data to confirm the existence of sociological profiling within the speech itself. In return, the interpretation of these profiles is difficult because the construction of this representation is purely algebraic. In contrast, if they rely at first on prominent words identified by the Chi-square distance, the grid of categories is developed on a semantic basis. Neglecting lexical units considered too circumstantial, other units contribute to one or other of the socio-semantic categories added. The construction of a corpus in which each word is replaced by its category, including non-categorized words represented by a unique arbitrary symbol, allows us to verify the stability of the results with the categorized lexicon. The axis system thus becomes easier to interpret because it is now based on a reduced number of vocabulary categories. This provides a very good tool with which to validate the construction of the socio-semantic categories grid. As a result, the most eccentric categories, including the most significant of our interpretative system – *appearance, addiction, costs, education, death,* and *soc-friend* – spread to the four cardinal points. In contrast, plan the trivial categories that constitute the common references of speech appear in the center of the plan.

15.7. Conclusion

Going beyond the descriptive observation and comment, this approach illustrates how interpretation can be based on transparent and explicit methodologies made possible by the use of computer programs. Thus, the construction of grids of categories can be based on statistical indices that take the overall corpus into account. The application of the grid gains in terms of systematicity and accuracy. Finally, through the combination of methods of analysis, we increase the reliability of conclusions by providing a means by which to corroborate or refute assumptions and conclusions.

15.8. Bibliography

[BOU 96] BOURQUE G., DUCHASTEL J., *L'Identité Fragmentée. Nation et Citoyenneté dans les Débats Constitutionnels Canadiens, 1941-1992*, with the collaboration of V. ARMONY, Fides, Montreal, 1996.

[DAO 05] DAOUST F., Système d'analyse de texte par ordinateur, ATO Center, UQAM, 2007, updated 2009. (Available at: http://www.ling.uqam.ca/sato/satoman-fr.html, accessed February 9, 2010.)

[DAO 06] DAOUST F., DOBROWOLSKI G., DUFRESNE M., GELINAS-CHEBAT C., "Analyse exploratoire d'entrevues de groupe: quand ALCESTE, DTM, LEXICO et SATO se donnent la main", *Actes des JADT-2006*, Les Cahiers de la MSH Ledoux, no. 3, vol. 1, p. 313-326, 2006.

[DUC 99a] DUCHASTEL J., LABERGE D., "La recherche comme espace de médiation interdisciplinaire", *Sociologie et Sociétés*, vol. 31, no. 1, p. 63-76, 1999.

[DUC 99b] DUCHASTEL J., LABERGE D., "Des interprétations locales aux interprétations globales: combler le hiatus", in N. RAMOGNINO and G. HOULE (eds.), *Sociologie et Normativité Scientifique*, Presses Universitaires du Mirail, Toulouse, p. 51-72, 1999.

[GEL 04] GELINAS-CHEBAT C., DAOUST F., DUFRESNE M., GALLOPEL K., LEBEL M., "Analyse exploratoire d'entrevues de groupe: les jeunes Français et le tabac", Le Poids des Mots, Acts of JADT-2004, vol. 1, p. 479-487, 2004.

[HAB 05] HABERT B., *Instruments et Resources Électroniques pour le Français*, Ophrys, Paris, 2005.

[LEB 05] LEBART L, *Data and Text Mining*, Ecole Nationale Supérieure des Télécommunications, Paris, 2005.

[MEU 90] MEUNIER J.-G., "Le traitement et l'analyse informatique des textes", *Management Gestion de l'Information Textuelle*, ICO, vol. 2, III, p. 9-18, 1990.

[REI 02] REINERT M., ALCESTE Manuel de Référence, CNRS, University of Saint-Quentin-en-Yvelines, 2002.

[SAL 03] SALEM A., LAMAILLE C., MARTINEZ W., FLEURY S., Lexico Manual 3, version 3.41, available at: http://www.cavi.univparis3.fr/Ilpga/ilpga/tal/lexicoWWW/manuels.htm.

PART VI

Distance Cooperation

Chapter 16

A Communicational and Documentary Theory of ICT

16.1. Introduction

In this chapter, we will propose a communicational and documentary analysis of information and communication technologies (ICT), itself contained in the framework of transactional analysis of the action. This documentary vision presents a double advantage according to us.

On the one hand, it allows us to de-center the technical vision to focus on the uses and practices which, for the most part, are in continuity with documentary uses and practices that predated computerization, even if the latter has significantly expanded and renewed them.

On the other hand, it allows us to take into account the recent developments marked by an even greater democratization and large-scale generalization of transactions mediated by textual and multimedia and textual documents (use of blogs and wikis, Web 2.0); transactions that we call "mediated docu" [ZAC 07b].

This documentary theory of ICT is itself being renewed by document researchers (for example [PED 06]) and is being extended by work that we are conducting with the Tech-CICO team at the University of Technology of Troyes on:

– documents for action [GAG 06, PAR 06, PAY 06, PRAT 06, ZAC 04, ZAC 06a];

Chapter written by Manuel ZACKLAD.

– the socio-semantic and hypertopic Web [CAH 04, SRI 06, ZAK 05, ZHO 06];

– annotation processes [LOR 06, ZAC 03, ZAC 06a, ZAC 07a];

– open information search [ZAH 06]; and

– community uses of ICT [SCH 06, ZAC 07b].

We stress in particular the importance of documents for digitized action for the mediation and regulation of collective activities in the current uses of ICT, not only within professional organizations but also in the general public, recreational, civic or associative activities. To understand the role of ICT, we believe it is necessary to rely on a theory of collective action that can reflect the great diversity of contexts in which it occurs:

– formal organization;

– community of practice with porous boundaries; or

– distributed collective practices whose spatio-socio-temporal extension can be very important [TUR 06].

This theory of action, is made by the transactional approach of action (ATA). ATA is a socio-psycho-economic approach that aims to reflect the value of creation processes associated with communicational activities and take into account the importance of material support in the mediation of any form of action. After defining the notion of transaction, we will introduce concepts of transactional flows and coordination levels within these flows, to account for sequences of complex and interdependent activities for which ICT are essential.

16.2. Transactional approach of action

The concept of transaction that we borrow from Dewey and Bentley [DEW 49], corresponds to productive meetings from which a medium and stakeholders are transformed [ZAC 06b, ZAC 06c]. The people involved in the transaction are in the position of director (or co-director) and beneficiary (or co-beneficiary), all configurations of symmetry and asymmetry between these positions being possible. An individual person may be involved in a transaction with him- or herself. The achievements of the transaction, as diverse as a conversation, a meal, the joint motion of a material object, are both its purpose and the necessary condition for interlinking people. Indeed, any transaction must be mediated, either by the actions or words of people involved, or "remotely" through perennial artifacts circulating from one body to another.

Drawing from Greimassian's semiotics [GRE 06] used in the area of organizational communication by Coreen [COR 99], we identify four stages in transactions [ZAC 06c]:

– virtualization (the potential beneficiary expresses the vision of a project to which the director adheres or *vice versa*);

– skills acquisition (the director or co-directors acquire the skills necessary to carry out the project);

– performance (more or less simultaneous transformation of the work and people creating the content of the project);

– the sanction (symbolic or tangible rewards for the director from the beneficiary).

In our approach, every human action is part of a transactional trial and action analysis, which involves identifying ongoing transactions.

16.3. Transactional flows: rhizomatic and machinery configuration

Action analysis from a transactional perspective is difficult because of the fragmented and distributed character of transactional trials.

Indeed, not only is a transaction distributed according to different spatio-socio-temporal parameters, but it is likely to create others more or less systematically derived from it. To resume the geological metaphors of Deleuze and Guattari [DEL 80], we assume that transactions are flows circulating at variable speeds, crossing each other at various locations. Any situation of effective action can thus be analyzed in terms of its "genealogy", which makes it part of both a direct and obvious transactional trial, but also puts it in terms of previous transactional trials to which participants in the situation do not necessary refer explicitly. Moreover, the same situation of activity can allow the simultaneous occurrence of multiple transactional phases belonging to different lines.

Thus, the free circulation of transactional flows is likely to create rhizomatic configurations [DEL 80][1], whose nodes are actual activities corresponding to different phases of one or several distributed transactions. The channels (or links) joining these nodes correspond to the genealogical relationships between different

---

1 "(…) underground stem that grows, from it, buds from outside – does not begin and end, it is always in the middle, between things, inter-being, intermezzo. The tree is affiliation, but the rhizome is alliance, only alliance. The tree imposes the verb to be, but the rhizome has as fabric the conjunction […]. There is in this conjunction enough strength to shake and uproot the verb to be." [DEL 80], p. 36.

activity situations. Figure 16.1 offers a graphical representation of a transactional flow that, from a generator activity situation, leads to the derivation of a multitude of activity situations [ZAC 06c]. In this figure, we distinguish two types of activity situations: propagator or sedimented. In a propagator activity situation, the commitments made by people are still sufficiently "active" to generate new activity situations (new encounters, productions, commitments, etc.). A sedimented activity situation is no longer the carrier of active commitments because the associated sub-transaction was closed at the end of the sanction phase. It is still in the memory of the transactions and has been documented, but now it plays an archive role and is likely to be the subject of a re-exploitation in the framework of a historical outcome but is unlikely to motivate a direct and precise personal action.

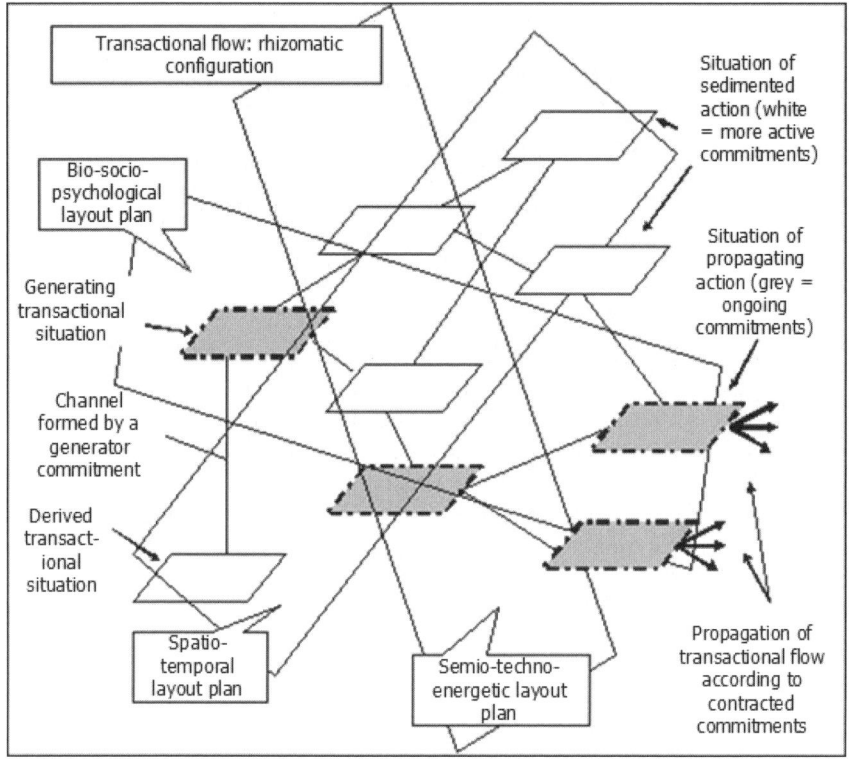

Figure 16.1. *Schematic of a rhizomatic transactional flow*

In some cases, transactional flows do not have the rhizomatic structure associated with spontaneous transactional diffusionism. The periodization and regionalization obey strong constraints (for example, locations and timings of prescribed activities). The intermediate productions are strictly standardized and relations between participants are formalized in the framework of rigid organizations. In this case, we believe that channeled transactional flows are part of a machinery configuration. In these configurations, actual situations of activity arranged as much as possible according to specific rules that reflect the social hierarchy and systematic organization of activity. The sequence of situations ensures repetitive cycles corresponding to mandatory periodizations (the year and the balance sheet, for example).

16.4. ICT and documents: transition operators between situations of activity within a transactional flow

Whether the configuration of the transactional flow is rhizomatic or machinery, a major reason that ICT will be sought is to ensure coordination between these distributed situations of activity that allow people to extend or revise ongoing transactions in the most coherent way possible. Our conceptualization of ICT is itself based on the concept of the document. In our approach to *documentarization* [ZAC 04, ZAC 06a], a document is a semiotic production transcribed or recorded on a perennial support equipped with extensive features allowing its re-operation. Semiotic productions are the product of symbolic communicational transactions characterized by their creative character and expressive dominant characteristic. Documentarization equips a perennial support with attributes that will facilitate its circulation in space, time and communities of interpretation (title, numbering, table of contents, metadata, ratings, etc.).

Under this approach, the document category not only includes written texts but also multimedia productions of all types: graphic work, photography, audio and video cassette, CD, DVD, video game, and even objects manufactured for arts purposes, liturgical objects, or even instruments essentially derived of their utilitarian function to serve a primarily semiotic function [BRI 51, BUC 97]. Whether the support used is ephemeral or perennial, it is a use of alternative mediations to overcome the direct perception of semiotic productions from the body of the author, as in the case of performing arts or face-to-face conversation [ZAC 04, ZAC 06a]. While the use of support of waves was a major characteristic of modernity in the 20th Century (in the extension of print), the "return" to digital documents, enriched by all possible media, seems to us and Pedauque [PED 06] to be a feature of the current century.

As we shall see below, digitization has made the emergence of new documentary practices particularly suited to the coordination of transactional flows that spatio-temporally distributed possible. These documentary practices seem to have been made possible by a new type of document, the document for action (DFA), which is increasingly essential for cooperative collective actions and corresponds to a growing number of collective uses of ICT in the context of finalized collective actions. DFAs correspond to very diverse systems, characterized by their durability, rapid circulation of the document, multi-authors, fragmentation and extended incompleteness [ZAC 03, ZAC 04, ZAC 06a]: text files or annotated drawings, forum systems, blog or wiki systems, messaging systems and others.

When they include productions corresponding to transactional situations that are partly independent but integrated in a single transactional flow, we are talking about "action files". Examples of the use of DFA use are very diverse and occur in many professional contexts. These are:

– documents of engineering design (mechanical, software or other);

– patient records in medicine;

– contract documents in a business context going from commercial proposal to a contract in due form [GAG 06];

– quality records, which are increasingly digitized;

– study reports in management consultancy; and

– exchange forums in the area of free software [RIP 06].

Before returning to the particular importance we attach to DFAs, we present a typology of different documents whose digitization allows the "computerization" of the information system.

16.5. Four classes of documents within the information system

We will rely on two systems of variables presented in the Tables 16.1 and 16.2. The first (Table 16.1) establishes a distinction between two essential components of ICT and more generally of the information system of organizations. The first variable of Table 16.1 is made, on the one hand, by computer and telecommunication networks that carry signals most often digitized, and on the other hand, by records or transcription support allowing the storage of these signals in a more or less perennial way (network-telecom-support system)[2]. The second dimension of Table 16.1 is made by the documentary system which, from the data

---

2 Note that a perennial record or transcription support becomes a document only after documentarization and transformation of the support to allow its re-use later (see above).

stream from the first dimension, documents the files to allow their re-use later[3]. The second variable guarantees the circulation of a continuous flow of information from one point to another, connected by the network, from the acquisition and processing of the signal to its restoration on adapted equipment. The second variable organizes the information content, its storage and access. The second variable establishes a classic but essential separation between two spreading modes that apply to both the network-telecom-support system and the documentary system: an indiscriminate distribution mode, that we find in audiovisual mass media, the distribution of documents (newspapers, books), and a directional and interactive distribution mode corresponding to the use of the telephone or video on one hand, and documents allowing interactive collective "writing" on the other (databases, messaging, annotated files).

General typology of ICT	Dominant mediatization: network-telecom-support system	Dominant documentation: documentary system
Indiscriminate distribution	Audio-visual types of media (radio, television, etc.)	Resource documents for the general public (books, newspapers, etc.)
Directional distribution and interactivity	Means of interpersonal communication (telephony, video, etc.)	Articulatory document or document for action (databases, annotated files, messaging, etc.)

Table 16.1. *Typology of the two sub-systems constituting ICT (networks and digital documents)*

The second system of variables (Table 16.2) crosses the level of codification of the document and frequency at which it is likely to be updated. The frequency of update makes a distinction between stabilized documents used as reference or resource without undergoing modification and documents used to "track" events occurring in the transactional flow. The degree of codification[4] itself refers to the degrees of standardization of transactions. Coded documents are produced in expressive transactional situations of routine nature that are most often brief,

---

3 Documentarization produces "visible" attributes for the user but also "invisible" technical attributes generated automatically and dependent on editing and viewing features (for example SGML tags). We can consult [MAR 94] on this matter for a broad view of digital document.
4 The term codification is not used as often as in knowledge management (explicit knowledge) but appears in the sense of a "system of symbols to represent information in a technical field" (http://atilf.atilf.fr/tlf.htm, accessed February 9, 2010).

repetitive and automated. The vocabulary used in documents is clearly defined in advance so that it is not subject to ambiguity.

Typology of documents	Coded document	Non coded document
Rare update ("frozen" or stable document)	**Coding document** • list of nomenclatures (rare update) • program file	**Resource document** • work fixed on a support: book, report, music recording, film, photo
Frequent update (insertion in an active transactional flow)	**Articulatory document** • paper files • coded computer databases called transactional • coordination mechanism	**Document for action** • text or computer aided design files collectively annotated • electronic messages • forum, blog or wiki system

Table 16.2. *Typology of documents according to the degree of codification and frequency of update*

As mentioned above, in our sense of the term, the codified information is less invested with subjectivity than semiotic productions made in the creative transactions whose meaning is open to interpretation. In documents carrying semiotic productions, rarely updated "resource documents" or frequently updated "documents for action", the attributes associated with documentarization come from the explanation of certain parameters of the transactional situation and of certain implicit links between fragments of semiotic production.

In the case of information transactions, codified and based on frequently updated "articulatory documents" for example, the value of attributes of the form is provided by the simple identification of already explained parameters of the standardized transactional situation (the fields) and by the selection of an item within a predetermined set. If the design of a coding document (nomenclature or computer program for example) requires a certain creativity that corresponds to the initial specification of transactional situation parameters and to coding to their possible values, their exploitation is most often routine or fully automated. Codified information transactions that cause the update of an articulatory document occur, for example, during the act of an automated purchase and produce codified information that cannot be treated as semiotic productions.

In computer science literature, it is often customary to refer to the codified information associated with the updating of articulatory documents by considering that it is "structured data" managed by classical automated database management, as opposed to "semi-structured" data conveyed by new information and communication technologies[5]. Of course, resource documents and DFA also contain codified information. We must understand this expression by considering that it implies that in classical automated management systems, the codified information tends to exhaust the meaning of the articulatory document while the codification concerns only small zones dedicated to referencing and explicit links between fragments in the non-codified document. By contrast, the characteristic of semiotic productions conveyed by ICT (carrying semi-structured information) that correspond to resource documents and DFA, is that they cannot be interpreted without considering the characteristics for variable parts of the transactional situation which is not, in the case of ICT, completely standardized.

Therefore, knowing the name of the director of the semiotic production is necessary for correct interpretation of the semiotic content. This information must thus be provided in the document to facilitate its external and internal use while they are not always in an entry form. In this case, the knowledge of the function of "the author" of the form can often be substituted, to some extent, by the precise knowledge of his/her identity.

16.6. ICT status in the coordination and regulation of transactional flows

We are now able to address the essential question of the role of ICT in the coordination of activity situations distributed within a transactional flow. The artifacts and methods for solving problems linked to the articulation of complex interrelated activities are coordinated, in the words of Schmidt and Simone [SCH 96], ensuring continuity between situations of activities belonging to the same transactional flow. Coordination problems arise within situations of activity (intra-situational), but also between different situations of activity within a transactional flow (inter-situational).

Inter-situational coordination problems also often reflect on intra-situational coordination. We distinguish three levels of coordination: through access to activity situations, through prior standardization of the primary transaction and through the use of regulating transactions in a situation.

---

5 The distinction between ICT and new information and communication technologies being decreasingly common, we will distinguish between them in the rest of the chapter.

16.6.1. *Coordination through access[6] to situations of activity and their ingredients*

The first level of coordination corresponds to material conditions allowing access to the locations, productions and people that compose the situations of activity. The pre-condition for exchange is to have common places and to refer to calendars for organizing meetings. The access also concerns:

– the current productions, artifacts with an expressive character (speeches, documents and images);

– material (objects and instruments);

– energetic (means of transport, repair services, etc).

These productions can be associated with a specific place where moving from one space to another by following the transactional flow and location of situations of activity.

Finally, access concerns people in their physical, psychological and social dimensions[7]. In terms of access, network-telecom-support equipment allows substitution of the direct presence of some attributes of the transaction body (voice, image) in a synchronous logic of use (telephone, videophone, instant messaging). They also offer a sustainable support to semiotic production for its recording or transcript in an asynchronous logic of use (storage and sending of a recording or a transcript offline).

16.6.2. *Coordination by prior standardization of the primary transaction*

For people involved in the transaction, coordination can present specific difficulties, themselves likely to generate a new transaction for resolution. In other words, coordination itself appears as a problem capable of causing a specific work. The purpose of these derived transactions, called regulating transactions here, is to coordinate the primary transaction by facilitating articulation between the different situations of activity. These transactions sometimes focus on people and sometimes on productions by following different "regulating paradigms" [ZAC 06b][8]. This distinction is similar to the distinction between cooperative work and articulation work of the cooperative work [SCH 96], for example. It also corresponds, partly, to

---

6 We use this term as a reference, particularly, in respect of Rifkin's book, *The Age of Access* [RIF 00].

7 In [ZAC 06c], we identify three dimensions of artifacts (material, energetic and physical) and three dimensions of people (biological, psychological and social).

8 There are *a priori* four paradigms: legal-psycho-managerial, human, techno-instrumental and epistemic [ZAC 06c].

the problem of "organization work" [DET 03], which extends the concept of independent regulation [RAY 89]. Among all of these authors, howver, regulations are essentially seen as dealing with "social relationships" between people (which reveals the objectives focused on people). In the transactional approach of action, regulations can also directly deal with the characteristics of artifacts in media.

The regulating work may be exercised in two quite radically different ways. In the first case, it takes place prior to the primary transaction that is to be regulated. The different intra- and intersituational interdependencies are determined as far in advance as possible. This early treatment of interdependencies leads to standardization of the conduct of the primary transaction and its results. This is because it is based on previously-defined resources by reducing the degrees of freedom of participants in the course of action. Indeed, standardization of the transactional process is more easily implemented when the products are themselves standardized.

The degree of standardization of coordination is variable but, in most social structures, particularly organizational ones, it is always present. Standardization concerns locations, their dimensions, their functional characteristics and periodization, repetitive use of time, duration of time slots, etc. When it aims to facilitate circulation of artifacts, it concerns possibilities of the integration of different product (material) components, interoperability of movements and procedures (energetic), or logical consistency of the different fields of a form (expressive). Finally, with people, standardization corresponds to standardized relations and predefined protocols of interactions accelerating transactions from both the director and the beneficiary's points of view.

The use of ICT in the context of standardization of the primary transaction corresponds to the use of articulatory documents. These have a different status as the primary transaction itself possesses a material or energetic (e.g. manufacturing of parts) dominant or expressive dominance (e.g. formulation of a diagnosis in the field of health). The first case (manufacturing of parts) corresponds to industrial information systems and has very codified coordination mechanisms [SCH 96][9].

The second case (diagnosis in healthcare), can also identify coordination mechanisms corresponding to the use of forms that are, for example, to prescribe the conduct of a path of care between different specialists. Depending on whether the

---

9 Among the examples cited is the Kaban system (in industrial time) that allows a cooperation, which we believe to be "passive", between the operators without involving direct communication, or examples of workflow systems (electronic processes), one of the flagship technologies of the Computer Supported Cooperative Work.

form is more or less "open", the decisions are likely to be annotated and to follow different paths, corresponding to an articulatory document or to a DFA (open form).

Dominant of the transaction / Level of coordination	Effects on people	Realization of expressive artifacts	Realization of material or energetic artifacts
Level of access to the activity situation: **network-telecom-support system (NTSS)**	NTSS for synchronous communication with direct effect on people and instantaneous adjustments	– Support allowing the re-transcript of semiotic expressions – NTSS allowing access to media	NTSS allowing remote monitoring of objects and processes and recording on perennial media
Level of prior standardization of the primary transaction: **coding and articulatory documents**	Delayed and standardized effects on people referring to relations that are themselves conventional. Interactivity is in the context of a predefined protocol	– Codified administrative DB[10] (in various sectors of activity: bank, medicine, distribution, civil status, etc.) – Administrative coordination mechanism (*workflow*)	– Codified technical database (tracking of production, inventory, etc.) – Computer program
Level transaction regulation in a situation: **resource documents and DFA**	Delayed but non-standardized effects on people. The effect is the result of a located process of interpretation. The response is also delayed but contextual and personalized	– Annotated documents, mail, forum, blog, etc. – Resource documents: report, book, audio-visual recordings, etc. – "Open" decision support or documentary DB	Technical documentation on an artifact and serving as support (accessory) to a transaction of material or energetic dominant

Table 16.3. *Role of ICT in the coordination and regulation of situations of activities in distributed transactional flows*

---

10 Databases.

16.6.3. *Coordination through the use of regulating transactions and resources in situation*

The third level of coordination is to develop systems that aim to facilitate the regulation work by people involved in the primary transaction during the course of action. At the second level, standardization aims to anticipate problems of coordination using a preliminary design of the articulation of work. The design is often implemented by people other than those involved in the primary transaction.

However, at this level of coordination, it is the participants in the primary transaction who must actively exploit the regulating resources to resolve the problems of coordination in a situation. These resources can also be designed by people other than those involved in the primary transaction, but this design must offer sufficient "hold" on the parties involved in to facilitate articulation in the course of action. This distinction between these two forms of coordination corresponds to the distinction between passive and active cooperation.

In the first case, that of coordination through standardization, the people in a situation are required to make local adjustments, but most of the management of interdependencies has been defined previously. In contrast, in the case of the use of regulating resources, we speak of active cooperation that involves participants in a continuous way to resolve problems of coordination that arise during the course of action.

In terms of regulating transactions, the digital documents used are no longer articulatory documents corresponding to coordination mechanisms but documents that need to be actively interpreted by people in the transaction in order to regulate situations of collective activity. These documents can be resource documents providing general principles under different regulating paradigms, or DFA updated during the activity's progress. We will call these documents "support documents of regulation", and we will distinguish them from documents serving as coordination mechanisms.

Among support documents of regulation, we will look at DFA in particular.

16.6.4. *Internal structuring of DFA according to the nature of the transaction*

DFA can be used in the context of hybrid transactions, where the document accompanies a material or energetic transaction, or in the context of transactions of an expressive dominant in which the document is the principal object of the activity. In the last case, we can establish a distinction between the content associated with primary transactions that corresponds to the main semiotic content of the document and the content associated with regulating transactions. The latter corresponds to the result of documentarization, allowing a re-exploitation of certain parts of the document in the context of a deferred party. These attributes constitute a "system of orientation and access"[11] (SOA) which performs functions of internal articulation of fragments of semiotic content (captions, notes, index) and external articulation of the document (title, reference number of document, keywords) in a documentary collection or within a folder. Even if the border is not always easy to establish, we will consider that the content associated with primary transactions occupies specific zones of support (primary zones) while the system or orientation and access dedicated to coordination occupies articulation zones that are themselves more or less codified.

In the context of creative transaction that interests us here, the DFA changes in two different ways:

– by contributing annotations that are new fragments of semiotic productions corresponding to the primary transaction;

– by the addition and enrichment of associative annotation in a more or less automatic way (often incorporated into contributing annotations) that enable us to articulate the new fragments with the rest of the semiotic content (footnote, citation of the previous message, date and automatic signature of an intervention posted on a site, etc.)[12].

A similar process occurs during the addition of small document candidates for incorporation into a folder or larger collection that we need to articulate within it. In both cases, the associative annotations provide a more or less explicit form of justification of the relevance of contributions to the cooperative work and will offer a means by which to re-exploit these contributions during new activity.

In this context, part of the regulation within the DFA is done by introducing associative supporting and signaling annotations. These annotations are themselves chosen within a SOA that can be radioactive. In the work that we conduct within the Tech-CICO laboratory, for several years we have defended the use of semiotic ontologies in the socio-semantic Web apprach[13], for the realization of these SOA (see, for example, [CAH 04, ZAC 05]).

Semiotic ontologies can be considered to lie at the intersection between two techniques of resource description used in information sciences and on the Web:

---

11 In part, SOA is a form of "interface" with the semiotic content associated with the primary transaction. This concept is probably close to that of Architext [JEAN 99].

12 See, for example [ZAC 06a], for an illustration in the context of a forum.

13 Which we distinguish from the computationally semantic Web [ZAC 05].

multi-facet indexation according to the different stakeholders involved in the transaction[14] and the ascending description through step-by-step association from terms suggested by users (as in methods of "tagging" (semantic markup) made fashionable in Web 2.0)[15]. Here we consider they contribute to the progression of regulating transactions allowing, coordination of work in situations of distributed activity within a transactional flow.

16.7. Conclusion: document for action and distributed transactions

In this chapter, we have shown how ICT offer new systems of coordination of transactional flows based on digital documents of the DFA type. DFA indeed appear to be perennial, fragmented and evolving, with media facilitating the conduct of creative documents despite the distribution of situations of activity within a transactional flow. They are both more weakly codified than articulatory documents and more easily updated than classical resource documents. For them to be effective, they must have systems of orientation and access that can be updated during the progression of mediated regulating transactions involved through the coordination of situations of activity.

Current Web developments offer increasingly integrated cognitive and social technologies, allowing the editing, visualization, navigation aid and open information search within DFA collections (relations between annotated files, blogs, messaging, forum, engines, bookmarks, etc.)[16]. Because DFA are weakly codified but possess a relevant semiotic organization thanks to the system of orientation and access that they have, they allow us to compensate for the loss of semioticity often associated with ICT use based on highly codified articulatory documents (codified administrative databases or closed-question forms). Because DFA can easily and remotely be updated by people, they can partly compensate for the loss of relationality associated with difficulties in updating resource documents.

DFA thus provide support for the development of new forms of distributed cooperation in the field of organizations, divided communities or distributed collective practices. The unprecedented proliferation of DFA within organizations and success of projects such as Wikipedia or open source software, that bring together large distributed communities, demonstrate the potential of these new documentary technologies that have enabled the regulating paradigms represented in these projects.

---

14 As the thesaurus used in documentology.
15 Communities sharing photos (flickr) or bookmarks (del.ic.ious).
16 The socio-semantic Web is typically a Web made up of constantly changing DFA, mediating relations between communities of distributed author-readers.

16.8. Bibliography

[BRI 51] BRIET S., *Qu'est-ce que la documentation ?*, EDIT, Paris, 1951.

[BUC 97] BUCKLAND M. K., "What is a "document"?", *Journal of the American Society for Information Science,* vol. 48, n° 9, p. 804-809, 1997.

[CAH 04] CAHIER J.-P., ZACKLAD M., MONCEAUX A., "Une application du Web Socio Sémantique à la définition d'un annuaire métier en ingénierie", in MATTAN N. (ed.), IC 2004 : *15es journées francophones d'ingénierie des connaissances*, Presses Universitaires de Grenoble, Grenoble, 2004.

[COO 99] COOREN F., *The Organizing Property of Communication*, John Benjamins Publishing Company, Amsterdam/Philadelphia, 1999.

[DEL 80] Deleuze G., Guattari F., *Capitalisme et schizophrénie 2. Mille Plateaux*, Les éditions de Minuit, Paris, 1980.

[DET 03] DE TERSSAC G. (ed.), *La théorie de la régulation sociale de Jean-Daniel Reynaud - Débats et prolongements* , La Découverte, Paris, 2003.

[DEW 49] DEWEY J., BENTLEY A.F., "Knowing and the known", in BOYDSTON J.A., *John Dewey: The later Works*, 1925-1953, vol. 16, p. 2-294, Southern Illinois University Press, Carbondale, 1989.

[GAG 06] GAGLIO G., ZACKLAD M., "La circulation documentaire en entreprise comme analyseur de pratiques professionnelles, une étude de cas", *Sciences de la Société*, n° 68, p. 93-109, 2006.

[GRE 96] GREIMAS A.J., *Sémantique structurale*, Larousse, Paris, 1966.

[JEA 99] JEANNERET Y., SOUCHIER E., "Pour une poétique de 'l'écrit d'écran'", *Xoana*, vol. 6, p. 97-107, 1999.

[LOR 06] LORTAL G., TODIRASCU-COURTIER A., LEWKOWICZ M., "Soutenir la coopération par l'indexation semi-automatique d'annotations", in LEWOKOWICZ M. (ed.), *17es journées Ingénierie des Connaissances*, Nantes, 28-30 June 2006.

[MAR 94] MARCOUX Y., "Les formats normalisés de documents électroniques", vol. 6, n° 1-2, p. 56-65, *ICO Québec*, 1994.

[PAR 06] PARFOURU S., GRASSAUD A., MAHE S., ZACKLAD M.,"Document pour l'Action comme media pour la Gestion de Connaissances", CIDE.9, *9e Colloque International sur le Document Electronique, SDN'06*, Fribourg, Swiss, 18-20 September 2006.

[PAY 06] PAYEUR C., ZACKLAD M., "Modèle d'accès multi-supports et multi-canaux aux documents d'actualité", CIDE.9, *9e Colloque International sur le Document Electronique, SDN'06*, Fribourg, Suisse, 18-20 September 2006.

[PED 06] PÉDAUQUE R.T., *Le document à la lumière du numérique*, C&F Editions, Caen, 2006.

[PRA 06] PRAT N., ZACKLAD M., "Analyse des dépendances entre fragments documentaires pour le management des communautés d'action médiatisées", *AIM 2006, 11ᵉ Colloque de l'Association information et management*, Luxembourg, 7-9 June 2006.

[REY 89] REYNAUD J.-D., *Les Règles du jeu : L'action collective et la régulation sociale*, Armand Colin, Paris, 1989.

[RIF 00] RIFKIN J., *L'âge de l'accès*, Maspéro, Paris, 2000.

[RIP 06] RIPOCHE G., SANSONNET J-P., "Experiences in Automating the Analysis of Linguistic Interactions for the Study of Distributed Collectives", *Computer Supported Cooperative Work*, vol. 15, 2006.

[SCH 96] SCHMIDT K., SIMONE C., "Coordination mechanisms: Towards a conceptual foundation of CSCW systems design", *CSCW Journal*, p. 155-200, 2-3 May 1996.

[SCH 06] SCHMIDT K., SIMONE C., LEWKOWICZ M., ZACKLAD M., "Beyond Electronic Patient File: Supporting Conversations in a Healthcare Network", dans Herrmann T. *et al.*, *6th International Conference on the Design of Cooperative Systems*, IOS Press, Carry Le Rouet, Pays Bas, 11-14 June 2006.

[SRI 06] SRITI M-F, EYNARD B., BOUTINAUD P., MATTA N., ZACKLAD M., "Towards a semantic-based platform to improve knowledge management in collaborative product development", *IPDMC 2006*, Milan, p. 1381-1394, 2006.

[TUR 06] TURNER W., BOWKER G., GASSER L., ZACKLAD M., "Information Infrastructures for Distributed Collective Practices", *Computer Supported Cooperative Work*, vol. 15, p. 93-110, 2-3 June 2006.

[ZAC 03] ZACKLAD M., LEWKOWICZ M., BOUJUT J.-F., DARSE F., DETIENNE F., "Formes et gestion des annotations numériques collectives en ingénierie collaborative", dans Dieng-Kunz R., *14es journées francophones d'Ingénierie des Connaissances*, 01-04 July 2003, Presses Universitaires de Grenoble, Laval, Grenoble, 2003.

[ZAC 04] ZACKLAD M., "Processus de documentarisation dans les Documents pour l'Action (DOPA) : statut des annotations et technologies de la coopération associées", *Le numérique : Impact sur le cycle de vie du document pour une analyse interdisciplinaire*, 13-15 octobre 2004, Montreal, 2004, http://archivesic.ccsd.cnrs.fr/.

[ZAC 05] ZACKLAD M., "Introduction aux ontologies sémiotiques dans le *web* Socio Sémantique", dans Jaulent M.-C., *16es journées francophones d'Ingénierie des Connaissances*, Grenoble, Presses Universitaires de Grenoble, 2005, http://archivesic.ccsd.cnrs.fr/.

[ZAC 06a] ZACKLAD M., "Documentarisation processes", *Documents for Action (DofA): The Status of Annotations and Associated Cooperation Technologies, Computer Supported Cooperative Work*, vol.15, p. 205-228, 2-3 June 2006.

[ZAC 06b] ZACKLAD M., Gestion du connaissant et du connu dans la théorie transactionnelle de l'action (TTA), Document de travail, 2006, http://archivesic.ccsd.cnrs.fr/.

[ZAC 06c] ZACKLAD M., Une approche communicationnelle et documentaire des TIC dans la coordination et la régulation des flux transactionnels, working document, 2006, http://archivesic.ccsd.cnrs.fr/.

[ZAC 07a] ZACKLAD M., "Annotation : attention, association, contribution", dans SALEMBIER P., ZACKLAD M. (ed.), *Annotation dans les documents pour l'action*, Hermès, Paris, 2007.

[ZAC 07b] ZACKLAD M., *Réseaux et communautés d'imaginaire documédiatisées*, 2007.

[ZAH 06] ZAHER H., CAHIER J.-P., ZACKLAD M., "Information Retrieval and E-Service: Towards Open Information Retrieval", *Proceedings of IEEE International Conference on Service Systems and Service Management IC SSSM, 2006*, Troyes, France, October 2006.

[ZHO 06] ZHOU C., LEJEUNE C., BENEL, A., "Towards a standard protocol for community-driven organizations of knowledge", *Proceedings of the Thirteenth International Conference on Concurrent Engineering*, Antibes, 18-22 September 2006.

Chapter 17

Knowledge Distributed by ICT: How do Communication Networks Modify Epistemic Networks?

17.1. Introduction

Much of what we know and what we learn is not first-hand knowledge but is obtained either directly by communicating with a person or a group, or indirectly by the use of an informational artifact. With Internet technology, the opposition between these two modes tends to fade as network technology combines communicational and computational properties. A loose coupling between communicative technology (e-mail, chat, list of spreading, forum, blog, wiki and collaborative platform) and active social cooperation is involved in the processes of knowledge distribution. It therefore becomes difficult to distinguish, in the distribution of informational statements, the coupling that occurs between people in groups and that between people and artifacts as the two are intertwined in a common technical and relational system.

Some technical systems of arrangement of messages related to ICT, such as the *Usenet*[1] system, are particularly interesting to study as they modify the structure of the interaction, the modes of cooperation and communication schemes. We will

---

Chapter written by Bernard CONEIN.
1 An analysis of a system of *request for comment* type, such as *Usenet*, is proposed by the *Free On Line Dictionary Of Computing*; see also [SMI 02] and [KOL 96]. The *Usenet* protocol, which supports the distribution most lists is like a mailing list where messages initiated by requests are exchanged without any intervention of a mediator.

present analysis proposals based on a description of some constraints exerted by threaded communications both on social relationships and on the recipient design inside an open-source community (debian.user.french)[2].

Previous studies of newsgroups in open-source software organizations [LAK 00, SAC 04, SMI 99] have constructed a new unit of analysis: threads as a communication network. The most interesting findings of these studies are when they investigate the effects of communication networks on social interactions. The richness of these studies[3] lies in the possibility of analyzing the structure of threads by estimating the degree of mutuality and collaboration of online relations [DOR 06]. Substructures of threads, such as local communication networks, give rise to various epistemic networks such as advice or discussion networks. Thus, some changes in the formal structure of these network communications can promote the formation of advice or discussion. It is, we believe, by analyzing these local relational structures that we will be able to understand how an epistemic[4] community emerges online as a form of distributed system of knowledge development [CON 04, CON 05]. According to this relational approach, an online community can become an epistemic community only when it works as a distributed system.

17.2. ICT and distributed cognition

Knowledge can be said to be *socially* distributed when it is communicated or produced through a relationship that coordinates an information seeker and an information provider. Knowledge can be said to be *ecologically* distributed when an agent is coupled with an equipped environment. When an artifact is the unique cognitive aid of the task of an agent, as in human-computer interaction, the distribution first begins between the external structure of the artifact and the internal operation of the person. An artifact contributes to the process of knowledge acquisition by propagating, storing and reformatting information in the same way as

---

2 The idea of treating discussion threads as networks owes much to the research of Remi Dorat (LIFL) and of Matthieu Latapy (LIAFA) as well as to the PERSI project (Research Program of Social Networks of Internet), see Chapter 8 of this book.
3 See ANR (Agence Nationale de la Recherche – French National Research Agency) Autograph draft.
4 Epistemic communities are communities of evolving knowledge production that distinguish from user groups who use Internet to acquire information. The term is used both in economics for the analysis of organizations [COH 05] and in *social epistemology* [LON 02] for the analysis of scientific production.

an expert can contribute to a discussion with other experts [CLA 98][5]. However, if we observe the distribution of this knowledge in a seminar or consultation with an expert, two cases where verbal communication predominates, cognition is distributed between the beliefs of group members. The social dimension of knowledge is therefore manifested at the relational level through an advice- [BLA 86] or discussion-based [LON 02] relationship.

Another way of knowledge distribution involves examining distribution at the level of a larger system [HUT 95] that not only includes social relationships but also informational artifacts (book, computer, speed indicator, microscope, calculation tool, etc.). Thus, an artifact, such as protocols of online queries, can act on communication relations by facilitating the coordination between sender and recipient but also by recording the messages, arranging them and by publishing them in a certain format. The request/reply format in digital media can, for example, simplify information gathered from agents by storing, spreading and pre-computing the information sought [SPE 01]. The distributed system becomes an intermediary between social and technical systems. The Internet thus seen as a network technology presents new properties to assist knowledge acquisition[6] by combining social and technical components. An important objective of current research on ICT is to find new models of knowledge propagation that take the properties of social interactions and that of artifacts into account simultaneously. Among the available methods, if the theory of distributed cognition is a more promising hypothesis than others, it is because it aims to analyze the coordination between all the components that contribute to knowledge enhancement.

We can, however, question the ability of distributed models to fully take the interaction between these two components into account. These models have difficulties accounting for the interaction between social and functional elements of the/a digital system. Researchers often view distributed cognition as mainly a socially distributed relational system [GOL 99] or as a system operating primarily through cognitive technologies [NOR 91]. However, if ICT dramatically changes the way cognition is distributed, it is because network technologies act on the arrangements between groups of people and artifacts. Thus, the relationship between people and artifact is neither social nor ecological as both components are intertwined and dependent upon each other.

---

5 A system is distributed when the elements it coordinates are strongly coupled: "the human body is linked to an external entity by mutual interaction, thus creating a coupled system which can be rightfully considered as a cognitive system. All the components in the system play an active role because they jointly govern behaviour in the same way as natural cognition" [CLA 98].

6 The term ICT underscores both the I-dimension as IT and the S-dimension as social communication technology.

Another reason that limits the application of ICT analysis to distributed cognition models is that they were most often developed to study very specific processes related either to the manipulation of tangible tools [CON 94, KIR 99] or to control processes from digital visual displays [HUT 95]. However, even if the artifacts used are different, these two contexts share common properties. They are situations where agents master their skills and seek primarily to reduce the cognitive cost by using visual cues (affordances) or tangible objects (direct manipulation). In most examples used in these studies [HUT 95, KIR 99, NOR 91] artifacts are conceived as "facilitators" and are used in contexts where the exploitation of skill knowledge by experts in routine tasks predominates (for example, flying an aircraft, cooking or using a personal computer).

In the example of Open Source, the constant development of an evolving and unstable knowledge and its scattering through ICT, predominates over routine exploitation of a standard knowledge [TUO 01]. In a situation of creative development of new knowledge, the coupling between external components takes a different form. Indeed, the more knowledge acquisition requires continuous elaboration, the more the coupling is provided by intensification of collaborative interactions between equal peers and the less direct manipulation of the artifact plays an important role[7].

If, therefore, in the cooperative uses of ICT, and in particular in Open Source, the model of uses differs, this means agents seek less to acquire routines than new knowledge. Among these skills, only a portion is fully controlled, the remainder being unstable. If Open Source is an alternative model to the user-centric model[8], it is mainly because its user model is relevant in the case of collaborative contexts for knowledge exploration. With cooperative uses of ICT, the relational dimension in knowledge acquisition plays a major role as it relies on dynamic coupling between a communicator and a receiver.

When social components related to the quality of message transmission contribute in a massive way to knowledge acquisition, the visual and motor interactions with the artifact and the weight of graphical representations appear to decline. By posting information on a mailing list, interface use will play a secondary role compared to that of communication, active diffusion and modification of expert knowledge. When the artifact is communicational and cognitive at the same time, its function is first to support a cognitive process of collaborative acquisition of

---

7 The perception/action coupling that are central in direct manipulation interfaces become less important.
8 User-centric models appeared before distributed cognition models, as it is a logical antecedent to these models. Authors such as Norman and Hutchins are all proponents of both the user-centred model and situated/distributed approach to cognition [CON 94].

knowledge [GOL 99]. In this context, coupling and distribution support is mainly organized through coupling between agents in communication networks.

17.3. Mailing lists as a distributed system

The mailing lists of users of free software can be considered as a distributed system because they present a case of original coordination between the two components of distribution. The system becomes relational because it gathers a large number of individuals as members of the user list, but the individuals are coordinated only through an artifact that connects each of them through the broadcast of queries. The artifact is therefore not only an epistemic aid to individuals' cognition, it allows them to coordinate and communicate among themselves. In this example, the artifact contributes more directly than other cognitive technologies to the social dimension of distributed cognition. *Usenet* technology defines a protocol for sending requests. *Usenet* shapes the transmission of knowledge between senders and receivers in conversational sequences initiated by a question (*posted request*), preceded by a title (subject, address, date). Each member on the list can reply to the entitled query and become involved by sending back a piece of advice or an idea.

Archive

To: debian-french@lists.debian.org

Subject: archive

From: kazem < kazem@planetepc.fr>

Date: Sun, 02 Sep 2001 15:08:21 + 0200

Hello,

Does anyone know where are archives of the mailing list debian-french@lists.debian.org Thank you

Figure 17.1. *Example of a message on Usenet*

How does the architecture of a mailing list, through the *Usenet* system, modify knowledge distribution within the list? This architecture first acts on the transmission and sequentialization of messages through *threads*. The thread shapes each message as queries or requests and as replies to queries through the "reply-to" function. The system then acts on the address design and on the sender/recipient coordination through the server by sending messages to the recipients asynchronously. The messages are displayed as request/reply conversational sequences (threads), but it is the server that regulates their order of publication. The coordination between sender and recipient is arranged in such a way that the sender of a request does not select any recipient individually. The system thus exercises a constraint that affects turn-taking, at the same time as message sequencing (request/reply format) and orienting towards the recipient in the first turn (the questioner addresses his/her request to the group without singular speaker selection).

On the other hand, the list as an artifact intervenes at another relational level by facilitating the acquisition and propagation of advice to all of the members of the list. When pieces of advice are distributed among a large number of individuals, threaded communication facilitates individual requests for advice by increasing distribution within the list to all members. Getting advice quickly to a large number of recipients increases both the amount of knowledge for the first recipient and that of the other members of the list, who benefit at the same time the advice is posted. The protocol of posted queries therefore facilitates the consultation of advice by facilitating communication between novices and experts.

The list can also be seen as a cognitive tool by which to acquire new knowledge. In the case of Open Source, experts are mainly interested in advancing the discussion on problems for which there are no recognized solutions, require a better formulation or even raise a critical debate.

This leads to two remarks: one concerning the specificity of the Open Source software as a model of distributed cognition [TUO 05] and; the other concerning the generic properties of any distributed system [CLA 98, CLA 02].

– Primarily, free software through mailing lists allows users within the same community to receive both old explicit knowledge and new evolutive knowledge.

– Secondly, Open Source communities differ from other knowledge communities by the fact that they combine distinct learning styles[9] and receive novices and experts within their number instead of putting them in two separate communities.

One of the best reviewers of free software, Tuomi [TUO 05], has stressed this specificity of groups of users of free software:

---

9 Knowledge diffusion based on *instructional learning* with an advisor cohabitates with knowledge production based on *collaborative learning* between equal peers [TOM 99].

"In the Open Source model, development depends essentially on users who are also developers of the technical system".

This means that regardless of *debian.user.french*, like any other open-source software list, the list provides a way of sharing ideas that not only brings together novices and experts but allows two modes of learning to interact: instructed learning between novice and expert and collaborative learning between experts.

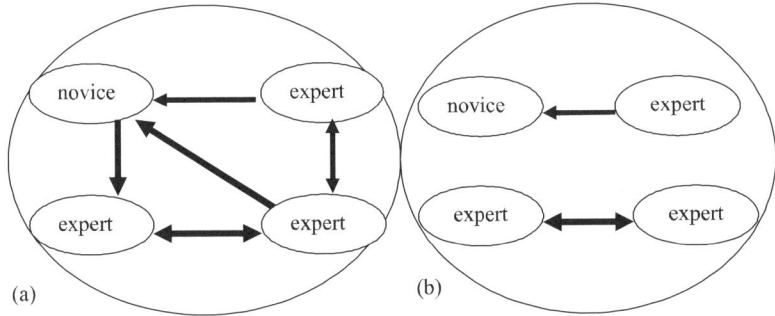

Figure 17.2. *Types of cognitive distribution: a) Highly distributed system (cohabitation); b) Lowly distributed system (intense interaction)*

The essential characteristic of the evolving dimension of the system is that it enables people with different levels of skills and learning to interact. This system of collective expertise without centralized control coordinates two populations: a group of recognized experts and a group of novices who are peripheral participants. It may seem preferable for the dynamics of the collective expertise system that the novices can receive expert advice on secondary matters and at the same time experts can discuss information among themselves to create new knowledge (development), or improve the security and quality of the software (maintenance). It can thus be assumed that for a massively distributed system to evolve, it should promote *interaction* and not only *cohabitation between the two knowledge exchange structures*. A learning logic of collaborative type based on a relative equality between the experts (collaborative learning) can coexist with learning through asking for advice (instructional learning) based on deference and trust [TOM 99]. It also means that the system will be scalable only if it allows the production two request and two advice formats:

– requests that focus on secondary issues of the field, and requests concerning fundamental questions on the architecture of the modules [GOL 02][10];

---

10 Goldman distinguishes, in a field of knowledge, the primary and secondary questions.

– pieces of advice that are recognized as reliable by all and advice that is either subject to discussion or cannot yet be answered because the problem has not been properly formulated.

17.4. Evolution of communication networks

Usenet modifies the processes of knowledge, not directly but through the thread as a communication framework. It is the communication networks, shaped as a posted request system, that answer requests for advice and their eventual transformation into a collaborative discussion. The thread as a sequence of replies is, as we have seen, always constrained by two elements: the selection of messages by the server[11] and posting of the request through a collective address [KOL 96]. A first way to take into account the form of the thread as a communication network is to look at the thread as a network whose nodes are either messages or messages labeled by authors [DOR 05]. These local networks can be of variable sizes because they may contain a single message (that is a request without answer), or form a dyad or extend to contain between 10 and 20 messages[12]. A significant number of threads contain only two messages (1,320 out of a total of 6,731 over one year within *debian.user.french*) and thus they are reduced to a non-mutual dyadic relation of advice[13]. When a thread extends beyond two messages, its structure can become complex and display new properties of coordination.

Threads do not only present variations in form in terms of size but also in terms of the way messages are addressed and the degree of reciprocity between senders and recipients. Mutual relations in the thread can evolve towards relations of reciprocity within the list only if the rate of interaction between a restricted number of participants increases[14]. If a thread is a network, it can also be considered partly as a group with a complex participation framework with several levels of address

---

11 The sequentiality of replies to a query is only partially under the control of the contributors because it is the server that distributes messages. The time lags between messages are not only due to the asynchronous nature of the sequences but also due to the path followed by a message in the digital network.

12 We examined, over one year, a flow of 26,682 messages grouped in 6,731 threads [DOR 05, DOR 06], which allowed us to examine how their structure evolved over a long period in terms of messages and agents that intervene.

13 [DOR 05] notices that these threads concentrate a smaller number of advisors (428) compared to the number of request-providers (750).

14 A mutual relation is reflected in the sequentiality of replies between two agents. Reciprocity can be postponed, according to a generalized-exchange system. An advisor can be advised on another subject by another advice giver. It means that non-mutuality at the thread level does not exclude reciprocity at the level of the list.

[GOF 79]. Gibson [GIB 05][15] has observed the behavior of agents in conversational groups in direct communication. He has shown that in a conversational group, singular speaker addresses and group addresses are combined. Each conversation with several participants therefore presents open options in the mode of co-orientation between senders and recipients. A group address has a direct implication on the mutuality of the relation. A "threaded consultation" seems thus to hinder the mutual structure of a dialogic consultation at first [BLA 86]. When A emits a request, it addresses the group; it does not select any singular recipient and it is the respondent who self-selects him/herself (*turn claiming*)[16]. In the case of lists, the system of requests favors this option at the turn where the request is posted because the server distributes the request to all and makes a consultation possible without any reply or with a single response that is automatically an answer to the questioner (the request-provider). It is only at the third turn that the orientation towards the recipient[17] may take a reciprocal character. The relational structure of the thread may then diverge and allow relations of reciprocity between sender and receiver. We then find ourselves facing two possible scenarios: *m2* and *m3* are answers to *m1*, or *m2* is an answer to *m1* and *m3* a reply to an answer *m2*.

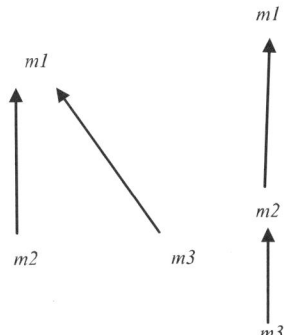

Figure 17.3. *Relational sub-structures*

These two threads show very distinct relational sub-structures. If in both cases the second message *m2* is always an answer to the request *m1* (except in the case of reformulation of the request by the questioner), it is only at the third turn (*m3*) that two options are open in the orientation of replies and the selection of the recipient.

---

15 [GIB 05] is inspired by the speaker selection preferences in the conversation analyzed by [SAC 74] but also by the notion of participation framework provided by Goffman [GOF 79].
16 Self-selection is an option in the conversation when the speaker in place does not mark up any sign of particular attention from one of the recipients.
17 The orientation of a reply is given in a thread by quoting the extract of the message to which it responds.

Depending on how the respondents co-orient [MON 99][18], a thread can thus generate two types of reply structure. The threads can either be a network of replies to a request or a network of discussion of responses whose graph takes a line-shaped form [DOR 05] (see Figure 17.4).

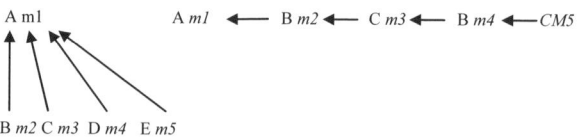

Figure 17.4. *Graphs of replies*

If we now consider the authors (A, B, C...) of messages *m1*, *m2*, *m3*, *m4* and *m5*, we can see that they can, at each contribution, select different respondent roles. When *m3* is sent by C, two configurations can arise:

– a configuration where C as a second respondent can, like the first respondent B, direct his/her own reply towards request *m1* and thus towards A, the author of the question;

– a configuration where the authors of messages at each turn respond to previous respondents by repeated dyadic mutual relations (B/C, B/C).

The second configuration presents interesting properties and it is useful to understand how a relationship of discussion can emerge within a flow of advice-giving activity. It can be said that the second advisor, the author of *m3*, has the possibility at the third turn of specifying, discussing or commenting on the first advice *m2*. If conversational groups, coordinated by an interface of *request for comment* type, are first arranged by answer-replies oriented towards the posted request, the system leaves open a second option, in the third turn, to construct replies oriented towards the previous respondent. The local structure of the thread can therefore evolve from a system of replies to a request towards other forms.

However, for this evolution to be possible, the respondents should be able to select privileged partners from distinct respondents. Indeed, if the structure does not evolve, then it is reduced to a weakly mutual system of two turns/two contributors: a request for advice and a production of advice[19]. For dialogical message pairs to be

---

18 Each response in the *Usenet* system indicates by the ">" sign the segment of the message to which it replies, which allows knowing who is the recipient.
19 This is because mutuality cannot be identified with reciprocity that an advisor may receive advice from another advisor later. This does not contradict the idea that a two-turn request/reply sequence is weakly mutual.

formed, it is necessary that the system of the list goes beyond an online consultation of advice with weak member commitment. The emergence of a dialogical pair with mutual co-orientation is only possible at the third turn, however.

We therefore assume that it is the existence of a dual structure of co-orientation of authors that allows sequential message organization and a sender/recipient relationship to evolve. If these two sub-structures are not equivalent, it might be because they do not produce the same effects on transmission and reception of messages and on the extension of the size of threads. The emergence of a repeated dialogical pair between respondents is only possible after a contributor replies to the first respondent, or when the questioner replies to the first advisor. The group address of the request posted to members of the list thus does not prohibit the ensuing selection of a single contributor, it just delays it at the third turn. When this branching appears, the advice relation tends to change. At the same time that the mechanism of partner selection changes, the structure of mutuality of relations also changes.

Once the rate of dialogical interaction increases between contributors there is a relative decrease in the number of participants and thus there are processes that promote the emergence of mutual repeated interactions in the thread and list. As it seems logical that experts prefer to exchange knowledge with experts, the transformation of the relationship of advice into a relationship of discussion must be a possibility offered by the network structure. Mutual co-orientation between respondents would then become the most direct way to gather the experts and to increase the amount and quality of knowledge production.

17.5. Epistemic networks and discussion networks

How do communication networks operate on the production of knowledge, and in particular on the transformation of the relation of advice into a relation of discussion? The distinction between advice and discussion is epistemic and indirectly related to knowledge transmission. One consequence is that communication networks cannot be identified with epistemic networks.

Indeed, if the form of the digital communication network affects the distribution of knowledge, at the same time there is no direct mapping between question/answer format and relation of advice, or between answer/reply to the answer and relation of discussion. The orientation options, for the first three turns, can be represented in the form of three situations, each with simple triadic structures. These situations combine relations between replies and relations between advice providers and advice seekers. They connect each participant to a particular role in relation to a reply type and to another participant.

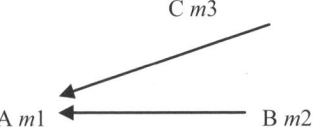

Figure 17.5. *Situation of advice: three agents and two answers to m1*

This first structure (Figure 17.5) expresses the structure of novice/two experts (N/E2), analyzed by Goldman [GOL 02][20]. This structure evolves in the case of digital communication, because the two answers are subject to constant peer evaluation by other list members. Any intervention can cause a counter-reply of a reader.

Figure 17.6. *Hybrid advice/discussion situation: three agents, one answer to m1 and a counter-response to m2*

This second structure (Figure 17.6) includes a branching in the supporting structure by allowing collaborative communication between the two advisors, B and C. The advice seeker cannot just compare two answers but become involved in discussion between its two advisors when C replies not to A but to the answer provided by B[21].

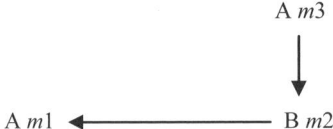

Figure 17.7. *Discussion situation: two agents, an answer and a counter-response of the questioner*

---

20 The problem of expertise for Goldman concerns the capacity of a novice to evaluate the contribution of an expert.

21 The graph of this network shows a thread-like structure in the sequencing of messages as each message replies to a past message. A dyadic conversational structure thus always takes the form of line-shaped graph.

This last configuration (Figure 17.7) represents a structure of discussion where a request for advice is transformed into a collaborative peer structure. In this case, the advice seeker has expertise and can assess the response of the first advisor by discussing the advice obtained. A question can lead to a collaborative and dialogical structure if the questioner can assess the additional knowledge provided by his/her informant by thus becoming his/her collaborator.

The relationship between communication network and epistemic network can become complex when the demand for advice is initiated by an expert. One example of this type can be found in a thread studied on the *debian.user* list where the demand for advice was posted by an active contributor who produced 124 messages in three months including 119 items of advice and re-intervened by discussing advice given to him. He interacts with a consultant who is also an active advisor of the list[22].

In summary, two elements seem to favor a relation of discussion: the degree of competence of the person posting the request and his/her ability to re-contribute to the thread. Generally, discussion sequences fit within a relation of advice, as in the hybrid situation. This means that threads are mainly structured as advice and hybrid threads rather than discussion threads[23].

The development of a discussion sequence within a consultation for advice often leads to expanding the size of the thread (number of messages)[24]. However, when the rate of interaction between respondents' increases, the number of authors tends to decrease since the increase in interaction rate is based on the occurrence of sequences of dialog between two respondents. When the size of a thread of discussion is extended, two phenomena can be combined or evolve distinctly:

– branching (hybrid situation) characterized by a change of orientation in the sequence of advice when some of those intervening address their replies to one of the respondents[25];

– increase of the rate of interaction that is characterized by a relative decrease in the number of participants compared to the number of messages, and which can result in the emergence of re-interventions by the same author.

---

22 We can assign to each contributor a rank by giving to him a rate of contribution (posted messages) and a rate of advice (posted responses).
23 If there is no pure relation of discussion without opening by a request, a relation of advice is possible without the sequence of discussion.
24 67% of threads have 0-2 responses, 14% 3-4 responses and 18% 5-15 responses.
25 Cases of extended files where all the interveners address their response to the sender of the request are thus few [DEL 02] compared to threads where interveners address responses according to the quality of the question or responses to the question.

The orientation towards the respondent at the third turn can generate a branching in the request system, as it is both a way to stimulate a discussion and a means by which to extend the size and density of social relations between contributors. Indeed, the more a thread permits a discussion, the stronger is relational density (degree of mutuality) becomes and the more agents tend to coordinate and control participation by reducing the number of participants.

17.6. Conclusion

Knowledge distribution within Open Source software user lists is a massively distributed system of collective expertise. We can observe the dynamics of distribution by looking at how the reply structure in the threads evolves over long periods of time. We have shown that the probability that a relation of advice can be transformed into a relation of discussion is a condition whereby network structures are transformed into epistemic networks, as the quality of knowledge relies on a dynamics that promote a collective expertise based on collaborative learning. The structure of threads can take two forms of co-orientation between contributors: an orientation of respondents towards the advice seeker and an orientation towards the previous respondent. A collective expertise works as a distributed system only when the dynamics of communication networks facilitate an interaction between two types of epistemic networks, those of advice and those of discussion.

If digital communities are distributed dynamic systems, this means that no centralized system controls either the size of the thread or the rate of interaction between contributors. There is a corrective mechanism in advice transmission that allows users to select qualified partners despite the fact that any posted request in *Usenet* is addressed to the group without expert selection. Social pursuit of knowledge is the result of both thread extension beyond the dyad and the emergence of a collaborative structure based on reciprocity. It is only through local mechanisms that emerge within and between threads that collaborative networks develop between privileged partners.

If current studies on cooperation in online communication networks remain largely at an exploratory phase, their progress is conditioned by a better conceptual integration between social and cognitive sciences. In terms of social sciences, the progress of research requires the analysis of social networks and the social quest for knowledge to be taken into account. The overlap between social networks and epistemic network analysis will also be an important future issue for these studies [ROT 05]. It should, however, be emphasized that many methodological and conceptual problems will be clarified when these approaches report relevant results.

17.7. Bibliography

[BLA 86] BLAU P., *Exchange and Power in Social Life*, Transaction Publishers, New Brunswick, 1986.

[CLA 02] CLARK A., "Mind, brain and tools", in CHAPIN H. (ed.), *Philosophy of Mental Representation*, Clarendon Press, Oxford, 2002.

[CLA 97] CLARK A., *Being There: Putting Brain, Body and World Together Again*, Mitt Press, Cambridge, MA, 1997.

[CLA 98] CLARK A., CHALMERS D., "Extended mind", *Analysis*, vol. 58, no. 1, p.7-19, 1998.

[COH 05] COHENDET P., "On knowing communities", *Advancing Knowledge on the Knowledge Economy*, National Academies, D.C., Washington, 10-11 January, 2005.

[CON 04a] CONEIN B., "Communauté épistémique et réseaux cognitifs: coopération et cognition distribuée", *Marché en Ligne et Communautés d'Agents, Revue d'Economie Politique*, p. 141-160, 2004.

[CON 04b] CONEIN B., "Relations de conseils et expertise collective : comment les experts choisissent leurs destinataires dans les listes de discussion?", *Connaissance et Relations Sociales, Recherches Sociologiques*, vol. 35, no. 3, p. 81-87, 2004.

[CON 94] CONEIN B., JACOPIN E., "Action située et cognition : le savoir en place", *Sociologie du travail*, vol. 36, no. 4, p. 475-500, 1994.

[DEL 02] DELSALLE S., Debian user as collective for a project, Cognitive interdependence and exchange regulation, DEA Paper, University of Lille III, 2002.

[DOR 05] DORAT R., Modélisation des Threads de Discussion dans une Liste de Diffusion, DEA Paper, June 2005.

[DOR 06] DORAT R., LATAPY M., CONEIN B., AURAY N., "Multi-level analysis of an interaction network between individuals in a mailing list ", *Annals of Telecommunications*, 2006.

[GIB 05] GIBSON D., "Taking turns and talking ties: networks and conversationnal interaction", *American Journal of Sociology*, vol. 110, no. 6, p. 1561-97, 2005.

[GOF 63] GOFFMAN E., *Encounters, Two studies in the Sociology of Interaction*, Mac Millan, New York, 1963.

[GOF 79] GOFFMAN E., "Footing", *Semiotica*, vol. 25, p. 1-29, 1979.

[GOL 02] GOLDMAN A., "Experts: Which one should we trust ?", *Pathways to Knowledge*, Oxford University Press, Oxford, p. 139-163, 2002.

[GOL 99] GOLDMAN A., *Knowledge in a Social World*, Clarendon Press, Oxford, 1999.

[HUT 00] HUTCHINS E., Distributed Cognition, Working paper, University of California at San Diego, 2000.

[HUT 95] HUTCHINS E., *Cognition in the Wild*, MIT Press, Cambridge, MA, 1995.

[KIR 99] KIRSH D., "Distributed cognition, coordination and environment design", *Proceedings of the European Cognitive Science Society*, available at: http://adrenaline.ucsd.edu/kirsh/articles/Italy/published.html, 1999.

[KOL 96] KOLLOCK P., SMITH M., "Managing the virtual commons: Cooperation and conflict in computer communities", in Herring K. (ed.)., *Computeur Mediated Communication*, John Benjamins, Amsterdam, 1996.

[LAK 00] LAKHANI K., VON HIPPEL E., *How Open Source Software Works*, MIT Sloan School, Working paper, 2000.

[LAZ 95] LAZEGA E., "Concurrence, coopération et flux de conseil dans un cabinet américain d'avocats d'affaires: Les échanges d'idées entre collègues", *Revue suisse de sociologie*, vol. 1, p. 61-84, 1995.

[LAZ 01] LAZEGA E., *The Collegial Phenomenon*, Oxford University Press, Oxford, 2001.

[LAZ 99] LAZEGA E., "Le phénomène collégial: une théorie structurale de l'action collective entre pairs", *Revue française de sociologie*, vol. 40, p. 639-670, 1999.

[LON 02] LONGINO H., *The Fate of Knowledge*, Princeton University Press, Princeton, 2002.

[MON 99] MONDADA L., "Formes de séquentialité dans les courriels et les forums de discussion : une approche conversationnelle de l'interaction sur Internet", *Recherche*, vol. 2, no. 1, p. 3-25, 1999.

[NOR 99] NORMAN D., "Cognitive artifacts", in CARROL J.M. (ed.), *Designing Interaction: Psychology at the Human-computer Interface*, Cambridge University Press, New York, (French translation, EHESS, Paris, 1993), 1999.

[ROT 05] ROTH C., BOURGINE P., "Epistemic Communities, description and categorization", *Mathematical Population Studies*, vol. 12, no. 2, p. 107-130, 2005.

[SAC 04] SACK W., DETIENNE F., BURKHARDT J.M., BARCELINI F., DUCHENAUT N., MAHENDRAN D., "A methodological framework for socio-cognitive analyses of collaborative of design Open-Source Software", *CSCW, Journal of Computer-Supported Cooperative Work*, Chicago, 6-10 November 2004.

[SAC 74] SACKS H., SCHEGLOFF E., JEFFERSON G., "A simplest systematics for the organization of turn-taking for conversation", *Language*, vol. 50, p. 696-735, 1974.

[SMI 01] SMITH M., FIORE A., "Visualization components for persistent conversations", *Proceedings of the SIGCHI Conference on Human Factors in Computing Systems*, pp. 136-143, 2001.

[SMI 99] SMITH M., "Invisible crowds in cyberspace: Mapping the social structure of the Usenet", *Communities in Cyberspace*, Routledge Press, London, 1999.

[SPE 01] SPERBER D., "L'individuel sous l'influence du collectif", *La Recherche*, vol. 344, p. 32-35, July/August 2001.

[TOM 99] TOMASELLO M., *The Cultural Origins of Human Cognition*, Harvard University Press, Cambridge, MA, 1999.

[TUO 01] TUOMI I., "Internet, innovation and Open Source", *First Monday, Peer Reviewed Journal on Internet,* vol. 6, no. 1, January 2001.

[TUO 05] TUOMI I., "The future of Open Source", in WYNANTS M. and CORNELIS M. (eds.), *How Open is the Future,* University Press, VUB Brussels, p. 429-459, 2005.

[VON 01] VON HIPPEL E., Open Source Software: Innovation by and for sers: No Manufacturer Required, Working paper 133, MIT Sloan School, 2001.

Chapter 18

Towards New Links between HSS and Computer Science: the *CoolDev* Project

18.1. Introduction

In 1991, Kuutti [KUT 91] started a mini revolution in the field of research related to computer-supported cooperative work (CSCW) by publishing *The Concept of Activity as a Basic Unit of Analysis for CSCW Research*. Wanting to address the problem of the definition CSCW itself, Kuutti proposed to use a theory from the works of human and social sciences (HSS) and, more specifically, the concept of activity from activity theory (AT). However, although this theory has been involved in CSCCW for more than 15 years, the question remains as how to best use it to develop groupware [HAL 02].

Many researchers have shown how AT provides a framework for analyzing and improving CSCW situations. Although these works are very interesting, the results obtained are strongly related to the case studies and, from the computer scientist's viewpoint, hardly reusable in another context. Starting from this observation, we have chosen to tackle the problem in reverse. The conceptual framework of AT describes generic mechanisms, that is elements common to all human activities. Our idea is to use this genericity to identify computer techniques that *a priori* support them. This chapter is based on our own experiment in the context of the creation of the *CoolDev* environment, a tool to support the cooperative development of software.

Chapter written by Grégory BOURGUIN and Arnaud LEWANDOWSKI.

18.1.1. *A new support to cooperative activities of software development*

Software development is an inherently cooperative activity. However, many studies have shown that this dimension is poorly supported in integrated development environments (IDEs) such as *Visual Studio Team System* [SRI 04] or *Eclipse* [IBM 06], which aim to integrate various tools (editor, debugger, code sharing and others) involved in development activities. Trying to fill this gap, we initiated the *CoolDev* project, a new IDE built as an extension of the *Eclipse* platform.

18.1.1.1. *The CoolDev project*

CoolDev (Cooperative Layer for software Development, see Figure 18.1) proposes to orchestrate a set of tools involved in development activities. The basic function of *CoolDev* is for each player in the process to recreate his/her work environment based on his/her role. In a very simplified way, when he/she starts *CoolDev*, a user is prompted to identify him/herself, allowing the distributed environment[1] to find what his/her role is (developer, project manager, or any other role having been defined) in the activity in progress. This has the effect of shaping his/her client workstation by loading (starting-up, positioning in the interface) and configuring (connections to dedicated servers, rights management, communication channels with the other players, etc.) the tools (editors, sharing documents, instant messaging) that he/she needs to fulfill his/her role.

In our example, the user *Arno* was identified as *developer*, which had the effect of loading the tool (*CooLDAView*) that allowed him for example:

– to see who the other participants involved were;

– to verify whether they were connected or not and in what role;

– to start the code editor connected to the CVS (Concurrent Version System) server of this activity, in the right project with the proper rights;

– to start the chat tool that allows him to discuss with the other participants;

– etc.

However, *CoolDev* does not simply configure the environment at start-up. Indeed, the actions of each participant can have repercussions in the global environment. For example, updating a portion of code in the CVS server by a developer can lead to the display of a message in the chat of all of the particpiants and modify the status of the role of *testers* in order to enable them to comment on the portion of software concerned.

---

1 The environment is both distributed in space (the actors are connected on remote machines) and in time (the actors are not necessarily all connected at the same time).

New Links between HSS and Computer Science   285

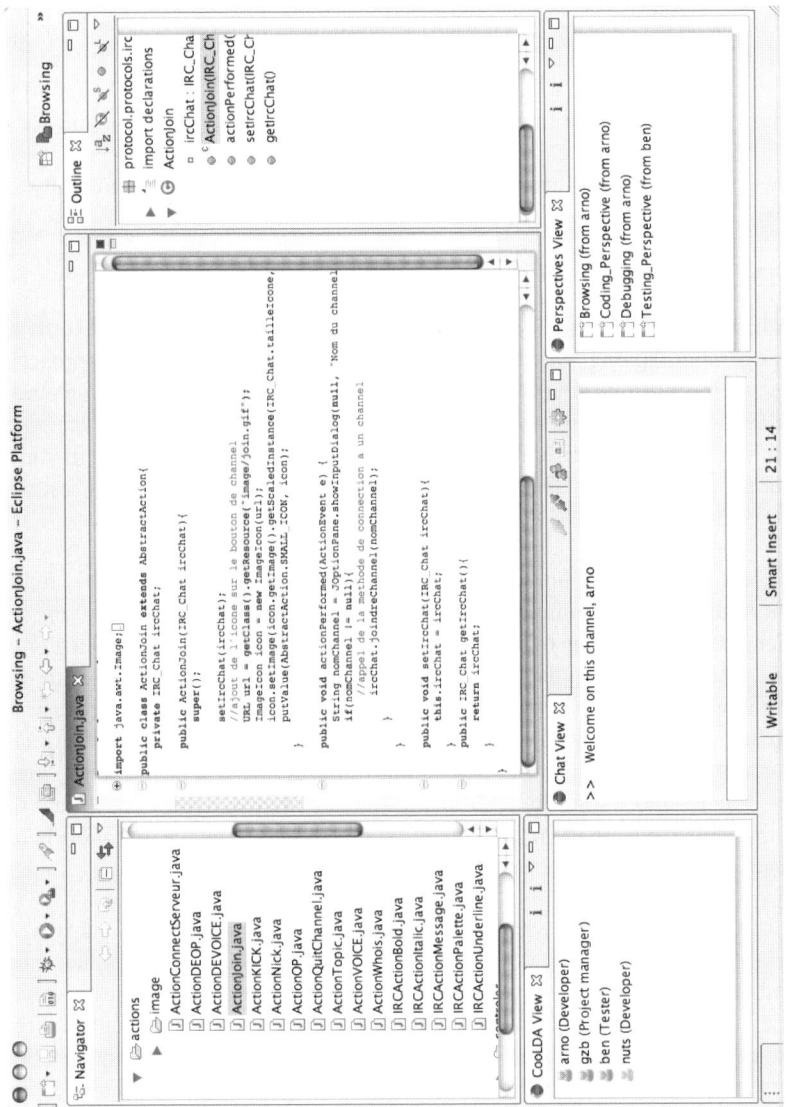

Figure 18.1. *An example of perspective for the role of developer in CoolDev*

We must note that the basic properties of *CoolDev* that are described above are elements that can be found in most environments for software development that are interested in collaboration. *CoolDev* wants to introduce this "classical" but missing collaborative dimension within the *Eclipse* IDE. However, our final objective is primarily to create an IDE that brings new properties offering a "better" support to collaboration. In order to define these new properties as well as the means to implement them, we became interested in the results obtained in other projects attempting to create better groupware, especially those dedicated to software development activities.

18.1.1.2. The role of human and social sciences in groupware design

Several proposals [BAR 02, KOR 02, SOU 03] have recently emerged to better support cooperation among software development tools. If we study the approaches underlying their creation, it is interesting to note that most of them implement theories developed in human and social sciences (HSS) – AT in the examples mentioned – in order to better understand the human cooperative activities in which the development is achieved. This type of approach is very characteristic of research in the CSCW field.

What holds our attention in these very characteristic examples is that the studies carried out highlight shortcomings related to our problem, but offer results that are very difficult to reuse in our own approach. This problem is illustrated in Figure 18.2. It is established that the creation of "good" groupware happens in the multidisciplinary encounter between HSS and computer science in the case of interest to us. The meeting points between these disciplines are generally made through one-off projects. Just like numerous publications on groupware, these studies are directed more towards HSS than towards computer science itself and the elements that come out of them are usually strongly linked to the situation being analyzed.

Thus, a question remains: what can the computer scientist gain in his way of designing *a priori*, of using and developing his own techniques? It is true that these work in situations. However, it is clear that the development of disciplines involved in a multidisciplinary approach is made, from this point of view, through indirect links, through specific works based on particular projects involved. The idea that we advocate is to imagine that we can today forge more generic links by proceeding directly to an analysis of elements that constitute the successive developments of various disciplines, often through one-off confrontations.

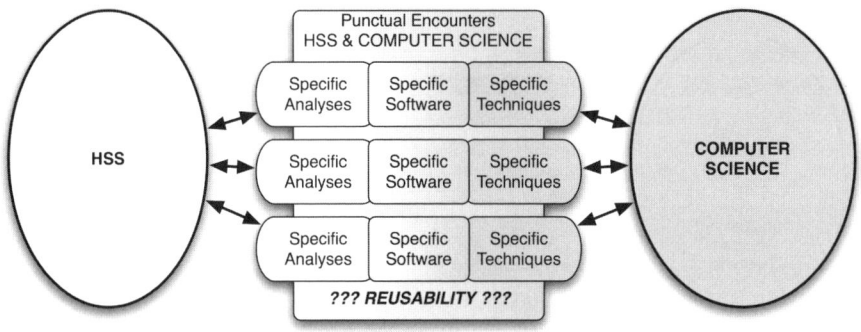

Figure 18.2. *The problem of the reuse of results provided by ad-hoc meetings between HSS and computer science*

18.2. Towards new links between HSS and computer science: application to AT

From our point of view, described in Figure 18.3, the generic link between HSS and computer science are revealed in their respective goals. HSS try to create theories, methods and tools (such as the structure of activity proposed by Engeström [ENG 87] or the checklist of Kaptelinin [KAP 99]) helping to understand human activities. Computer science tries, on its part, to develop theories, methods and tools for creating software that support them.

It appears that the interface between these disciplines is the *human activity* in its generic form, that is, the description of points common to all human activities. The questions that then arise are: what are the elements linked to the generic properties of human activity that were identified in each discipline, and what links can be built up between these elements? The goal of any system being to support human activities, we are convinced that these generic links identified can easily be reused in the creation of particular systems.

To our knowledge, only one team of researchers close to our area of interest has attempted to forge such a type of link. These are the works initiated by Dourish and Button in what they call technomethodology [DOU 98]. However, after a leading article discussing the principle of this approach, we did not find any more advanced developments. It should also be noted that this proposition has its roots in the ethnomethodology while our own works focus on AT.

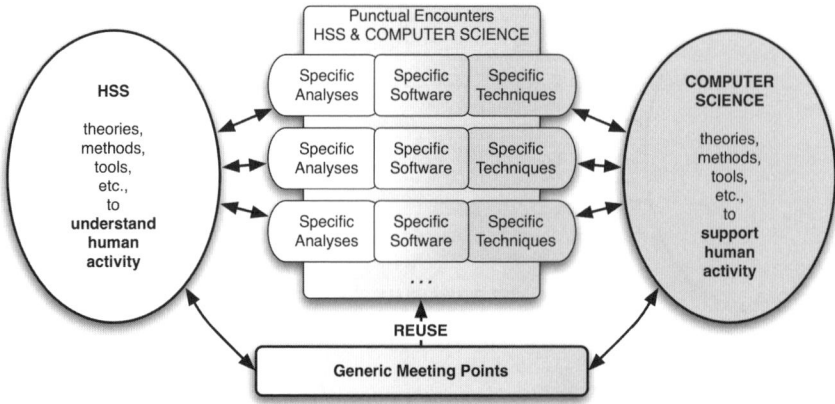

Figure 18.3. *A generic approach to multidisciplinarity*

18.2.1. *Activity theory*

AT [BED 97, KUT 91, NAR 96] originates from the Soviet historical and cultural school of psychology founded by Vygotski. Over the years, it has focused on mediation through the tool or instrument and has proved to be a body of concepts whose aim is to unify the understanding of human activity by providing bridges towards other approaches coming from human sciences, in particular approaches in social sciences and those of behavior. The fundamental analysis unit of AT is human activity which is defined as a coherent system of internal mental processes, of an external behavior and motivational processes that are combined and directed to achieve conscious goals.

Engeström [ENG 87] defined the "basic structure of an activity", which is the most classical reference to AT. This model expresses the relation between the subject and the object of the activity. This relationship is reciprocal: the subject realizes the object of the activity, but at the same time, the properties of the object transform the subject by increasing its experience. It is furthermore mediated by the concept of the tool representing everything that is used in the realization process of the object.

The tool both allows and limits: it allows the subject to realize the object of its activity, but limits by masking some of the potential of transformation. On the one hand, the tool assists in the realization of the object by subjects that use it because it carries with it implicit goals that were set by its developers. On the other hand, it is itself transformed and (re)constructed during the activity. It is often modified or adapted by subjects in a reflexive process and in response to contradictions that may emerge between elements participating in the activity, depending on their emerging needs, their goals, and experience. Thus, this subjects' experience crystallizes in the tool that thus carries in it the cultural heritage of the situation. Mediation by the tool corresponds then to a means of transmitting a certain culture and experience. Moreover, the individual is not isolated but forms part of a community that share the same object of activity. Community-subject and community-object relations are mediated by concepts of rules and division of labor also containing the cultural heritage of the situation. As with the tool, these mediators are open to new developments. More detailed descriptions on our understanding of AT can be found in [BOU 05].

18.2.2. *In search of generic tools for human activities*

The AT has often been criticized [HAL 02], in particular by HSS researchers, because its approach is sometimes considered to be too general not to allow a certain level of detail in the description and understanding of analyzed activities to be attained. Paradoxically, it also seems to be the most approved theory by computer technologists, hence its success in CSCW. Indeed, beyond the creation of computer supports for particular activities, one of the major motivations that guides computer scientists is genericity. The software creator tries in most cases to create a tool aimed to support a more or less generic task that can be instantiated for various users involved in similar activities. For example, in the case of *CoolDev*, our goal is to support various activities of software development so that the tool we create is potentially useful to a maximum of users. In the same way, computer science itself tries to address the generic problems of support of human activities. Thus, part of what the computer scientist seeks in the answers of HSS is a generic description of properties of the activity he tries to support. From our point of view, this corresponds to what provides AT by proposing a model that describes the components and mechanisms common to all activities.

A typical example of this search for generic solution in HSS by creators of groupware is their extensive use of the basic structure of activity. Many works claim being inspired from AT because they propose a conceptual model inspired by the structure of Engeström, very useful for the conceptualization of a generic support for cooperative activities, as this corresponds to what it describes. We ourselves have used this structure to conceptualize *CoolDev* [LEW 05]. The concepts of role, tools, etc. described in section 16.1 are to be directly related to the entities described in AT (see Figure 18.4). However, although this example shows that the structure of Engeström allows the information to use AT directly in the design, we do not think that it is a real generic link between foundations of activity and computer science. On the contrary we are convinced that these links can be identified, not only by considering this structure, but also by examining the properties and mechanisms that are associated to it in AT.

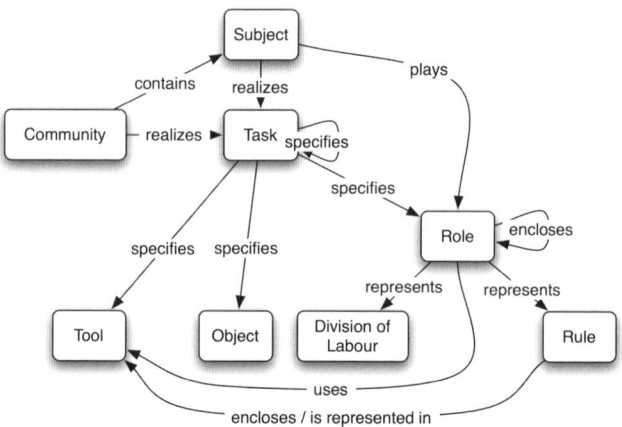

Figure 18.4. *Elements of an ontological model based on AT*

18.3. Generic links between AT and computer science: application in *CoolDev*

Every software is a tool whose purpose is to equip a human activity. Thus, in AT, the description of the mechanisms surrounding the tool, mediator of activity, and the properties it is supposed to present can help the computer scientist on the fundamental properties of software tools he attempts to create. If we consider that certain properties of software come from the very foundations that were used in their creation, that is, techniques that constitute computer science, we then therefore create generic links between computer science and HSS, and hope that techniques supporting generic properties of tools in human activity would allow producing better software. We will now exemplify this subject by highlighting a few particular points from the application of this approach in *CoolDev*.

18.3.1. *Link between generic properties of AT and computer science techniques*

We have identified certain connections between the properties of the human activity described in AT and computer techniques that allow supporting them. These techniques have been implemented in *CoolDev* following its theoretical analysis (as opposed to an empirical analysis) as part of AT. The outlines of this analysis are the following: *CoolDev*, as a *tool* to support development activities, may be changed by its users (the *subjects*) according to their emerging needs and their *experience* in development. This *experience* should be able to *crystallize* within *CoolDev* so that it can then be transmitted to other users.

Finally, *CoolDev*, as a groupware, contains a representation of other elements of the activity as the roles of actors who reflect certain *rules* and *division of labor*. These elements being also subject to development and to the crystallization of the experience, their computerized representation should reflect these properties. The techniques implemented to support these generic properties are listed in Table 18.1.

Activity Theory	Computer techniques
Reflective attitude	Introspection, causal relation
Understanding by the subject	Framework, meta-modeling
Evolution, crystallization of the experience	Intercession, customization, integration, extension
Development of experience	Prototyping
Transmission of experience	Model-driven engineering components

Table 18.1. *A connection between properties from AT and computer techniques*

To enable subjects to understand the functioning of their tool in the event of contradictions, *CoolDev* uses *introspection* [MAE 87] that allows the system to provide a representation of its own functioning, during the execution. *CoolDev* proposes for example a point of view on the scenarios, the actions available for the roles, etc.

However, these elements correspond to computer objects that are hard to understand for non-computer scientists. This is why techniques of *framework* and *meta-modeling* are very useful. A framework provides a set of computer elements that guide the application towards a particular area.

In our case, the framework of *CoolDev* is a set of objects (in terms of object oriented languages) that participates in the support of development activities. Meta-modeling allows defining one or more languages oriented towards subjects that will allow them to manipulate elements defined in the framework. In *CoolDev*, the meta-model defines for example what a role is, what a tool is, and how these entities can be composed to create a particular activity support. At runtime, *CoolDev* performs the translation between the computer entities of the framework and entities of the language defined by the meta-model.

This translation is made in both directions: from the framework to the meta-model to provide a representation of the functioning of a specific activity support understandable by the subjects; from the meta-model to the framework to make effective within the system the transformation of specifications of an element by the subjects (see Figure 18.5). The maintenance of this link between the representation provided to subjects and the computer representation that allows the application to run corresponds to a *causal relation*, as defined in computer reflective systems [MAE 87].

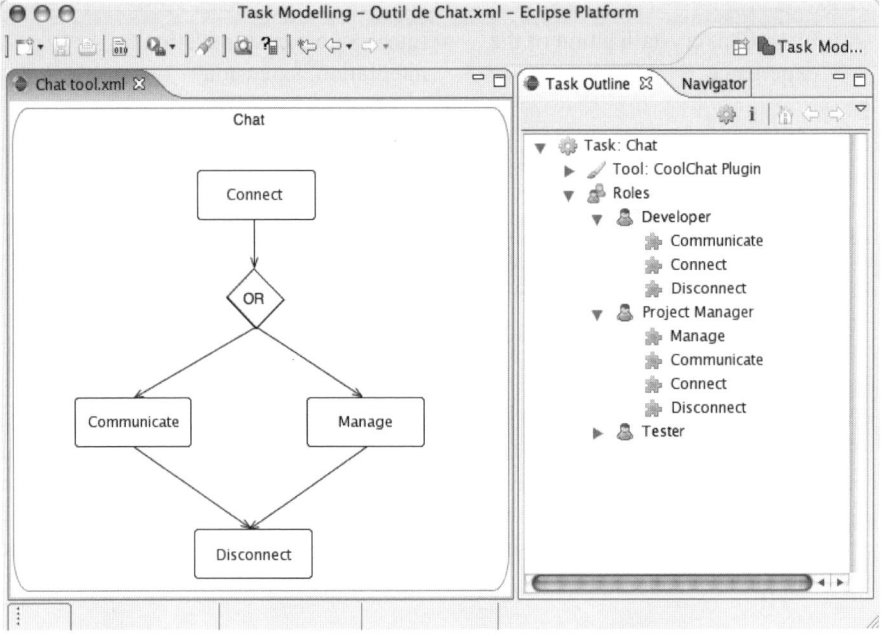

Figure 18.5. *An example of access to (re)definition: the sub-activity of Chat (explanation of actions (as tasks) available for the roles)*

With these techniques, subjects have the opportunity to reflect on their activity support and to identify elements causing problems. The intercession technique [MAE 87] makes it possible for this support to evolve, that is, the mediators of the activity, by transforming elements that constitute it, under the framework, through a meta-model and during the execution of the system. This intercession can take different forms whose complexity is inversely proportional to the power of transformation. These are the techniques of *customization, integration* and *extension* described by A. Morch [MOR 97].

An example of customization in *CoolDev* is the possibility offered to subjects, according to their role, to choose a perspective (as the arrangement shown in Figure 18.1) on their activity. *CoolDev* being built as an extension of the Eclipse platform, the concept of perspective corresponds to that of the same name in Eclipse: it is an arrangement of tools such as an editor, a chat or other. The integration technique allows subjects to introduce in the activity support new tools in the environment.

In *CoolDev*, these tools correspond to any Eclipse plugin[2] [IBM 06] available on the Internet. Finally, the extension mechanism of *CoolDev* allows for example (re)specifying the rights of roles of actors on tools involved in this activity, this through the metamodel and within the framework.

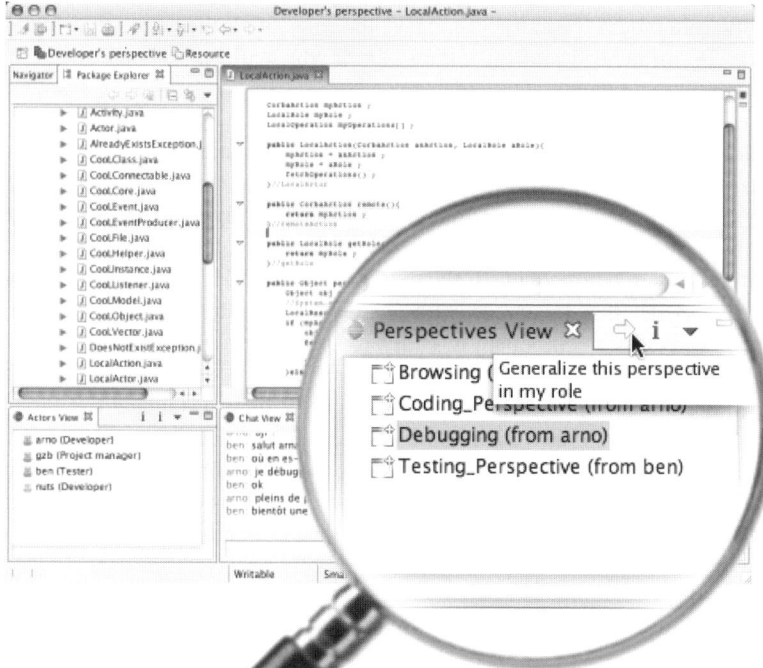

Figure 18.6. *Experience crystallization of by generalization of perspective in the models*

---

2 Integral component extending the functionalities of the platform.

All these transformations of the environment that reflect the reflexive attitude of subjects are carried out according to their experience. However, every transformation is not necessarily the reflection of an experience suited to crystallize within the tool. Indeed, subjects may need to carry out tests before wanting to share their experience.

For example, in *CoolDev*, each actor has a role, an instance of a shared model. We have implemented prototyping mechanisms allowing instances of roles to detach from their model in order to allow particular experiments, such as the choice of arrangement of tools, developed during the activity. This mechanism corresponds to a controlled release of the causal relation between the role model and its instances. Thus, in our example, each developer has the opportunity to integrate and arrange new tools in his environment. Hence, all the developers do not have exactly the same perspective, even if they share the same role. However, each instance of role functions as a prototype that represents the preferences but also the experience of each actor in his role. Finally, when a perspective or an instance of role represents an interesting experience, this instance can be crystallized, that is generalized at the level of the role model (Figure 18.6). Thus, the roles of all the actors that follow this model can be influenced and the future actors who will be assigned this role will directly benefit from the crystallized experience.

Finally, we have identified several computer techniques that indeed allow crystallizing and sharing the experience developed by subjects. When compared to the techniques that we just mentioned, the *component*[3] approaches and techniques related to *model-driven engineering* allow the sharing of experience. Indeed, the very purpose of a component is its reuse by others. However, any component contains experience that was recorded by its developers. In a similar approach, each model can be seen as a component intended to specialize a generic structure. Thus in *CoolDev*, a particular scenario of software development activity, transformed and developed during its implementation, is a model that has crystallized the experience developed by those who have used it and caused it to evolve. This model can then be reinstantiated in the platform in order to support a different (involving for example other actors), but similar (starting with the same definitions of roles, the same tools) activity.

18.4. Discussion: towards new developments

Faced with the problem of the reuse of results from ad hoc confrontations between HSS and computer science, we have proposed a first comparison between

---

3 Computer components are software elements defined in order to facilitate their reuse since they offer a particular interface hiding the complexity of their implementation.

the generic properties of human activity described in AT and the techniques of computer science that allow supporting them. It can be noted that these computer techniques have, at least in some cases, already been implemented in particular situations. Introspection, intercession and approach by models have already been used to alleviate problems of rigidity of workflows [KAM 00] and intercession techniques have been used to address the problem of malleability [MOR 97]. This is not surprising since, as we have pointed out, computer science and HSS fed on their respective experiences.

However, the fact of addressing these comparisons in a generic way allows us to approach tool development in a more theoretical way by directly identifying the generic properties of human activity that we want to support and techniques that have emerged from computer science and that allow to do so. Moreover, this approach could allow defects to be revealed in each discipline. For example, in computer science, component approaches have existed for many years. However, even if technical solutions exist for their implementation, they are much less well defined from a semantic point of view.

Considering a component in terms of human activity highlights the fact that it is today very difficult to understand what will be its place in the activity that wants to integrate it. Only experienced and motivated computer scientists are able to really reuse most computer components. However, as we saw previously, this technique proves to be very beneficial for the reuse of experience *in situ* and by end users. This problem of semantics is due to the fact that with current techniques, it is necessary to almost completely rebuild the task model of a component to be able to contextualize it in a particular activity.

Indeed, its task is diluted in the code and the documentation means generally provided, as the Javadoc, do not really allow understanding its underlying overall logic. This is why, in order to facilitate the dynamic integration of components, we are working today on a new approach inspired by works on AT and task modeling from the field of HCI [LEW 06]. This example demonstrates how the fact of considering computer science from the point of view of HSS can lead to transformations in its techniques. Conversely, the study of computer science, crystallizing the experience developed by computer scientists, can certainly also feed the HSS. It is still too early for us to really measure the impact of this approach. However, we hope that it will lead in the future to a better exchange between these disciplines.

18.5. Bibliography

[BAR 02] BARTHELMESS P., ANDERSON K.M., "A view of software development environments based on activity theory", *Computer Supported Coop. Work 11(1-2)*, p.13-37, 2002.

[BED 97] BEDNY G., MEISTER D., *The Russian Theory of Activity, Current Applications to Design and Learning*, Lawrence Erlbaum Associates, Mahwah, 1997.

[BOU 05] BOURGUIN G., DERYCKE A., "Systèmes Interactifs en Co-évolution, Réflexions sur les apports de la Théorie de l'Activité au support des Pratiques Collectives Distribuées", *Revue d'Interation Homme-Machine (HCIR)*, vol.6-1, AFIHM Europia, p. 1-31, June 2005.

[DOU 98] DOURISH P., BUTTON G., "On "Technomethodology": foundational relationships between ethnomethodology and system design", *Human-Computer Interaction*, vol. 13, Lawrence Erlbaum Associates, p. 395-432, 1998.

[ENG 87] ENGESTRÖM Y., *Learning by Expanding*, Orientakonsultit, Helsinki, 1987.

[HAL 02] HALVERSON C., "Activity Theory and Distributed Cognition: Or What Does CSCW Need to do with Theories?", *Computer Supported Cooperative Work (CSCW)*, vol. 11, n° 1-2, p. 243-267, 2002.

[IBM 06] IBM Corp., Eclipse Platform Technical Overview, http://www.eclipse.org/articles/, 2006.

[KAM 00] KAMMER P.J., BOLCER G.A., TAYLOR R.N., HITOMI A.S., BERGMAN M., "Techniques for supporting Dynamic and Adaptive Workflows", *Computer Supported Cooperative Work*, 9(3-4), p. 269-292, 2000.

[KAP 99] KAPTELININ V., NARDI B., MACAULAY C., "The Activity Checklist: a Tool for Representing the "Space" of Context", *Journal of Interactions*, p. 27-39, July/August 1999.

[KOR 02] KORPELA M., MURSU A., SORIYAN H.A., "Information Systems Development as an Activity", *Computer Supported Cooperative Work*, 11(1-2), p.111-128, 2002.

[KUT 91] KUUTTI K., "The concept of activity as a basic unit of analysis for CSCW research", *Proceeding of the Second ECSCW'91 Conference*, Kluwers Academics Publishers, p. 249-264, 1991.

[LEW 05] LEWANDOWSKI A., BOURGUIN G., "Inter-activities management for supporting cooperative software development", *Proceedings of the Fourteenth International Conference on Information Systems Development (ISD'2005)*, Karlstad, Sweden, 15-17 August 2005.

[LEW 06] LEWANDOWSKI A., BOURGUIN G., TARBY J-C., "Vers des Composants Logiciels Orientés Tâches", *Actes de la 18es conférence francophone sur l'Interaction Homme Machine (HCI'06)*, Montreal, Canada, 18-21 April 2006.

[MAE 87] MAES P., Computational Reflection, Thesis, V.U.B, Brussels, 1987.

[MOR 97] MORCH A., Method and Tools for Tailoring of Object-oriented Applications: An Evolving Artifacts Approach, part 1, Thesis, 241, University of Oslo, 1997.

[NAR 96] NARDI B.A., *Context and Consciousness: Activity Theory and Human-Computer Interaction*, MIT Press, Cambridge, MA, 1996.

[SOU 03] DE SOUZA C.R.B., REDMILES D.F., "Opportunities for Extending Activity Theory for Studying Collaborative Software Development", *Workshop on Applying Activity Theory to CSCW Research and Practice, in conjunction with ECSCW 2003*, Helsinki, 2003.

[SRI 04] SRIDHARAN P., Visual Studio 2005 Team System: Overview, http:// msdn.microsoft.com/library/default.asp?url=/library/en-us/dnvsent/html/vsts-over.asp.

PART VII

Towards Renewed Political Life and Citizenship

Chapter 19

Electronic Voting and Computer Security

19.1. Introduction

Works on electronic voting go back to the late 1990s, thanks to a strong impetus given by the European Commission through its research program for the Information Society Technologies (IST). Since then, Europe has recognized the importance of fostering e-democracy and has called for a number of projects on this topic. In the 2000s, only the US and, to a lesser extent, the UK could claim to have significant experience of "remote elections". Among them, we can cite in particular the primaries of the Democratic Party in Arizona that were held in 1996. However, their level of security was very inadequate and did not stand up to scrutiny.

Within the above-mentioned European program, various projects of electronic democracy [MON 02b] have come to fill these gaps. The most famous are CyberVote, headed by EADS, whose work is presented here; e-Poll, led by Siemens; and the EU-StudentVote[1], under the auspices of the organization of the same name. These experiments have allowed Europe and France to push forward a certain number of ideas on the problems of security and protection of personal data.

Chapter written by Stéphan BRUNESSAUX.
1 The European Commission's website on the CyberVote project, an innovative system of online voting from fixed and mobile terminals can be found at http://www.ist-world.org/ProjectDetails.aspx?ProjectId=677844cb40d240ea8a96b548c8011ab9. The e-Poll project, an electronic polling system for remote voting operations is at, http://www.e-poll-project.net/. The EU-StudentVote, presentation can be found at: http://www.europe2020.org/?lang=en. Websites accessed February 11, 2010.

19.2. Motivations of electronic voting

The advances in information and communication technologies (ICT), and in particular in security techniques, have rapidly led politicians and engineers specializing in software development to promote electronic voting under a certain number of expectations.

19.2.1. *Reducing costs and time mobilized for elections*

Setting up an election mobilizes staff and resources. According to our personal information, the average cost of an election would is approximately 3 Euros per voter. This figure is generally accepted both by the electronic voting solution editors and by representatives of the state in charge of elections. However, in addition to these expenses, we should add all of the hidden costs corresponding to the hours volunteers put in, the premises used for polling and counting, catering costs of staff involved, etc.

Compared to traditional voting, voting by electronic means allows us to reduce the cost of voting due to the absence of paper and a reduction in the number of people (president, assessors), reduction of time during which they were mobilized and generates instant vote counting. Thus, the necessary initial investments required in the IT field (in terms of materials or software) may be depreciated over several years, even during a much shorter time for elections whose periodicity is more important.

19.2.2. *Improving the participation rate*

Electronic voting has always attracted statesmen who, from the beginning, have seen a way to increase the participation rate by simplifying the procedure, the latter being justified in particular by abolishing the need to go to polling stations.

At the same time, the initial results of studies [MON 02a, VED 03] conducted on the impact of electronic voting showed that on ballots with low participation rate, electronic voting may not increase participation. This is very understandable. If the poll does not present any challenge in the eyes of the voter, the fact of being able to vote electronically is not necessarily a sufficient argument to mobilize the voter. This solution has nevertheless appeared very attractive in some countries like Switzerland, where the principle of a referendum leads voters to vote regularly on particular decisions.

19.2.3. *Meeting requirements of mobility*

If there is a population that can benefit from this system, it is those with mobility problems. The CyberVote project by the European Union was launched with the particular intention of increasing participation by those who would have difficulty travelling to polling stations. It aimed at developing a highly secure online voting system that enables a voter to vote from any family or office computer or from a third-generation terminal, such as a smart phone or a PDA.

Thus, the project targeted all people who were hospitalized, disabled, suffering from mobility difficulties and also people geographically distant from their polling station on polling day: business men or women on the move, people on paid leave, those who cannot take time off work or simply away for the weekend with friends. Following the abolition of voting by mail for French political elections, the use of proxy voting was reviewed but it is limited to certain categories of voters and is subject to highly regulated steps [PER 01]. In particular, the voter may give a proxy to a representative (a voter who votes in his/her place) but only if the representative is registered in the same village. In addition, the proxy is only valid for a determined election and must be renewed for other elections. It is established by an empowered authority and must be sent to the town hall early enough to enable the representative to vote. While the right to vote is a fundamental right, it appears that there is a certain inequality in the chance to use this vote depending on whether the voter resides near to his/her polling station and can move with or without difficulty.

These benefits, largely sullied at the beginning of the digital era, are being replaced in the context of a growing and steady progression of Internet use in France to communicate, work, play, buy, sell, perform banking transactions, carry out administrative procedures[2] or tele-declare online taxes.

Given these different expectations, critics of electronic voting[3] have put forward the argument of digital division, emphasizing the fact that those who are well off have the means to procure computer materials necessary for remote voting. On the other hand, the existance of the polling booth is often mentioned as a republican ritual essential to democracy [CHE 03]. Finally, entrusting such an important operation of machines or software whose function remains obscure to uninitiated people raises the problem of transparency and trust required for the voting operations thus carried out. All these considerations are the basis of our ideas on the implementation of a secure electronic voting system [FDI 03] that is proposed as an

---

2 For example, Denmark is one of the first European countries to have made e-government mandatory [GUI 06].
3 See Pour une Ethique du Vote Automatisé (Vote Electronique) http://www.poureva.be and http://www.ordinateurs-de-vote.org/, accessed February 11, 2010.

alternative to mail or proxy voting. Since its early design, our approach was to offer to the voter similar or even higher guarantees: the guarantee the ballot will be delivered and secrecy of the ballot (compared to proxy voting). We are going to detail these points after defining the different approaches of electronic voting.

19.3. The different modes of electronic voting

The term "electronic voting" in reality encompasses different definitions. We support in part the terminology proposed by the European project CyberVote [VAN 01] and the Centre National de la Recherche Scientifique [MON 02b]. Thus, according to these two sources, electronic voting systems can be divided into two distinct categories depending on whether the system is online or offline.

In offline voting systems, the computers used represent a set of independent machines without an interconnection or network. It is in this category that we put together different voting machines, such as those used in Florida in November 2000 or in Brest in March 2004 [LEF 04]. In France, the use of such machines has been accepted since the Electoral Code was decreed on May 10, 1969 (Decree no. 69-419). With this type of system, the voter continues going to the polling station. Once in the polling booth, he/she expresses his/her vote electronically using the voting machine. The counting of votes is facilitated by the absence of paper.

In online voting systems, the computers used are connected to each other via a computer network. This network can be local (not accessible from a machine outside the network – as in the case of the confidential zone of a corporate network for example) or, conversely, open, that is, connected to the Internet. In the first case, we will talk about voting on the Intranet, while in the second we will use the term "Internet voting".

Voting on the Intranet will be particularly suitable for private elections, such as for university councils, staff representatives, etc. Internet voting can, itself, be used as an alternative to postal voting. On the network, we will distinguish server machines from client machines. The server machines host the voting application and all elements necessary for the authentication of voters, while client machines offer access to the voting application through the use of a Web browser or dedicated application.

The voter may have to visit a polling station to use a client machine in order to express his/her vote. Here we will then talk about online voting from a polling station (for example, in the case of a voter from city A who would go to the station in city B to vote). Conversely, the voter may be asked to use any computer (home, polling station, cybercafe, etc.) as a client machine. In this case, we talk of remote online voting or Internet voting. It is this access that we have prioritized.

19.4. Prerequisites for the establishment of an Internet voting system

The different studies conducted around electronic voting systems [CAL 00, IPI 01, VAN 01], have highlighted the requirements such a system must satisfy. They are of different orders:

– *Non-discrimination*: each voter must have equal access to the electronic voting system. This implies on one hand the use of a computer to vote and, on the other hand, the absence of specific knowledge to use the computer system. In general, we propose the introduction of computers in public areas for people without computer skills. Although the developers of such systems are redoubling their efforts to make them increasingly user-friendly, the use of electronic voting requires some computer skills that excludes the part of the population that is not familiar with this type of tool.

– *Secrecy*: nobody should be able to link a ballot to the voter who has cast it. The different electronic voting systems are based on different technical approaches that guarantee the secrecy of the vote. The so-called blind signature or homomorphic encryption (see further) approaches offer maximum secrecy. In all cases, it is essential to ensure the secrecy of the vote, to respect the right of the voter to cast an anonymous vote and to avoid abuse of the system (paid or manipulated voting).

– *Freedom*: the voter must be able to vote freely. The design of electronic voting systems must ensure candidates are presented in a neutral manner and remove links to professions of faith, propaganda or advertising sites. The absence of the voting booth, however, makes it impossible to totally avoid external pressure (family threats, coercion, etc.) on the voter at the time of his/her vote. Remote Internet voting is, in this case, comparable to mail voting where the person concerned is not immune to such pressures when he/she inserts a ballot into the envelope.

– *Integrity*: a cast vote cannot be changed. One of the major points of electronic voting systems is to ensure the integrity of the ballot. The implications are numerous. First, it means that the choices made from the voting terminal (the voter's computer for example) are correctly coded and correctly transmitted to the voting server. Thus, if the voter chooses candidate A, it is important that his/her choice is faithfully reproduced in his/her ballot and that the latter is not falsified during uploading to the virtual ballot box. No change should be made by a third person (hacker on the network for example) during the delivery of the ballot to computers containing the ballot box. Then, once the ballot is placed in "the ballot box", it must not be subject to any change. To certify this, security mechanisms, such as "signature" or "evidence", as well as computer-based and physical (access to dedicated computer rooms) access control systems are set up.

In France, the French data protection authority (CNIL - Commission Nationale Informatique and Libertés) has opened the use of an Internet voting system with its recommendation of the July 1, 2003 related to the security of on-site or remote electronic voting systems, in particular over the Internet. It is therefore necessary, when setting up an online election, to verify the compliance of the adopted system to this recommendation [CNI 03], which lists the prerequisites for the implementation of such a system:

– those specific to the election;

– those applicable during the vote; and

– finally those related to post-vote control operations when requested by the election judge.

Thus, the CNIL requires, among other things, access to the source code of the solution to be able to perform an audit. From a computer science point of view, it is essential that there is a physical separation between the hardware or software systems responsible for managing the votes and those responsible for the register in order to ensure that the link between the voter and his/her vote is indeed broken. The use of physical (access control, identification of people entitled to act on the system, etc.) and logical (firewall, access protection to applications, etc.) measures are required in order to maintain the security and confidentiality of personal data, in particular against intrusions from outside. Similarly, the establishment of a sealing process for sensitive voting system files is compulsory. Sealing helps to identify and list any modification made to the system, whether it is voluntary in the case of the implementation of a patch or as a result of external intervention (hacker for example). Other measures that complete this recommendation include the establishment of an emergency system in case the primary server fails.

19.5. Operation of an Internet voting system

19.5.1. *Authentication of the voter*

The authentication of the voter is a crucial point when using an Internet voting system. The absence of a polling station with a person in charge of verifying the identity of the voter requires the use of computer authentication. There are different approaches and each has respective advantages and disadvantages:

– *Authentication by electronic certificate*: a digital or electronic certificate is a small file stored on the computer of the voter or on a smart card. It enables the voter to be identified, provided that the latter has not given another person the password protecting the certificate. It is this type of authentication that is used today in France for teledeclaration of taxes, for example. If the use of certificates guarantees a strong authentication, this technique nevertheless represents a deployment cost higher than other forms of authentication and can rapidly become an economical hindrance. In France, the CNIL recommends the use of digital certificates for political elections [GEN 03].

– *Authentication by login and password*: this is by far the simplest and cheapest authentication system to implement. The voter receives a password in the form of a number combining numbers or letters (for example, a voter registration number, consular identification number, social security number, etc.) and a password that he/she is the only one to know. With these elements, he/she will identify him/herself on the computer system before he/she can vote. One of the key elements of this technique is the means of communication used to deliver to the identification elements to the voter. There may be hand delivery, by registered post with or without the use of a masking process, such as a scratch card to reveal some or all of the elements. In France, the CNIL recommends using two separate channels for the delivery of identification elements when these are of the "login plus password" type. Thus, the first element of identification may be sent by post while the second will be sent electronically.

– *Biometric authentication*: the idea in this case consists of using one or several measures physically characterizing the voter in order to identify him or her. The most widespread biometric authentication system today is the fingerprint reader. It is supposed to guarantee the physical presence of the person. However, DCSSI experts [WOL 03] (the French authority in charge of central management of information system security) are questioning the use of biometrics as an authentication process by highlighting the possibilities of fingerprint falsification using gelatine as well as possible vulnerabilities in software security. Furthermore, its cost is currently crippling and its use can only be considered on a large scale when computer manufacturers decide to equip it in a standard configuration all their materials.

These authentication techniques can be supplemented by the use of a challenge. In this case, it consists of asking the voter a supplementary question (for example, what is your date of birth, your birth town?) and checking whether the answer is correct. The choice of question will depend on the quality of the challenge, knowing that if the answer is publicly available on a database or an administrative service (such as the date of birth when it concerns a company director[4]), the challenge will then only bring a sense of superficial security. Whatever the authentication process used, the problem of the voting booth remains. In the case of remote voting (Internet

---

4 See the websites of information on companies and their directors such as http://www.societe.com, accessed February 11, 2010.

voting or mail voting), it is not possible to guarantee the physical isolation of a person, which again raises the problem of the secrecy of the vote.

With these authentication elements previously received, the voter is prompted to log on to a specific Website on election day. The verification process of his/her ballot concerns two main points: the correct registration on the electoral roll and the absence of a signature that would indicate that the person has already voted. Where appropriate, access is given to the list of candidates or resolutions for which the voter is required to vote.

19.5.2. *The choice of candidates*

Once correctly authenticated, the voter views the list of candidates corresponding to the election. In France, French law requires that the order of candidate registration determines their position on the Webpage. The voter can make his/her decision by ticking the boxes associated with the names of the candidates for whom he or she wants to vote. The electronic voting systems organize all of the outlines of election, majority voting, proportional voting, with mix, without mix, etc. The voter is offered the chance to send a blank ballot. However, electronic voting systems are generally not designed to send a void or invalid ballot, the latter being counted as blank.

19.5.3. *The vote*

When the voter is sure of his/her choice, he or she is invited to confirm his/her vote. This step may eventually include entering a secret code. Following this assumption, the voter enters the secret code that was sent to him/her by post or email. In case of input error, the vote will not be validated and the voter must repeat the authentication steps and choice of candidates. Once confirmed, the vote is digitally signed and encrypted and then sent via the Internet to server machines to carry out operations of signing and insertion of the ballot into the ballot box. Signing and encryption operations allow us to guarantee the validity of the ballot and the secret of the voter's choices. Once the vote is acknowledged by one or many server machines, the system carries out verification operations of the ballot in order to ensure its validity and then proceeds to update the list of signature and insertion in the electronic ballot box.

19.5.4. *Acknowledgement*

The vote ends with the sending or display of an acknowledgement indicating that the voting ballot was properly processed by the server machines. This step, relatively transparent to the voter, is fundamental because it reassures the voter that the voting routine was properly conducted and was not modified during the "transport".

19.5.5. *Counting of votes*

If there is an undeniable advantage of electronic voting on other voting systems, it is the speed of tallying. This last operation, which will lead to the proclamation of results, is indeed greatly facilitated by the automation of the election. The tallying of the ballots can indeed be done in a few minutes, at most a few dozen minutes, and this, relatively independent of the number of voting ballots received, from a few thousand to several million. The procedure of tallying an election organized electronically takes on the same principles as for a traditional election. The qualified persons, members of the electoral committee or assessors, are gathered to proceed with the steps of initialization and launching of the calculation of results. Authenticated minutes are then prepared, to which printouts of the results are attached. In the case of Internet voting completed by mail voting, a preliminary step is to verify the absence of double voting. This step is facilitated by reading a barcode affixed prior to the back of the envelope containing the ballot. It is therefore appropriate to process all ballots received by post before starting to count electronic votes. In the case where Internet voting complements voting in a station, it may be wise to close the period of online voting a few days before voting in stations in order to have the lists of signing on available containing information of previous signing via the Internet.

19.6. Prevention of threats

19.6.1. *Threats*

Like any computer program, an Internet voting system is subject to different threats that can be divided into two main categories:

– *accidental threats*: threats of an accidental nature are related to natural accidents such as:

- climatic phenomenon (earthquake, flood, lightning, etc.) or external phenomena (fire, water damage),

- malfunctions of the direct environment of the system (electrical power failure, malfunction of air conditioning, network problem, etc.),

- accidental failures or malfunctions of system components (hardware failures or software problems),

    – unintentional errors of the personnel directly or indirectly involved in the implementation of the system (input error or error in location of a backup media for example);

    – *computer abuse*: this is the result of intentional acts motivated by greed, challenge, demonstration, nuisance or revenge. These acts may be committed by hackers, technical journalists, dismissed or corrupt personnel, rival companies, protesters or political activists, criminal groups, terrorist organizations or foreign secret services. These attacks could be carried out with or without the complicity of operators of the system, either directly on the system or on elements necessary for its operation (network access for example).

19.6.2. *Prevention*

Faced with these different threats, there are different preventive measures, of technical nature or not, to prevent their occurrence. Some measures apply directly to the computer architecture by setting up a system of hardware and software redundancy to take over in case the main system is unavailable from a geographically remote site and this, without disruption of the service or data loss. The implementation of logical (administrative rights, passwords) and physical (secure computer rooms, badges, surveillance cameras, recording systems, etc,) protection systems allows us to verify the rights and actions of those who need to intervene on the computer system. To cope with possible external attacks, intrusion detection systems, firewalls and other anti-viruses deal with the majority of these issues. Finally, the use of protection techniques for secret protection will complete these procedures.

It is appropriate at this point to recall recent developments in the law adapting the information society. Thus, in France, the automated processing of personal data must be declared to the CNIL under the "computer science and freedoms" law of 1978. This measure requires those responsible for computer processing to secure personal data. Thus, if a hacker gains access to personal data about a voter, the latter may file a complaint[5] against the officer responsible for the security of personal data. At the same time, hackers face jail sentences[6] for fraudulent access, an obstacle to the functioning of the system or damage to data.

---

5 Article 34 of the Data Protection Act. In case of recognized negligence, the one responsible of processing faces a penalty of five years imprisonment and a fine of 300,000 Euros.
6 The penalties are two years for simple fraudulent access and five years in the two other cases, with fines ranging from 30,000 to 75,000 Euros.

19.7. The different technical approaches

Different technical approaches have been developed to set up a secure voting protocol satisfying the prerequisites set out above. We will briefly summarize them. The common feature of all these approaches is the principle of asymmetric encryption. Unlike symmetric encryption, where we use the same key to encrypt or decrypt a plaintext, the asymmetric encryption requires the use of a couple of keys, one called public key and the other called private key or secret key. The public key, known to all, will encrypt the message to be protected (the voting ballot) while the private key will decrypt the message. Thus, if Catherine wants to ensure that the message that she will send to John can be read by him only, she will use the public key to encrypt the message to John. The message can thus be encrypted with this public key and sent to John without any risk of "leakage". John will then be able to reveal the content of the message by decrypting with his private key, which he is the only one to have. In the case of electoral operations, it is recommended to break the private key into different pieces that will be handed to different people. In this way, the trust normally placed in the private key holder will be spread over several people who would have to be gathered to rebuild it.

Three approaches based on these principles of asymmetric encryption have been developed to achieve Internet voting systems. They are mix-net, blind signature and homomorphic encryption.

19.7.1. *The mix-net*

The mix-net (mixer network) is a principle that aims to modify the order of arrival of voting ballots in order to avoid linking the voter and his/her ballot paper by affecting their logical distribution on the server. In other words, it intends to prevent the establishment of a connection between the ballot paper and the voter who has sent it by comparing the dates appearing in the logs of the computer system (for example, signature at 08h01'43" and receipt of the ballot paper at the same time within a few milliseconds). The process consists of inserting intermediate server machines – the mixers – to which the ballot papers are sent. These ballots are then sent after the final server machine. This technical approach has been adopted by the European project e-Poll.

19.7.2. *The blind signature*

The blind signature [CHA 84] is a protocol for separating a vote from the voter. This protocol is based on the use of a central authority that will authenticate the voter and then guarantee the authenticity of his/her vote without viewing the content. The principle is similar to the use of a carbon paper that the voter would slip with his/her ballot in an envelope for postal vote. This envelope is then presented to the authority who signs at the back of the envelope to certify the content after having verified that the voter has not already voted. Thanks to the carbon paper present in the envelope, this allows signing off on the ballot paper without having seen the content. The envelope can then be sent. This protocol however raises the problem of trust that we place in the central authority. The e-Poll solution combines the approach of blind signature with the mix-net approach.

19.7.3. *The homomorphic encryption*

Homomorphic encryption [SCH 00] is based on an extremely interesting mathematical property: homomorphism. This principle of elementary algebra has a significant application because with this technique the product of encrypted ballot papers is equal to the encrypted sum of the ballot papers.

Thus, during the courting of votes, the encrypted result will be obtained by simply multiplying all the different ballots that were encrypted using public key between themselves. A single decryption operation is then applied, using the private key, to obtain the result of the vote by decrypting the product obtained. The result of the election is thus calculated and made public without having decrypted a single ballot paper! This approach is extremely elegant and offers interesting properties, such as verifiability of the election. It is however only applicable for closed elections where the candidates are fully defined in advance. This approach has been adopted by the European project CyberVote.

19.8. Two examples of realizations in France

The two pioneering examples of realization of Internet voting, described below, were both realized with the CyberVote solution, produced by EADS [GRI 06] as a result of the European project of the same name. For a long time, France was behind countries such as Estonia [MAA 04], Belgium [DEL 05], Switzerland [BRA 04], Spain [RIE 04] or even Brazil, in approving through legislation in 2004 the renewal of Chambers of Commerce and Industry members via the Internet. It then continued in 2006 with the first political election in the context of the partial renewal of their Assembly by French nationals abroad.

19.8.1. *The Chambers of Commerce and Industry vote*

The Assembly of French Chambers of Commerce and Industry[7] (ACFCI) and five partner chambers: Alençon, Bordeaux, Grenoble, Nice and Paris have offered, for the first time in France the possibility to vote by the Internet in the context of a mixed election (Internet voting and postal voting). Voting that was held from October 13 till November 3 2004 [ACF 04], was open to 300,000 voters – more than a quarter of traders and company directors in France. These voters, all heads of enterprise, participated in the election of their Chamber of Commerce and Industry by connecting to the Internet with a simple Web browser that supports secure connections of type 128 bits SSL[8]. The system, used in accordance with CNIL recommendations, had been the subject of an expertise under the control of CLUSIF[9] (Club of the Security of French Information Systems). It was also controlled by an independent committee made up of specialists in security, under the auspices of the ACFCI. With his/her personal identifiers received in a secure envelope, the voter could vote in three steps: identification, selection of candidates, confirmation and cast of the vote. An acknowledgement of receipt guaranteed the users ballot would be included. This election was the first French Internet election set by a legislative decree. It was also the first time that so many people, with such diversity on the national territory, were called to vote in an election of legal value.

19.8.2. *French nationals voting from abroad*

It is under the supervision of the Ministry of Foreign Affairs that this election took place in June 2006. Voters, all French expatriated in Europe, Asia and the Levant, were called to vote to renew part of the Assembly of French nationals abroad[10] (AFE). The 380,000 eligible people had the choice of politically express themselves electronically, by postal vote or by going to a polling station.

The voting process used was subdivided into two distinct phases: a registration phase from May 30 to June 6 and a voting phase from June6 to 12, 2006. The registration phase was subject to a number of verification techniques (compatibility of the computer and Web browser) to guarantee the voter access to the election. To

---

7 Site of the Assembly of French Chambers of Commerce and Industry http://www.acfci.cci.fr/, accessed February 11, 2010.
8 SSL (Secure Socket Layer) is the protocol used to establish secure connections between a computer and a website (authentication of computers, integrity and confidentiality of data exchanged). There are mainly two versions with encryption keys of 40 bits and 128 bits. The latter offers greater guarantees.
9 https://www.clusif.asso.fr/index.asp, accessed February 11, 2010.
10 http://www.assemblee-afe.fr, accessed February 11, 2010.

meet the CNIL requirements, registration was done using two identifiers: the consular identification number (NUMIC) that the voter had received by postal mail and the email address that the voter had previously given the elections office. Once this registration was done, the voter received his\her password, necessary to access the voting site. The voting phase started with a new prior authentication step during which the voter had to enter the identifier sent to him/her by post, the password that he/she received by email after the registration as well as his/her date of birth. The choice of one or more candidates being completed, voter confirmed his/her selection by entering the secret code previously sent by post. His/her vote was then encrypted using the principle of homomorphic encryption described earlier, then sent to the voting server. This encryption principle ensured the voter anonymity during all phases of the operation, including during the tallying phase on June 18 2006.

Technical success was achieved thanks to a seamless security and a totally respected anonymity of votes, however questions were raised due to the complexity of the identification process. Therefore, the different stakeholders (instructing party, national commission for computer science and freedoms, industrial project manager, representatives of voters, security specialists, etc.) were brought together to analyze a way to simplify the process.

This first large-scale political election was renewed in 2009 with once again the option to cast ballots via Internet. There is a long road leading towards the generalized usage of Internet for political election. But we can still expect that overseas French nationals will one day vote over the Internet in French presidential elections, which may lead to a subsequent generalization to all French citizens.

19.9. Bibliography

[ACF 04] ASSEMBLÉE DES CHAMBRES FRANÇAISES DE COMMERCE ET D'INDUSTRIE, Questions-réponses sur l'expérimentation de vote par Internet pour les élections des membres des CCI, http://www.acfci.cci.fr/elections/Documents/Questions%20vote%20electronique.pdf, octobre 2004.

[BON 06] BONNET H., MONGON A., PIPUNIC N., "2006 European IST Prize Winners (year 2005)", http://www.ist-prize.org/winners/, Euro-CASE, mars 2006.

[BRA 04] BRAUN N., "E-Voting: Switzerland's projects and their legal framework – in a european context", *Proceedings of the Workshop on Electronic Voting in Europe – Technology, Law, Politics and Society*, p. 43-52, http://prodman.wu-wien.ac.at/~prosser/p47.pdf, juillet 2004.

[CHA 84] CHAUM D., "Blind signature system", *Actes de la conférence CRYPTO '83*, P. Press editor, New York, 1984.

[CHE 03] CHEVRET C., "Le vote électronique par Internet : du déplacement du rituel électoral à la perte de la symbolique républicaine", *Esprit critique*, vol. 5, n° 4, Autumn 2003.

[CAL 00] CALIFORNIA INTERNET VOTING TASK, A Report on the Feasibility of Internet Voting, http://www.ss.ca.gov/executive/ivote/, January2000.

[DEL 05] DELWIT P., KULAHCI E., PILET J.B., "Electronic voting in Belgium: A legitimised choice?", *Politics*, vol. 25(3), p. 153-164, http://www.ulb.ac.be/soco/cevipol/chercheurs/ponl_240.pdf, 2005.

[FDI 03] LE FORUM DES DROITS SUR INTERNET, "Recommandation – Quel avenir pour le vote électronique en France ?", http://www.forumInternet.org/telechargement/documents/reco-evote-20030926.htm, 26 September 2003.

[GEN 03] GENTOT M., Délibération n° 03-036 du 1er juillet 2003 portant adoption d'une recommandation relative à la sécurité des systèmes de vote électronique, http://www.cnil.fr/index.php?id=1356, July 2003.

[GRI 06] GRILHÈRES B., BEAUCÉ C., BRUNESSAUX S., "CyberVote : le vote par Internet : du rêve à la réalité industrielle", *ICSSEA 2006, 19es journées internationales génie logiciel et ingénierie de systèmes et leurs applications*, Cnam, Paris, December 2006.

[GUI 06] GUILLAUD H., "Danemark : l'administration électronique obligatoire", *Internet Actu n° 106*, http://www.Internetactu.net/, mars 2006.

[IPI 01] INTERNET POLICY INSTITUTE, Report of the national workshop on internet voting: issues and research agenda, sponsored by the National Science Foundation, conducted in cooperation with the University of Maryland and hosted by the Freedom Forum, mars 2001.

[LEF 04] LE FAUCHEUR JONCOUR E., Elections de mars 2004, les brestois inaugurent la machine à voter, http://www.a-brest.net/article430.html, February 2004.

[MAA 04] MAATEN, E., "Towards remote e-voting: Estonian case", *Proceedings of the Workshop on Electronic Voting in Europe – Technology, Law, Politics and Society*, p. 82-90, http://prodman.wu-wien.ac.at/~prosser/p47.pdf, July 2004.

[MON 02a] MONNOYER-SMITH L., MAIGRET E., "Electronic voting and Internet Campaigning: State of the art in Europe and Remaining questions", and TRAUNMULLER R. and LENK K. (ed.), *Electronic Government*, Springer, Berlin, p. 280-284, 2002.

[MON 02b] MONNOYER-SMITH L., MAIGRET E., "Le vote en ligne : nouvelles techniques, nouveaux citoyens ?", *Réseaux*, n°114, 2002.

[PER 01] PERRINEAU P., REYNIE D., *Dictionnaire du vote*, PUF, Paris, 2001.

[RIE 04] RIERA A., CERVELLÓ G., "Experimentation on Secure Internet Voting in Spain", *Proceedings of the workshop on Electronic Voting in Europe – Technology, Law, Politics and Society*, p. 91-100, http://prodman.wu-wien.ac.at/~prosser/p47.pdf, July 2004.

[SCH 00] SCHOENMAKERS B., "Fully Auditable Electronic Secret-Ballot elections", *XOOTIC magazine*, vol. 8, n°1, 2000.

[VAN 01] VAN OUDENHOVE B., SCHOENMAKERS B., BRUNESSAUX S., LAIGNEAU A., SCHLICHTING K., OHLIN T., "D4 vol.1 Report on electronic democracy projects, legal issues of Internet voting and users (i.e. voters and authorities representatives) requirements analysis", CyberVote project, http://www.eucybervote.org/ reports.html, May 2001.

[VED 03] VEDEL T., "L'idée de démocratie électronique origines, visions, questions", *Le désenchantement démocratique*, in PERRINEAU P. (ed.), Editions de l'Aube, La Tour d'Aigues, 2003.

[WOL 03] WOLF P., "De l'authentification biométrique", *Sécurité Informatique*, n° 46, October 2003.

Chapter 20

Politicization of Socio-technical Spaces of Collective Cognition: the Practice of Public Wikis

20.1. Introduction

We define *politicization* by the fact that social participants inscribe an explicit political dimension on a place or object that apparently did not have any, or that were not previously identified, as political. This politicization generally occurs by opening a debate. As highlighted in particular by Brossaud [BRO 05], this debate – from the Athenian agora to the bourgeois public space – has been the very foundations of the idea of democracy. It has gradually established a society of communication through discussions, consensus and contradictions. Our intention here is to highlight elements in the practices of "collective cognition" that bring politicization to the use of some collaborative systems, particularly wikis.

We anchor our analyses in the model of *distributed cognition*, supposing that in the context of uncertainty, participants rely on both technical and social resources to carry out their actions [HUT 95]. These resources would be made available in their environment, according to an organization that they learn to identify and use. Studies on communities of practice and epistemic communities have shown that the Web lends itself particularly well to the dynamics of distributed cognition [CON 04, LAV 91].

Chapter written by Serge PROULX and Anne GOLDENBERG.

However, these socio-technical spaces have structures of a different nature that are more or less flexible and adaptable. We also make the assumption that some socio-technical spaces lend themselves better than others to politicization, by the debate between heterogenous participants in particular. Thus, contrary to established plans of action in the control cabin of a ship [HUT 95] or in an airplane cockpit [HUT 96], cognitive schemes implemented in more flexible collaborative spaces (such as wikis) would cause particular forms of deliberation and debate to emerge in the research into conventional support for action [DOD 93].

A wiki is server software that allows users to create and edit Webpages via a browser. The first wiki was created in 1994 by Cunningham to support the collaborative creation of a knowledge base related to software engineering solutions. It has been designed to promote a rapid collective edition (*WikiWiki* means "very fast" in Hawaiian). Wikis today serve as support to the dynamic and collective edition of a multitude of projects built around the collaborative construction of knowledge. This tool has the distinction of being open and flexible, allowing free intervention but also the rewriting of texts, and sometimes, dialog between the authors[1].

We have tried to understand how the practice of wikis could contribute to a different way of thinking about policy in the context of the use of a writing device. Are we witnessing, through collective participation and negotiation of knowledge, a renewal of the communicational reasoning, conditional to the constitution of an Habermassian [HAB 78] public space? Does the (technical in particular) mediatization of these practices extend the elitist dimension of this model of political participation (debates between spokespersons and experts from rational arguments)? Or would we rather be facing the emergence of hybrid forums (between heterogenous individuals, experts and laymen) – a term suggested by Callon and his team [CAL 01] – that would guarantee the establishment of a new type of "cognitive democracy"?

After presenting different thoughts on the politicization of cognitive technologies, we will describe how the practice of politics is partially renewed around the production and definition of information and communication technologies (ICT). We will then propose the analysis of a wiki, whose principles of (political) management and (cognitive) understanding are based on agreements not yet established and partly left open for discussion.

---

1 We are particularly interested in public wikis that keep these features open.

20.2. Forms of ICT politicization

20.2.1. *Approaching politicization of cognitive technologies*

The study of the political dimensions associated with the spreading of media communication technologies developed significantly after World War II. Propaganda in the 1920s and modern advertising in the 1930s created the emergence of a critical analysis of media content, along with ideological criticism of the media themselves. In the context of industrial production of information by commercial, political and scientific elite, the socio-political analysis of ICT was first oriented towards the interests of cultural industries.

The critical theory of these industries has been especially developed by philosophers of the Frankfurt school, including Horkheimer, Adorno and Marcuse. As a member of the second generation of this school, Habermas analyzed the emergence of bourgeois public spaces as a political scene of discussions acting as counterweights to absolutist powers [HAB 73, HAB 78]. In cafes and other public places, the upper class (literate citizens) are engaged in debates that start to socially and politically make sense, i.e. their meanings in view of a democratic logic.

Starting in the 1980s, the French tradition of the sociology of uses, certain Anglo-American currents of Cultural Studies and constructivist approaches to technology provide a more nuanced, less deterministic understanding of technology by turning their analytical focus to the social reception of media and meanings of technology use. These approaches have led to a finer intelligence of what was between technology and sociology by exploring the abilities of users to rebuild and divert the meanings of technologies and media in daily life [PRO 02, PRO 05]. However, according to some observers, this orientation of analyses towards the micro-sociological pole of uses often resulted in overly descriptive approaches, at the risk neutralizing macro-sociological constraints that weight on the practices of users [JOU 00].

Coming back to the assumption of possible politicization of the technique, the philosopher Feenberg attempted to re-question the difficult relationship between critical theory and constructivist perspectives: "How to involve technology in the modern age in a context of democratization in light of the analyses made by the critical theory and constructivist proposals of analysts of the technique?" [FEE 04] He proposed to explore how participants invest the political dimension of a society marked by the technique.

Following the same line of thinking, Longford proposed to approach the practices of users to capture the emergence of a *technician citizenship*, assuming that the social participants could fully take part in debates and negotiations concerning technological trends in contemporary societies [LON 05].

These two authors invite us to pass from a critical theory of technical phenomenon to a sociological consideration of the *reflective* appropriation of IT by participants. This approach helps us to understand the social construction of a critical sensibility rooted in the everyday life of users. We could thus speak of *politicization of practices* when the *user* acquires the position of *participant* by taking effecting means of social control over his/her tools.

Pragmatic sociology has also contributed to the formulation of a political theory of technical activities. Shunning the emancipator imperatives of critical theory, it proposed to "follow the participants" to understand how they invest meaning in their actions. It is about describing how the latter restore political spaces of debates and conventions by managing the operation of technical devices. Responding to the problem of monopoly of scientific and technical expertise as identified by Habermas [HAB 73], but refusing the dualist approach (experts/laymen) of public consultation, Callon [CAL 01] suggests the implementation of hybrid forums that could allow a co-construction of knowledge and know-how, based on exchanges between heterogenous individuals with diverse statuses.

We have observed practices of co-construction of knowledge involved in open socio-technical spaces on wikis. Under what conditions can we speak of a "cognitive democracy" in this type of space? We assume that the presence of *controversies* and *conventions* are both indicators of politicization of practices of collective cognition. The *controversies* are defined by Latour as "discussions having partly as object scientific or technical knowledge that are not yet in place" [LAT 05]. The conventions are defined by Dodier as the common principles of justice and truth called for or established during the action. He distinguishes three approaches to the problem of conventions in social sciences [DOD 93]:

– An *interactionist* approach focusing on conventions that are made and unmade at the microsociological level of interactions through the incessant adjustment of participants with each other, subject to the contingency of circumstances. This is a characteristic approach to pioneering works in ethno-methodology [GAR 67] on the one hand, and those of Goffman, on another [GOF 69].

– A *culturalist* approach based on the idea of plurality of communities with regimes of respective conventions exists, i.e. that principles of truth and justice exist on their own. This approach assumes the existence of localized forms of conventions in space and time. This is the approach developed by Boltanski and Thévenot [BOL 89, BOL 91].

– A *universalist* approach that aims at establishing a general coordination model based on the implementation of common conventions brought by all human beings. This is the approach of Habermas [HAB 87].

The theoretical challenge is to articulate these two forms of indicators (controversies, conventions) to understand the politicization at work in practices of negotiated collective cognition on wikis. This is observable in the growing generality of arguments debated in controversies, and in the establishment of conventions governing collective action.

20.2.2. *Reflexivity and politicization of cooperation practices*

Reflexivity of participants in relation to their own practices and tools is presented by some individuals as a form of politicization. An associative individual from Brussels who uses wikis explained in an interview that: "Now, a website written in ASP[3], I won't read it anymore. It is the way that the site works that counts. It has the same importance as writing on recycled paper".

Politicization concerns less the characterization of traditional democratic procedures, such as electronic voting for example, than understanding social experiences online and how individuals take part in public debates about their own practices. We assume that the Habermassian public space has mutated and is found in other forms in cyberspace.

Some authors agree that an emblematic figure of a "technical citizenship" is that of the hacker, an amateur enthusiast with computing prowess who promotes a mutual aid economy based on the free circulation of codes and information [ESC 04, HIM 01, RAY 00]. The free software battles were initially about the *technical specification* of software. The term "free software" refers to a freedom of execution, study, distribution and improvement of a computer program. These freedoms are guaranteed by various free licenses.

A common example is the GPL (*General Public License*), which adds a viral character to these freedoms: any derivation of free software must retain the same rights of use. For some individuals – those we call "activists of the code" [PRO 06] – intellectual privatization (patenting software) is now the subject of political struggle.

The politicization of a technical device has hit areas outside those of computing and is now covers alternative media in open publishing mode, or open systems of

---

3 ASP (*Active Server Pages*) is a programming language developed by Microsoft.

scientific publication. There are practices whereby "digital knowledge" is spread. There is a *culture of freedom and gratuitousness* (access, use, exchange, sharing), but we would be wrong to consider this phenomenon only in terms of spreading. It is with respect to practices of innovation, expression and creation that significant changes seem to occur. ICT users have gradually become equipped with technologies promoting cooperation and collaboration. Thus, we are witnessing a global expansion of individual and collective participation in the circulation of information. The presence of these collective publishing technologies has led to an internal politicization of management practices in terms of publication. We are witnessing the construction of new conventions of use and to the emergence of a "management model" of collective expression but also of the diversity of viewpoints.

For an increasing number of communities of editing practice, content management systems (CMS) have become vital supports [JEA 06]. Technically mediated procedures are adopted, depending on the context and political habits of participants, with the entry into force of open publication management. *Moderation* (pre- and post-editing evaluation) then becomes the integral activity of a validation committee or the general readers. We are witnessing a logic of democratization and politicization of editing and publication which, through technical appropriation, sensitizes readers and contributors to the link between device and information type. In the next section we will explore a particular tool where moderation is not delegated but distributed to all contributors: wikis, and particularly the case of *Wikipedia* encyclopedia.

20.3. Writing on wikis: a practice of deliberative cognition

20.3.1. The plans of action of public wikis

20.3.1.1. Collaboration as principle of joint activity

Wikis have a simplified syntax that facilitates the creation of new pages and links between pages. They make a collective structuring of contributions and contents possible, with generally, maintenance of modification history and access to a discussion space on the text for some. They propose a particular framework of joint activity with the following main characteristics.

Cunningham spoke at the Wiki Symposium of San Diego in 2005 on a paper entitled "The crucible of collaboration" in which he made the distinction between collaboration practices and cooperation practices. It is crucial, according to him, to understand what distinguishes them in order to capture the specificity of wiki uses. His analysis puts the practices of the Ebay community into perspective with those of wiki users.

Cooperation refers to an action conducted jointly by a group of participants around a common goal and whose benefits must be carefully identified for the purpose of mutual and distinct recognition. Cooperation thus allows the benefits to be enhanced, i.e. a distinct value is assigned to different operations, opening onto a form of economic exchange.

Collaboration refers to an action conducted jointly by a group of participants whose contributions do not need identification because the success of the collective project seems to be the value of individual contribution. This is how we see discrete individuals appearing, sometimes called "WikiGnomes" (or "gardeners"), who complete the common project through minor contributions, such as spelling corrections. We deviate here from a model of "donation to donation" to a model of common good, where the contribution to collective good seems to take a value greater than individual performance [SMI 99]. Thus, in this mode of collective construction of knowledge, the signature of writing acts is not required[4]. As in a collaboration towards the construction of a common project, in many wikis, including Wikipedia, nothing requires the contributor to identify him/herself[5]. Thus, any password or identification procedure is considered as a further obstacle to the visitor who, therefore, will be reluctant to be a contributor. Writing in anonymity is innovative with regard to our habits of editorial responsibility. Unlike a logic based on the signature of an author rewarded for his/her contribution, the individuality of each counts less than the collective effort. While some write long pieces of text, others are just "gardening": for example, cleaning grammar, correcting typos, taking care of details without being known through these actions. Communities of public wiki users often rely on the principle of *soft security* to manage vandalism. Supported by monitoring tools, contributors are able to view and reset the vandalized pages in an instant. Thus, looting is deterred by the ease of countering it. The responsibility for managing the wiki then becomes collective and the mass of active contributors is the power of the wiki.

This logic for collaboration proposes is a shared responsibility that could be emblematic of system of collaborative public space. The legitimacy of knowledge produced by this collaborative principle is an emblematic debate of the politicization of socio-technical spaces of collective cognition.

---

4 "The open system of Wikibooks allowing everyone to modify pages, we cannot guarantee the permanent validity of the content. Please also note that all changes are archived in the history of the page as well as the Internet address of the unregistered contributors. Without name identification, it is the [Internet] address that appears in the thread of the last changes".

5 We will later see that controversy has led to the amendment of this rule on the English Wikipedia.

20.3.1.2. *The user, a public participant: a democratic wager?*

Many contributors and supporters see in the principle of open editing wikis a challenge for democratization of the use of the Web because this software tool encourages the contributions of non-expert users. Wikis offer an editing interface that allows the user to edit the page that he/she runs through to create a new page: the recording of his/her intervention then automatically becomes a version of a page of the site. This editing interface trivializes intervention on a Web space – until then only open to modifications by experts – to the action of the user redefined as an *empowered actor*.

Cooperation practices and involvement of the free software movement seems to have dissolved the boundary between users and developers. It seems that wiki systems encourage users to be contributors rather than just visitors. We could see a form of digital citizenship in it where, without being consulted, the members participate in the collective building of the project. The push to change what is given to us to read comes from an innovative logic in the history of websites, but also in text history. If a material support such as the book does not allow cumulative and permanent editing of the text, it is a major leap to authorize, as in the Wikipedia encyclopedia, every reader to contribute, correct or argue in the context of a writing project with universal vocation. It is also a democratic challenge to leave the writing of public knowledge open to anyone. Beyond that, this challenge also seems to reside in advertising, the "traceability" of the writing process and debates concerning it. Via the conservation of the history of changes, the memory of plural writing of the wiki is in the public domain and is a resource that can be used for discussions.

20.3.2. *Setting discussion of a collective cognition process*

Cognitive technologies have been developed by computer sciences. Their networking on the Web has given rise to new ways to construct knowledge, as well as to the emergence of coordination forms on a large scale. The most enthusiastic theorists have predicted, from the proper use of these techniques, the possible advent of an "era of collective intelligence". Thus, for Lévy, "collective intelligence is based, primarily, on a strong principle: everyone knows something" [LEV 97]. We can then consider how to organize the identification of the skills and knowledge of each participant creates the dynamics of cognitive interdependence making a renewed humanism possible. We remain mitigated on the perspectives of collective intelligence. Indeed, if the use of knowledge-building technologies has helped forms of collective cognition, it seems more appropriate for us to first question the conditions of the democratization of this process.

At a *micro-sociological* level, writing in a wiki can be considered under terms of activity of the language, as formulated by Austin[6] [AUS 70], if we understand that the statement on the Web is mostly written. In these cases is writing on a wiki is more interactive than performing? The technical structure allows, and even calls for, the modification of text by other readers. At this level, negotiation is first about the contents. In addition to the already-mentioned discussion pages, many wikis offer the chance to subscribe to pages whose contributor wants to follow the evolution, which facilitates group monitoring on distributed activities in the site. The validation of contents is then made via a process of reshaping or deliberation that extends over time, and never completely ends. There are always more readers than contributors, however, and the transition from reading to writing is made if the user finds that he/she has enough legitimacy to intervene.

At this level, it is the *cultural* characteristics, interest and knowledge of local contributors, that influence the nature of contents[7], the size and form of deliberations. It is at this level that rules governing use and local management of the wiki are negotiated. The rules of practice have, in Wikipedia, a space for deliberation and constitution dedicated to the community of contributors. The requirements of practice are also included in the technical areas on which discussion spaces are proposed.

Access to more global discussions is done in particular by advertising debates. Controversies about the system indeed become fully fledged when they are advertised outside the page being debated. The legitimacy of a collaborative and open encyclopedia is what seems to structure the majority of public controversies on Wikipedia. Let us take, for example, the instance of the case made by SlashDot accusing Wikipedia of censorship[8]. The censored pages related to Brian Peppers, an individual living with disability and a deformed physiognomy who was accused of sexual assault. "This case made him famous enough to enter the cyberculture", the authors of his biography argued on Wikipedia. "Disability and sexual aggression are not enough to make this man a very important person who may have his place in an encyclopedia", says Jimmy Wales, founder of the Wikipedia project, who prohibited the creation of a page on him until September 2007. The controversy not only concerns the legitimacy of the content and fact that the page has been deleted and banned, but also that on three occasions, the history of the discussions was also

---

6 *Quand dire ç'est faire*. The word is described as a performance, with power equal to a physical action.
7 See for example the difference of nature between the (conceptual) English and (historical) French article on the concept of collaboration http://fr.wikipedia.org/wiki/ Collaboration, http://en.wikipedia.org/wiki/Collaboration (websites accessed February 11, 2010).
8 http://slashdot.org/articles/04/09/05/1339219.shtml?tid=146&tid=1 (websites accessed February 11, 2010).

deleted. Shielding from advertising on behalf of the founder authority was seen as an autocratic gesture by protesting contributors, who started a site for the publication of censored articles. The issue of control and legitimacy of this so-called "free" encyclopedia has been publicly raised in this case[9].

As well as its content, the underpinning of the legitimacy of the collective encyclopedia project continues to be formulated, challenged, denied and reformulated in the public square. We must remember that the project is carried out by more than a million contributors who have filed or fought for a "fragment of knowledge". For now, the publicity of discussions around the collective construction of knowledge seems to be a symbolic arena for the defense of legitimacy of the project and, more broadly, for the protection of the principles of free and collective writing in a system of wiki type. The relative anonymity of contributors must allow equal treatment of interventions, distributing control to the majority and not to individuals recognized by their signature. This logic "collectivizes" the responsibility of progress anderrors, prompting the participant to increase his or her contribution according to a primarily deliberative logic.

20.4. Conclusion: the stakes of a cognitive democracy

Many studies were able to show that systems of free software production and use, based on the opening of the source code for a freedom of appropriation and modification, allowed the observation of an enlarged and politicized form of discussions (among technologies, expert actors and secular actors) [ESC 04, LON 05, COU 06]. Others have analyzed how the use of open CMS for publishing activists or scientific content has also politicized the understanding of systems and publishing policies among expert participants [COL 04, MIL 05, JEA 06]. Wikis would more particularly allow the practice of a cognition negotiated among heterogenous actors.

---

9 Going backwards, we find precedents to this controversy including that of the forged biography of a US politician. In September 2005, Jimmy Wales was charged with having initiated a project deemed *irresponsible* when *USA Today* published a text denouncing the fact that an *unsigned* article linked the name of J. Seigenthaler Sr., journalist, writer and US politician, to the assassination of J. F. Kennedy. This article was in the public domain for 132 days and J. Seigenthaler Sr. finally filed a complaint denouncing the irresponsible character of *Wikipedia* not being able to find the author guilty of this error. Following this case, Wales was ordered by the court to impose the signing of articles on the English encyclopaedia. Did this event cause a loss of legitimacy of the project? At the same time, *Nature* magazine published a study showing that there are on average an equal number of errors in both the *Britannica* and *Wikipedia* encyclopedias.

Editing conflicts on Wikipedia, becoming places where the truth of information is justified, illustrates the contemporary mutation of public writing policies very well. These practices appear to reshape traditional systems of legitimization. The credibility of information is now dependent on public debate. From a cognitive point of view, readers and publishers are confronted with managing an abundance of information and must acquire new tools to recognize the relevance and credibility of sources. The conditions of a cognitive democracy would thus lie particularly in the possibility to politicize (debate, discuss, contradict) information and how they organize themselves. Our observations have shown that the technical devices only allow formatting. Politicization is driven by a reflective practice on the nature of these supports, such as those of the circulating information. Wiki practices seem to create a world of knowledge in permanent redefinition with standard rules not yet stabilized, but for which it is possible, in principle, to trace the history of transformations. Our hypothesis is that there will be politicization of uses of a technology when a debate appears around three elements:

– a willingness for democratization promoting the opening and social appropriation of this technology;

– a problem prompted by controversies around the technical and symbolic characterization of this technology during its development;

– the creation of a negotiation space for the establishment of agreements to ensure the operation of this technical tool.

20.5. Bibliography

[AUS 70] AUSTIN J., *Quand Dire ç'est Faire,* Seuil, Paris, 1970.

[BOL 89] BOLTANSKI L., THEVENOT L., *Justesse et justice dans le travail*, (Cahiers du Centre d'Études de l'Emploino. 33), PUF, Paris, 1989.

[BOL 91] BOLTANSKI L., THEVENOT L., *De la Justification. Les Economies de la Grandeur*, Gallimard, Paris, 1991.

[BRO 05] BROSSAUD C., TIC, sociétés et espaces urbains. Bilan et perspective de la recherche francophone en sciences sociales (1996-2004), Maison des sciences de l'Homme "Villes et territoires", University of Tours, June 2005.

[CAL 01] CALLON M., LASCOUMES P., BARTHES Y., *Agir dans un Monde Incertain. Essay on Technical Democracy,* Seuil, Paris, 2001.

[CON 04] CONEIN B., "Cognition distribuée, groupe social et technologie cognitive", *Réseaux*, vol. 124, p. 52-79, 2004.

[COU 06] COUTURE S., "La construction des modèles du libre", *Congress of ACFAS "Free software as a sociotechnical model of innovation"*, University McGill, Montreal, 16 May 2006.

[DOD 93] DODIER N., "Les appuis conventionnels de l'action. Eléments de pragmatique sociologique", *Réseaux*, no. 62, p. 63-85, 1993.

[ESC 04] ESCHER T., Political Motives of Developers for Collaboration in GNU/Linux, Dissertation submitted for the degree of MA (Globalization and Communications), University of Leicester, 2004. http://www.chronovault.net/websites/tobi/research ID/results,

[FEE 04] FEENBERG A., *(Re)penser la Technique. Vers une Technologie Démocratique*, La Découverte, Paris, 2004.

[GAR 67] GARFINKEL H., *Studies in Ethnomethodology*, Prentice-Hall, New York, 1967.

[GOF 69] GOFFMAN E., *Strategic Interaction*, University of Pennsylvania Press, Philadelphia, 1969.

[HAB 73] HABERMAS J., *La Technique et la Science comme Idéologie*, Denoël, Paris, (original German edition, 1963), 1973.

[HAB 78] HABERMAS J., *L'Espace Public. Archéologie de la Publicité comme Dimension Constitutive de la Société Bourgeoise,* Payot, Paris, (original German edition, 1962), 1978.

[HAB 87] HABERMAS J., "L'espace du politique" in *Théorie de l'Agir Communicationnel*, Fayard, Paris, (original German edition, 1981) 1987.

[HIM 01] HIMANEN P., *L'Ethique Hacker et l'Esprit de l'ère de l'Information*, Exils, Paris, 2001.

[HUT 95] HUTCHINS E., *Cognition in the Wild*, MIT Press, Cambridge, 1995.

[HUT 96] HUTCHINS E., KLAUSEN T., "Distributed cognition in an airline cockpit" in ENGESTROM, Y., MIDDLETON, D. (ed.), *Cognition and Communication at Work*, Cambridge, University Press, New York, 1996.

[JEA 06] JEANNE-PERRIER V., "Des outils aux pouvoirs exorbitants?", *Réseaux*, no. 137, p. 97-131, 2006.

[JOU 00] JOUËT J., "Retour critique sur la sociologie des usages", *Réseaux*, no. 100, p. 487-521, 2000.

[LAT 05] LATOUR B., Description of controversies, online lecture notes, http://www.bruno-latour.fr/cours/index.html, version February 2010.

[LAV 91] LAVE J., WENGER E., *Situated Learning. Legitimate Peripheral Participation,* Cambridge University Press, New York, 1991.

[LEV 97] LEVY P., *L'Intelligence Collective. Pour une Anthropologie du Cyberespace*, La Découverte, Paris, 1997.

[LON 05] LONGFORD G., "Pedagogies of digital citizenship and the politics of code" *Techné: Research in Philosophy and Technology*, vol. 9 no. 1, p. 68-96, 2005.

[PRO 02] PROULX S., "Trajectoires d'usages des technologies de communication: les formes d'appropriation d'une culture numérique comme enjeu d'une société du savoir", *Annals of Telecommunications*, vol. 57, no. 3-4, p. 180-189, 2002.

[PRO 05] PROULX S., "Penser les usages des TIC aujourd'hui: enjeux, modèles, tendances" in Vieira L. and Pinède N., (ed.), *Enjeux et usages des TIC: Aspects Sociaux et Culturels*, Volume 1, Bordeaux University Press, Bordeaux, p. 7-20, 2005.

[PRO 06] PROULX S, "Les militants du code: la construction d'une culture technique alternative", *Congress of ACFAS*", *Le Logiciel Libre en tant que Modèle d'Innovation Sociotechnique"*, University McGill, Montreal, 16 May2006.

[RAY 00] RAYMOND E., "The Early Hackers", *A Brief History of Hackerdom,* 2000 (available at: http://oreilly.com/catalog/opensources/book/raymond.html, accessed February 11, 2010).

[SMI 99] SMITH M., KOLLOCK P. (ed.), *Communities in Cyberspace*, Routledge, London, 1999.

Chapter 21

Liaising using a Multi-agent System

21.1. Introduction

The purpose of this chapter is to understand the technological spirit of tele-liaising system manufacturing, that is, an interactive computer that supports collective and discussed decision-making [MOR 05]. Such a system aims to facilitate collaboration among participants working remotely and asynchronously. The experiment presented here was developed under the ADNT project (decision and negotiation support for land management) co-financed by the Rhône-Alpes region. Initiated in 2001, it aimed to develop methods and tools for decision and negotiation support that would be useful to local development individuals involved in projects related to territory. This work was conducted in parallel under the ICA program (Incentive Concerted action) "City" (CNRS/Ministry of Research). Aimed at promoting fundamental and multidisciplinary research, the ICA "City" has opened up perspectives mixing ICT and city sciences and applying them to major urban issues today. In the first part of this chapter, we will discuss our approach (section 21.2). In the second part, we will set it in relation to game theory (section 21.3).

The principle of the system proposed here will be presented in section 21.4. It is based on the analytic hierarchy process and adapts it to a multi-participant context, allowing each of them to associate with a software entity, called an agent. Each agent assists a participant and represents him/her in automatic dialogs. We will

---

Chapter written by Maxime MORGE.

describe the assistant multi-agent system in section 21.5. This provides an aid to consultation.

21.2. Motivations

In the real world, most decisions involve a wide range of individuals. Their success depends on their adherence to a consultation process. The resulting decision must thus convince them that everyone's point of view is fairly represented. This observation is the origin of a change of perspective in the development of democratic and technological procedures. The decisions must be collective and discussed [CAL 01].

In the area of governance, we can indeed distinguish two modes of forming a general consensus: representative democracy and dialogic democracy. Representative democracy is an aggregation process of individual preferences through which laymen and individuals delegate their power to elected representatives and experts. Dialogic democracy is a participatory process of composition of perspectives and interests through which civil society debates and deliberates. Hybrid forums, consensus conferences and discussion groups, whether they focus on sociotechnical controversies (BSE, gene therapy, mobile telephony, etc.) or territorial projects (motorway, nuclear power plant, water management, etc.), are dialogic democratic experiments.

Our work contributes to the last modality. It is about providing a computer support tool for liaising so that participants can collaborate remotely and asynchronously. We asked as a prerequisite that the decisions previously taken in camera, whose justification remains obscure, could become open and transparent via the computer tool. This change of perspective allows us, from a social and political point of view, to take advantage of the expansion of the consultation circle in order to enhance expertise and avoid potential bottlenecks. From the point of view of rational decision-making, it supports the idea of exploiting the creativity of different individuals during the formulation of the problem.

The system [GOR 96, GOR 01] is therefore an intelligent computer tool that mediates debates in the context of a dialogic democracy to facilitate the identification and resolution of conflicts rationally, efficiently and fairly [MCB 01].

21.3. Game theory

Game theory was introduced in 1940 by Morgenstem and von Neumann [MOR 40]. Based on an economic model of reasoning, game theory views collective decision-making as an aggregation process of preferences from a theoretical viewpoint. The rationality of an agent is then defined using a gain function to evaluate his/her satisfaction compared to the alternatives considered. This theory provides criteria by which to measure the quality of collective decision-making processes. We may in particular evaluate the performance of the result obtained at the end of such processes.

In order to propose a system of tele-consultation based on a multi-agent system, a first approach is to associate an agent who represents each participant in automatic negotiations. According to this hypothesis, the individual delegates part of the decision process. The agent signals to the participant that he/she is assisting in the alternative recommended at the end of the negotiations. This approach thus allows the cognitive schema and the value system of each actor to be respected. The vision and understanding of the same problem are subject to subjectivity. However, Arrow's theorem [ARR 63] indicates that it is not possible to build a satisfactory social choice function when we have more than three alternatives. According to this theorem, if we consider a group of agents each equipped with its own preference relation and a set of more than three alternatives, there is no social choice function that satisfies the following properties:

– universality, i.e. the social choice function must be defined whatever the preferences of the agents are;

– non-dictatorship, i.e. no agent should be able to impose his/her preferences, regardless of the preferences of others;

– unanimity, i.e. when all of the agents have the same preference, the social choice function must associate the same preferences with society;

– indifference of irrelevant alternatives, i.e. the relative ranking of the two alternatives should depend solely on their relative position for the agents and not on the ranking of third-party alternatives.

Economists call preferences a complete pre-order or a complete order, that is a reflective, transitive and total relation. In the case of an order, we then speak of strict preferences: the relation is antisymmetric. The property according to which the social choice function should be indifferent to non-relevant alternatives implies in particular that if we consider only one subset of alternatives, the social choice function must not lead to another ranking of this subset. The result of the second round of Presidential elections held on April 21, 2002 is one example of this. The inability to guarantee the existence of a quality process of collective decision-making was mathematically demonstrated by Arrow [ARR 63]. It only confirms the necessity to go beyond game theory and offers to assist in the debate (act as a consultant to facilitate the search for an agreement but leave the final decision to the participants).

21.4. The principles

By supporting and not replacing human judgment, users are at the heart of the problem.

The system represented in Figure 21.1 is based on the negotiation support system proposed by Ito and Shintani of Japan [ITO 97]. It is based on a multi-agent system, where each software entity assists a user and interacts with the other agents in the system. We propose functions for the collaborative development of argumentative schemas on one hand, and to clarify consistencies and inconsistencies among the preferences of the participants in order to detect consensus and conflicts on the other.

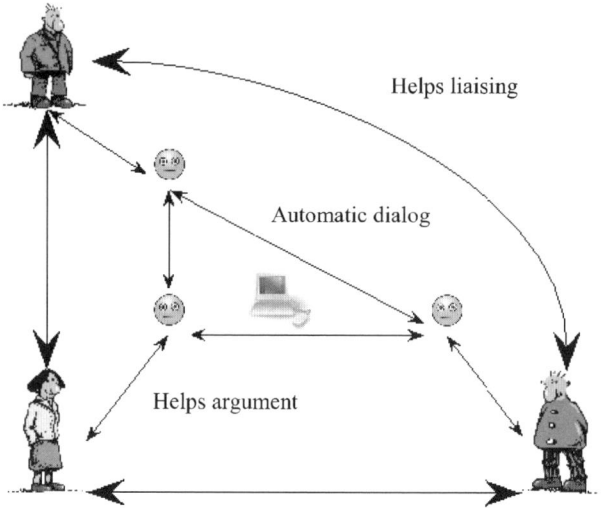

Figure 21.1. *Principle of the consultation support system*

In the field of territorial management, decision support tools used by participants are mostly based on multi-criteria analysis methods. These techniques are dedicated to clarify the understanding of a decision problem and its resolution. They become multi-criteria when the problem has several conflicting objectives.

We were particularly interested to the *Analytic Hierarchy Process* (AHP) proposed by Saaty [SAA 01]. It is a powerful and flexible decision process that facilitates the expression of preferences and allows decision-making using the judgments of the decision-maker, whether they are qualitative or quantitative.

This method is divided into three steps: the construction of a representation of the problem, the expression of preferences and the summary of judgments. Our goal will be to adapt this method to a multi-participant context.

To illustrate our point, we will now consider the problem of the location of International Thermonuclear Experimental Reactor (ITER), the first experimental installation of a thermal nuclear fusion plant.

21.4.1. *Representation of the problem*

First, the analytic hierarchy process (AHP) allows the development of an argumentative outline [GOR 97, RIT 73] to obtain a representation of the problem. In order to gather precise knowledge, our mind structures the complex reality into various components, divides them in turn and so on, in a hierarchical way. In a complex situation, we build taxonomy of evaluation criteria to structure decision-making.

AHP builds the problem into a decision hierarchy consisting of elements called "activities". The aim of the problem is identified and divided into sub-problems called "evaluation criteria". These allow a comparison of alternatives as being possible solutions to the problem to be elucidated. The criteria are in turn divided into sub-criteria and so on until we obtain leaves of taxonomy of criteria related to each other by an inheritance relationship.

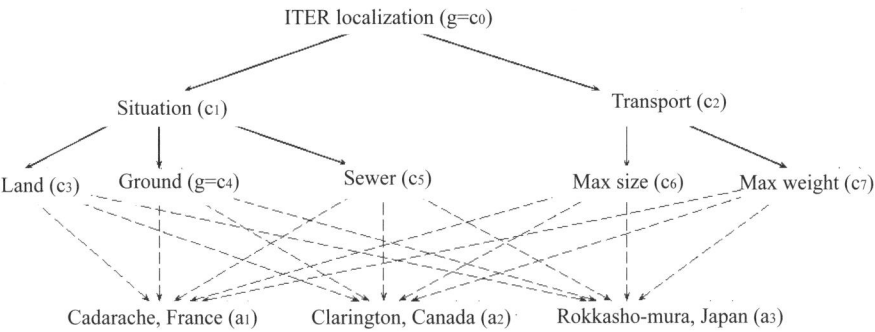

Figure 21.2. *Decision hierarchy for the location of ITER*

Figure 21.2 illustrates this representation of the problem through the example of the location of ITER. The goal is to select the right location from three alternatives (in France, Canada or Japan), three implantation sites that are evaluated according to criteria such as quality of the ground (c_4) organized in taxonomy. This goal is broken down into two criteria: the quality of the site noted c_1 and its accessibility noted c_2; c_1 and c_2 are broken down into sub-criteria c_3, c_4, c_5, c_6 and c_7.

It is from this representation of the problem that preferences are expressed.

21.4.2. *Expression of preferences*

AHP is based on feelings and intuitive judgments. We describe the importance of an activity in relation to another:

– absolute (9);

– attested (7);

– determining (5);

– low (3); or

– equal (1).

We can estimate the relative weight of an activity *i* as opposed to an activity *j* given an evaluation criteria c_k in the field of definition corresponding to qualitative judgments of comparison (1, 3, 5, 7, 9) to their intermediate values (2, 4, 6, 8) but also to their reciprocal (1/9, 1/8, 1/7, 1/6, 1/5, 1/4, 1/3, 1/2). The set of distinctions among similar activities on the basis of the parent activity can be summarized in a matrix of pair-wise comparisons.

Let us consider the decision hierarchy of the previous example. Criterion c_1 that corresponds to the situation is divided into three sub-criteria:

– the quality of the land c_3;

– the quality of ground c_4; and

– the quality of sewers c_5.

Criterion c_4, the quality of the ground, is four times more important than criterion c_3, the quality of the land, and four times more important than criterion c_5, the quality of sewers. Thus, criteria related to the quality of the ground and sewers are considered of equal importance.

The matrix of pairwise comparison on the quality of the land is represented in Table 21.1. The mathematical calculation of the eigenvector (W) associated with the maximum eigenvalue allows us to deduce the priorities of each of these criteria.

In summary, the method consists of comparing couples of similar activities of the same level on the basis of the parent activity and to establish distinctions between the two members of a couple by assessing the intensity of preferences of one compared to the other. Judgments thus expressed must be synthesized. Their consistency should also be checked.

A	c_3	c_4	c_5	W
c_3	1	1/4	1	1/6
c_4	4	1	4	2/3
c_5	1	1/4	1	1/6

Table 21.1. *Matrix of binary comparison*

21.4.3. *Summary of judgments*

Preferences are not necessarily consistent. The ratio of consistency allows us to evaluate this consistency. The more important the consistency ratio is, the less important preferences are. The threshold of acceptability is generally set at 10%. Within the limits of consistency thus defined, we can determine the relative priorities of the various activities.

A simple calculation allows us to determine the priority of an alternative compared to a criterion and deduce the optimal alternative as opposed to this criterion (the priority alternative). In the previous example, the respective weights of criteria c_3, c_4 and c_5 are 1/6, 2/3 and 1/6. Preferences are perfectly consistent: criterion c_4 is four times more important than criterion c_3 and four times more important than criterion c_5.

Preferences have been expressed using pair-wise comparisons among similar activities of the same level. The priorities of different activities as opposed to the parent activity are represented in Figure 21.3. It is from this information that the priority of alternatives is calculated. From this figure, we deduce by transitivity that the French site is optimal in relation to the goal.

In summary, this methodology of decision support allows a single individual to represent a problem, express his/her preferences from this representation and to synthesize its findings by measuring their consistency. Having defined the method used, in the next section we will present the elements of the underlying computer system that allows us to adapt this methodology to a multi-participant context through exchanges of information.

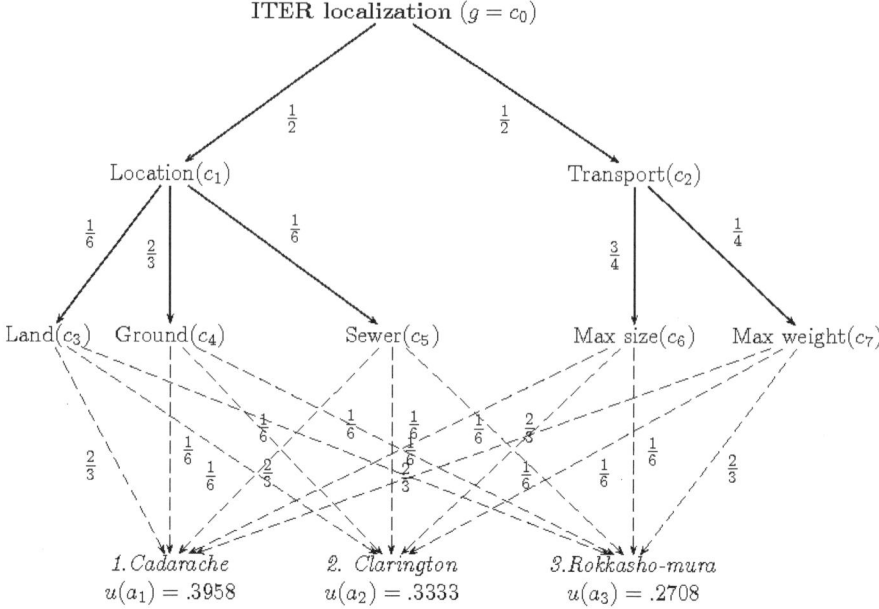

Figure 21.3. *Valued decision hierarchy for the location of ITER*

21.5. Multi-wizard system

The consultation support system proposed here is based on a multi-agent system, i.e. a set of software entities. Each agent assists a user and represents this individual in automatic dialogs. This allows participants to share their representation of the problem and compare their position. The participants may collaboratively develop their argumentative scheme and examine the consistencies and inconsistencies among their preferences.

21.5.1. *Joint elaboration of an argumentative scheme*

As with the HERMES [KAR 01] and ZENO [GOR 97] systems, our system allows us to jointly develop an argumentative scheme. Given that expertise is distributed, the process of sharing activities offers the possibility to create a common decision hierarchy while respecting the cognitive pattern of participants. When an agent has a new criterion in its decision hierarchy, it updates its preferences and spreads this activity throughout the system. The agent that receives such a proposal can confirm that it already has this criterion. Otherwise, it suggests this new criterion to its user. If the latter decides not to take this criterion into account, the agent declines this proposal. Otherwise, the agent carries out this proposal.

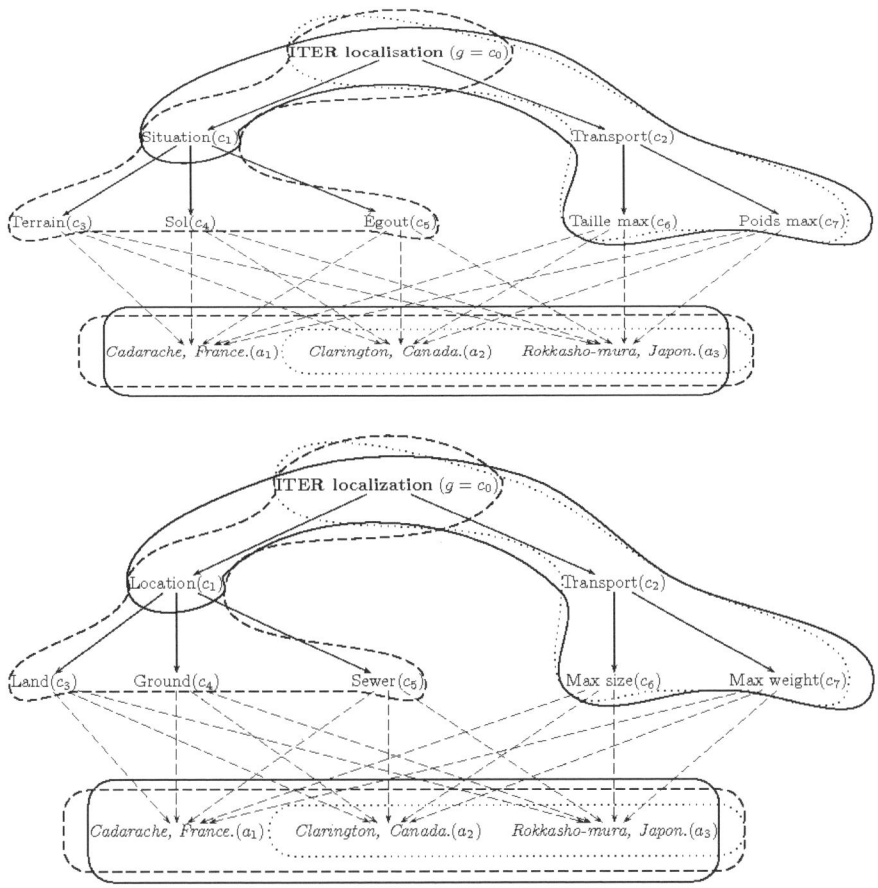

Figure 21.4. *Joint decision hierarchy for the location of ITER*

The process of sharing alternatives is very similar. A new alternative is suggested to the participants through the system. Users who integrate a new activity in their argumentative scheme must assess it. Similarly, participants can dissociate themselves from an activity. This consultation support system allows users to negotiate a common representation of the problem. All agents share the same goal but each has its own set of activities: alternatives and criteria. The set of activities can expand or shrink during the debate.

Let us consider three participants who wish to discuss the location of ITER. The joint decision hierarchy is represented in Figure 21.4. The common decision hierarchy initially consists of the goal and two Japanese and Canadian implantation sites. The first participant takes the common criterion c_0 and criteria c_2, c_6 and c_7 into account. He/she evaluates the two common alternatives: a_2 and a_3. The second participant considers the common goal g as well as two sub-criteria c_1 and c_2. He/she also takes into account two sub-criteria of c_2: c_6 and c_7. The third participant takes into account criteria c_0, c_1, c_3, c_4 and c_5. He/she evaluates three alternatives.

It is among activities shared by participants that conflicts and consensus are detected.

21.5.2. *Detection of conflicts and consensus*

Given that judgments are subjective, the system provides features that allow us to examine the consistencies and inconsistencies among the preferences of users.

Let us consider two assistant agents. A consensual criterion is a criterion that they share and for which one of the optimal alternatives is common. Inversely, a conflict criterion is a criterion that they share and for which the optimal alternatives are all different. A dialog between these two agents allows us to identify the major conflict or consensus that they share. A dialog ends when either one of the consensual criteria among the most general ones, or a conflict criterion among the most specific ones, is reached.

21.6. Conclusion

By adapting the AHP to a multi-participant context, the consultation support system proposed here can be based on a multi-agent system. Each agent assists a user and interacts with the other agents in the system.

On the one hand, the system allows us to represent a problem, expressing preferences from this representation and synthesizing these findings by measuring their coherence.

On the other hand, the system provides features for the collaborative development of argumentative schemes and enables us to elucidate consistencies and inconsistencies among the preferences of participants and thus detect consensus and conflicts. To evaluate the uses of such a tool and its possible diversions, this proposal must be subject to empirical validation. The use of such a system furthermore not only requires the use of a computer by all participants in the project considered and the absence of specific knowledge prior to implementation, but also greater transparency of debates on the part of all of the decision makers.

21.7. Bibliography

[ARR 63] ARROW K.J., *Social Choice and Individual Values*, Wiley, New York, 1963.

[CAL 01] CALLON M., LASCOUMES P., BARTHE Y., *Agir dans un Monde Incertain*, Seuil, Paris, 2001.

[GOR 01] GORDON T., MARKER O., "Mediation systems", in L. M. HILTY and P.W. GILGEN, (eds.), *15th International Symposium Informatics for Environment Protection*, Metropolis Verlag, p. 737-742, 2001.

[GOR 96] GORDON T., "Computational dialectics", in HOSCHKA P. (ed.), *Computer as Assistants – A New Generation of Support Systems*, Editions Lawrence Erlbaum Associates, Mahwah, NJ, p.186-203, 1996.

[GOR 97] GORDON T., KARAPILIDIS N., "The Zeno argumentation framework", *Proceedings of the 6th International Conference on Artificial Intelligence and Law*, ACM Press, New York, p.10-18, 1997.

[ITO 97] ITO T., TORAMATSU S. "Persuasion among agents: An approach to implementing a group decision support system based on multi-agent negotiation", *Proceedings of the 5th International Joint Conference on Artificial Joint Conference on Artificial Intelligence*, Morgan Kauffmann, 1997.

[KAR 01] KARACAPILIDIS N., PAPADIAS D., "Computer supported argumentation and collaborative decision making", *Information Systems*, vol. 26, no.4, p. 259-277, 2001.

[MCB 01] MCBURNEY P., PARSONS S., "Intelligent systems to support deliberative democracy in environmental regulation", *Information and Communications Technology Law*, vol. 10, no. 1, p. 33-43, 2001.

[MOR 44] MORGENSTERN O., VON NEUMANN J., *The Theory of Game and Economic Behaviour*, Princeton University Press, Princeton, NJ, 1944.

[MOR 05] MORGE M., Système dialectique multi-agents pour l'aide à la concertation, PhD thesis, Ecole des Mines, Saint-Etienne, 2005.

[RIT 73] RITTEL H., WEBBER M., "Dilemmas in a general theory of planning", *Policy Science*, vol. 4, p. 155-169, 1973

[SAA 01] SAATY T., *Decisions Making for Leaders; the Analytic Hierarchy Process for Decisions in a Complex World*, RWS Publications, Pittsburgh, USA, 1982, reprint 2001.

Part VIII

Is "Socio-informatics" Possible?

Chapter 22

The Interdisciplinary Dialog of Social Informatics

22.1. Introduction

What are the social consequences of using new information and communication technologies (ICT)? If we understood them better, could we design ICT more efficiently and effectively? These questions help to define the problems in the area termed "social informatics". These problems are not new. Computers began a mass penetration of the workplace in the early 1970s and since then, the social uses of computing have become a much debated subject. It is even more so now, in the 21st Century, with the global information infrastructure (GII) that is characterized by the growing convergence of Internet, the World Wide Web and satellite, mobile and wireless telecommunications. Increasing computing power is being incorporated into everyday objects that are capable of communicating with one another and with the people that use them through the GII. What will this mean for the organization of social activity?

Often we get the impression that the theoretical and empirical framework to answer this type of question is lacking. Beliefs, fears and enthusiasm seem to dominate thinking about the relationship between technology and social change, rather than the findings of solid academic research. This observation largely motivated researchers working in social informatics when they named their field in the mid 1990s [KLI 00]: they wanted to open an arena for an interdisciplinary dialog

Chapter written by William TURNER.

on how to collectively analyze the social uses of computing, and how to exploit the results of this research when designing new computer applications.

"Social informatics" brings research originating in computer science (primarily from software development, requirements engineering and artificial intelligence) and in the social sciences (primarily from sociology and anthropology) together under the same heading. Its specificity lies in the way that the dialog between these sciences is organized. We will first attempt to identify and describe the dynamics affecting this dialog and then show how understanding these dynamics can be used to strengthen the foundations of the social informatics field.

Our argument is that these foundations are changing. They used to be anchored in the belief that collective action can be objectively described by answering questions relating to such things as "who is doing what, where, how, with what resources and under what circumstances, with what results and what consequences?" Now, however, collective action is represented as being more the result of a learning process. Nothing in that process can be taken for granted, even if there are mechanisms that lock people into particular courses of action over time. Among these mechanisms are such things as getting to know one another, mutual trust and a willingness to help one another out when needed. That said, cooperation is now considered to be an accomplishment [QUE 93, p. 79]: relationships can go sour; people can drift apart; and confidence can quickly fade in the idea of achieving mutual respect and responsiveness. So the important question is no longer that of modeling a stable system of social interactions with the goal of improving system performance through the design of an appropriate computer application. Instead, it is to understand how the steady increase in the availability and amount of computing power in everyday life affects the skills and know-how required not only for learning how to do things together, but also for stabilizing the social system produced through on-going collective action.

22.2. Identifying procedures for configuring collective action

The interdisciplinary dialog of social informatics constitutes a crystal clear example of an unstable zone in the self-organizing dynamics leading to either cooperation or dissension. By exploring this zone and making an attempt to formalize procedures for working together despite differences in disciplinary perspectives, researchers in social informatics engage in the type of action needed to reflexively understand the skills and know-how at play when doing things together. They are no longer situated outside the socio-technical system, as observers of an objective reality that they are collectively trying to construct in order to improve efficiency, effectiveness and management, etc. On the contrary, these different rationalizations of why computing power is used in social situations are ad hoc, after the fact discourses, serving to justify a particular course of action. With respect to the course of action itself, however, the direction it takes depends upon the language games played by participants when attempting to position themselves in a particular socio-technical configuration.

The concept of a "socio-technical configuration" plays the same role in social informatics as the concept of "life forms" in Wittgenstein's discussion of language games [WIT 61]. For Wittgenstein, behavior is organized around social conventions: for example, if a chess player does not play by the rules of the game, he is not playing chess; he is playing another game. The same holds for language: people are able to anticipate the moves of others in the language games they play to the extent that they understand the rules of the game. However, a collectively shared understanding of these rules is virtually impossible to achieve, given the complexity of life forms. These are composed of a wide range of human, material, symbolic and other resources that can be combined in a potentially infinite number of ways, thereby producing heterogeneous structures of relationships that can be described from multiple viewpoints. So given the open-endedness of what can be represented and attributed to a life form, how do people set themselves up on common ground? Are there any rules? Wittgenstein answered this question negatively by pointing to what he called the paradox of the rule: defining the rule does not provide any guidelines on how that definition should be applied when doing things in a given situation. He argued that given this paradox, "whatever I do will, according to some interpretation, be reconcilable with the rule" [WIT 61, section 198].

An alternative to a rule-based interpretation of behavior is to consider that people interact with one another by anticipating the respect of an institutionalized code of behavior. They learn to get to know one another because they have a general idea of the game they want to play and they accept that in order to play it some rules have to be applied. When these two minimum conditions are met, people will consider themselves as being in a common "socio-technical configuration". They will feel that they know where they are and where they want to go, even if their perception of how to get there is fuzzy and confirmation of their project's doability is required.

These two conditions are met in the ambient intelligence research field. We will use this term to characterize three directions of research. The first aims at producing software agents that are mobile or incorporated into everyday objects and that will provide us with computing power wherever we are. The second concerns the construction of multi-modal interfaces in order to more naturally communicate with these agents using our senses, such as touch, hearing and vision instead of a computer keyboard. The last direction seeks to increase the cognitive autonomy of these agents so that they can function by themselves in order to find the resources necessary for carrying out their tasks.

When taken together, these different directions of research should make computing power more available, accessible and easy to use because software agents will learn how to anticipate the needs of users and will know how to make appropriate recommendations for assisting in specific tasks. The downside to this research lies in what is called user profiling. Technically, user profiling demands a set of clearly demarcated bins into which things represented symbolically in the system will neatly and uniquely fit. These bins can be organized hierarchically, as independent elements in a sequentially constructed list, or according to some logical function. Their very existence contributes to sowing just about everything into the social fabric a standardized representation: animals, human races, books, pharmaceutical products, taxes, jobs and diseases. As Bowker and Star say, these categories take on life in the daily practices of industry, medicine, science, education and government. They produce "iron cages" hemming in the lives of people in the modern world (to paraphrase Weber's remark on bureaucracy). When research in the field of ambient intelligence is seen as producing these effects, the specific socio-technical configurations that it produces need to be treated with caution [BOW 99]. Clearly, the production of user profiles for improving system performance is creating a social and technical fabric that demands critical analysis.

Here then we have a typical debate about the social uses of computing couched in moral terms about whether or not technological evolution is good or bad. It is an area where an interdisciplinary dialog seems hard to engage because of what has been called a great divide [BOW 97]: on one side is the complex, political and emotion-laden reaction to the idea of seeing social behavior categorized and put into bins; on the other side is the equally complex, but rational argument that system performance requires standardization. Under these conditions, it would seem that the best we can hope for when engaging in an interdisciplinary dialog is to chart out a middle road between contradictions in order to place socio-technical configurations in some kind of functional equilibrium.

We disagree with the idea that the outcome of an interdisciplinary dialog will be to strike some sort of functional balance between standardization and change. We consider instead that standardization is part of a social learning process serving to dynamically adjust to change. It is implemented through adopting evaluation procedures. These procedures are institutional devices designed to provide people with incentives and rewards for mobilizing their skills, know-how and resources for doing things in a specific way. Their adoption initiates something similar to what Wittgenstein had in mind when talking about engaging in a game of chess. They specify the forms of behavior that are required of people in order to be recognized as skilful players in a particular type of language game. Procedures fix the game rules of interacting but, at the same time, efforts engaged to align a person's behavior upon an understanding of these rules leads to the adoption of new procedures because of the paradox of the rule defined earlier. When we try to show others that we know how to competently take part in a social activity, we are acting upon an interpretation of what is expected from us, and this interpretation will inevitably enter into conflict with other interpretations. But conflict does not block cooperation. On the contrary, it will strengthen it as long as there is confidence in the idea that people make an honest attempt to overcome their misunderstandings. New procedures emerge out of language games to the extent that people develop this confidence.

In summary, then, procedures open the playing field to a wide variety of participants but at the same time limit the game rules to those needed for mobilizing a particular set of skills, knowledge and resources. Procedural relevancy is measured not only by the intensity of the language games that each procedure initiates individually, but also by the ease with which the results of these language games can be adjusted and articulated with one another in a dynamic, confidence-building, learning process. That said, we will not discuss these measurement problems in this chapter. More information on the natural language processing techniques used to evaluate the intensity of language games and visualize the dynamics of inter-procedural articulation can be found in [TUR 09]. Instead, our goal here is to illustrate areas where an interdisciplinary dialog can be usefully engaged around the social uses of ambient intelligence.

22.3. Analyzing socio-technical change

GII is a configurable and programmable infrastructure capable of managing both information and communications. Services are being developed through the GII to knit, maintain and develop social contacts and networks, and to ease the cognitive charge of calculating appropriate courses of action. Take the fact that objects are increasingly being equipped with radio tags so that they are becoming progressively addressable and intelligent. Imagine, for example, young children whose house keys have RFID[1] tags on them so that once they get home, an RFID reader located someplace in the household can send a message to their mothers and fathers at work informing them that their children are safely home after school. Or take work to help the handicapped, such as that underway on the production of a virtual white cane. Now that objects are radio tagged, the cane of a blind person can be used to detect those objects and to provide a vocal description of the things that are in the person's immediate environment [JAC 04, JAC 06]. Finally, take efforts aimed at producing document management techniques capable of computer-supporting distributed collective practices [TUR 06]. Software agents are becoming mediators of social interactions by using the intelligence they acquire on the structure of socio-cognitive networks through the application of data-mining techniques [RIP 06a].

---

1 RFID: radio frequency identification.

That said, conclusions as to whether these types of projects are socially relevant require the adoption of evaluation procedures. Social informatics has addressed this issue using four rules, namely that computer applications:

– be easy to understand and use "intuitively", without explanation;

– allow people to anticipate the moves they should make in the language games they play in order to gain social recognition for their individual skills and know-how;

– help people identify additional skills, knowledge and resources for improving their capacity to participate in these language games;

– meet the specific needs of the task to be carried out.

The language games set in motion by these different evaluation objectives are played by both social scientists and engineers. The first concerns the need to collectively build standards to measure what is ergonomic and affordable, because a computer application that is intuitively easy to use will be both. Just as the shape of door knobs or handles invites a person to use them for opening a door, a computer application that is affordable – which incites people to act in a specific and appropriate way in a given situation – can be expected to have a greater impact on human behavior than one that is hard to understand and needs explanation. Intuitive interaction is the goal of multi-modal interface design, as we have seen. The virtual white cane example shows that an interdisciplinary dialog will concern the relationship between perceptions of stimuli, reactions to what has been perceived, and the sensorial channel through which information is communicated between humans and machines (hearing, touch, proprio-kinesthetics, etc.).

Representing these relationships requires the adoption of a model of human-machine interactions. One model that has often been used is based upon the assumption that a stimulus first has to be coded and then decoded in order to be understood, and that the efficiency of a communication channel can be measured according to the amount of additional noise that it introduces into the decoding process. Social scientists have long criticized this model. Wolton, for example, argues that the goal of developing information processing techniques for reducing noise completely misses the political and cultural signification of having noise in a social context. For him, the borderline between what is noise and what is not is an on-going subject of debate and conflict. It is where battles are fought to determine what will be remembered and what will be forgotten. It is there that collective memory practices are engaged and take form [WOL 99]. Other models of communication get around this criticism. For example, in our own work we are studying the practices of language games using the ideas of Grice, as developed by Sperber and Wilson [FER 07]. Our communication model is one of mutual manifestness, which holds that a stimulus has to produce not one but two effects in order to be perceived correctly. Take language stimuli, for example. In order to be understood, messages have to be informative in the sense of informing others about something that is both relevant to them and that they can understand cognitively with a minimum effort (it is affordable). However, at the same time, a message has to capture the attention of those people to whom it is addressed. Affordability and information content are not enough to communicate; awareness that people want to communicate something is equally important. Looking at a second area where an interdisciplinary dialog is being engaged will help to clarify this point.

This area concerns the question of how technology contributes to the construction of social capital. Broadly speaking, social capital refers to the resources accumulated through the relationships among people (COL 1990). It refers to being part of a network, having other people there when needed, to provide help in a variety of forms from the simple exchange of information and advice to material support and services. Social capital is grounded in norms of mutual respect, trust and reciprocity [PUT 95]; three norms that can be quickly undermined in situations of suspicion. Suspicion casts doubt upon the willingness of people to engage with others in a supportive way. It leads to uncertainty about what can be expected from cooperation and points to the inherent incapacity of people to calculate the return they can expect from investing in social capital construction. For unlike investing in physical capital, such as a car, that people can appropriate and use personally as they see fit, accumulating social capital requires the responsiveness of others.

So are we confident about the possibility of doing things together? Misunderstandings will cause suspicion and lead to us to answer this question negatively. They greatly diminish the vitality of language games because they prevent people from correctly anticipating the moves they should make in order to gain social recognition for cooperating. Augmenting the computing power distributed throughout our environment could help diminish these misunderstandings. Software agents could act as intelligent mediators of social interactions.

How can we represent the idea of using computers to help people in overcoming their misunderstandings? One approach lies in the possibility of encoding speakers' intentions into explicit categories. This approach moves in the direction of the language/action perspective introduced into computer science in the 1980s by Winograd and Flores [WIN 86]. The idea is that misunderstandings originate in the use of language.

Take, for example, the shopping list that we use when we go to the store for provisions. It will contain a variety of entries, such as "jam", "milk", "chocolate", etc., which serve to remind us of what is lacking at home. The mere sight of the items on the list is enough to trigger the appropriate actions for doing the shopping correctly. Moreover, the list can easily be transmitted to others because each word has a meaning which is (in general) understood because we are trained to use these words in our daily life. But what happens when we do not know how to name an object we are searching for? For example, when some cows went mad in the late 1980s, we did not know what the responsible agent was, how the disease was spreading, or whether it would infect people. So how do you put a name on something that you do not understand? Callon suggests that once again constructing a list is useful. By listing the opinions of different participants, their concerns, interests, options and identities are made visible; the constructed inventory captures the intentions of human actions [CAL 01]. It paints a picture of reality that is driven by interests, motivations and power relationships, rather than by relations of cause and effect. Consequently, it helps to make a situation intelligible because it clarifies why people want to communicate, that is, what they want to defend and with whom. An inventory provides a context for evaluating the veracity of different affirmations. Each new item recorded on the list constitutes an invitation to talk about, to examine how it reinforces the coherence of elements already gathered together in the interpretative categories organizing collective action or, conversely, how it calls these categories into question.

Computers manage lists. Software agents are able to assist people in doing a particular task to the extent that procedures exist for combining words and intentions in appropriate ways to inform a suitable course of action. As we will see in a moment, ontologies encode these procedures and producing these ontologies is currently a mainstream activity of the computer science community in connection with construction of the semantic Web. Our work in this area sees efforts aimed at standardizing language practices as a contribution to overcoming misunderstandings in language games [MAU 99]. The following story will serve to illustrate our approach. Suppose, first, that a software agent puts "jam" on a list because it is a word that it does not know but that word is used in the household environment where the agent is located. Suppose, second, that the agent has been given an algorithm that allows it to formulate a question combining new words such as "jam" with the tasks of the household that it has been assigned to carry out, such as stocking food provisions. Suppose that after launching the question using an Internet browser, the agent gets back a list of sentences containing the word "jam". The agent could then use a learning algorithm to class those responses in a way that would serve to identify different categories of jams, such as "organic", "non-organic", "strawberry jam", "marmalade", "low sugar" or "high sugar content", etc. After having done that the agent could ask members of the household to tell it which types of jam they prefer and then communicate their preferences to another software agent located in the neighborhood grocery store. This agent would then read the RFID tags on jam jars in the store in order to select the type of jam requested for delivery to the household.

Of course, this story of research in ambient intelligence will meet with criticism. For example, Suchman severely criticized the language/action perspective as a foundation for system design because it carries with it an agenda of discipline and control. For her, categorization systems are ordering devices, used to organize people, settings, events or activities by whom they are employed or to which they refer. She went on to argue that social reality is shaped by non-compliance with these ordering schemes [SUC 94], and this is our point too. Both software and human agents are capable of categorizing. When human agents encounter resistance to the way in which they have ordered their world, we call this a misunderstanding. They perceive that something unpredictable has happened that requires adjustment. We are using resistance to computer-generated classification schemes in the same way, as a means of improving the relevancy of learning algorithms under development to enable software agents to competently interact with humans [RIP 06b, SAN 03]. We maintain that this goal – the constant, on-going rationalization of list management techniques in order to take full advantage of the computing power incorporated in the objects surrounding us – will reduce misunderstandings, encourage people to initiate, develop and consolidate language games leading to mutually supportive actions. It will, consequently, contribute to increasing cooperation, which is the basis of social capital accumulation.

A third area of interdisciplinary dialog concerns the semantic Web[2]. The concept of a "semantic Web" designates a vast effort underway in computer science to make information available on the Web intelligible to machines through the construction of ontologies. Ontologies explicitly represent different areas of human activity in ways that prescribe an authorized interpretation of what takes place in that area, *a priori*, thereby permitting human-machine cooperation. Defenders of the semantic Web say that without the assistance of machines, people will have to support constantly increasing cognitive costs in using the vast amount of information available through the Internet. Opponents, on the other hand, react fiercely to the idea that engineers are seeking to build standardized, *a priori*, representations of human activity. The comments above about the "iron cages" created by classification systems illustrate this reaction. However, when we go beyond the rhetoric and look at the technical work involved in building the semantic Web, what is at stake seems easier to accept. Take the example of Berners-Lee [BER 01] who imagined a story in which an agent was assigned the task of organizing the transportation of Pete's mother to a hospital for medical treatment. In order to accomplish this task, the agent needed to understand:

– that hospitals are places, that places have addresses;

– that Pete wanted to find a hospital whose address was within 20 miles of his own house;

---

2 See Chapter 13.

– that his house was also a place;

– that it too had an address;

– that the two addresses are separated by a distance;

– that the number of kilometres separating the two addresses is a property of that distance; and that

– finally, 20 kilometers (or less) is the value that must be respected to help Pete.

Standardizing information processing procedures to enable a machine to help Pete help his mother does not seem to raise an "iron cage" problem. To the contrary, it appears helpful and socially quite useful to try and generalize the proposed taxonomy in order to carry out similar types of calculations in other situations.

In conclusion, then, the high importance accorded in this chapter to maintaining and reinforcing the language games of social informatics lies in the fact that through the interdisciplinary dialog, a line of demarcation is established between what is socially acceptable and what is not when standardizing for reducing the cognitive costs of working with technology. The criteria determining social acceptability depends on us showing that additional computing power will efficiently and effectively contribute to human and social capital development. However, the procedures for applying those criteria are open-ended and require debate. Through debate, the "iron cages" of socio-technical infrastructures are identified and new spaces for enlarging and enriching our social interactions are opened.

22.4. Improving the design of computer applications

Social informatics is the name given to an interdisciplinary research field uniting efforts from both the computer and social sciences in an attempt to better understand the social uses of technology, and to use that understanding for designing open, convivial interaction spaces for supporting collective action. The methodological challenge lies in what some authors have called the "radical indeterminacy" of the unfolding course of human interaction [SUC 94]. In this chapter, we have accepted the idea of unpredictability, but have argued that people successfully manage uncertainty by adopting procedures to reduce it. Although there is indeterminacy, it is not radical and that is precisely why the idea of "socio-technical configurations" is a useful conceptual device for design.

This concept is not used in this chapter to designate the outcome of a classical requirements study; it does not represent the agreed-to result that a design team will seek to achieve after having specified the social and technical constraints of their work together. The starting point for an interdisciplinary dialog is not an agreement upon a blueprint for action, it is the action itself. Language games originate in conflicting interpretations and the need to adopt appropriate procedures for doing things together. Collectively building a socio-technical configuration consequently needs to be considered as an achievement. It manifests a willingness to cooperate. It shows that through their interdisciplinary dialog, computer and social scientists are gaining a better understanding of how to distribute computing power over the objects of our environment. It links technology and social change by reducing the social and cognitive costs of information processing while, at the same time, contributing to the dynamics of human and social capital formation and development.

22.5. Bibliography

[BER 01] BERNERS-LEE T., HENDLER J., LASSILA O., "The semantic Web", *Scientific American*, vol. 284, no. 5, pp. 34-43, May 2001.

[BOW 97] BOWKER G., STAR L., TURNER W.A., GASSER L. (eds.), *Social Science, Technical Systems and Cooperative Work, Beyond the Great Divide*, Lawrence Erlbaum Associates, Mahwah, NJ, 1997.

[BOW 99] BOWKER G., STAR L *Sorting Things Out: Classification and its Consequences.* Cambridge, MA: MIT Press, 1999

[CAL 01] CALLON M., LASCOUMES P., BARTHES Y., *Agir dans un Monde Incertain, Essai sur la démocratitic technique*, Seuil, Paris, 2001.

[COL 00] COLEMAN, J. *Foundations of Social Theory*, Belknap Press of Harvard University, Cambridge, MA, 1990.

[FER 07] FÉREY N., TOFFANO-NIOCHE, C., MATTE-TAILLEZ, O., GHERBI, R., TURNER, W., "Using virtual reality for constructing specialized vocabularies", *Actes du Colloque IC 2007*, Toulouse: Cépaduès Editions, 2007, pp. 193-204

[JAC 04] JACQUET C., BELLIK Y., FARCHEY R., BOURDEA Y., "Aides électroniques pour le déplacement des personnes non voyantes, vue d'ensemble et perspectives", *Actes de l'IHM 2004*, pp. 93-100, ACM Press, 2004.

[JAC 06] JACQUET C., BOURDEA Y., BELLIK Y., "A component-based platform for accessing context in ubiquitous computing applications", *Journal of Ubiquitous Computing and Intelligence*, pp. 163-173, 2006.

[KLI 00] KLING R., CRAWFORD H., ROSENBAUM H., SAWYER S., WEISBAND S., *Learning from Social Informatics: Information and Communication Technologies in Human Contexts*, University of Indiana, Social Informatics, 2000. (Available at: http://rkcsi.indiana.edu, accessed February 11, 2010.)

[MAU 99] MAUNOURY M.-T., DE RECONDO A.M., TURNER W.A., "Observer la science en action ou, comment les sciences de l'information permettent de suivre l'évolution et la convergence des concepts de prion et d'hérédité non mendélienne dans la littérature", *M/S Médecine Sciences,* vol. 15, no. 4, p. 577-582, April 1999.

[QUÉ 93] QUÉRÉ L., "Languge de l'action et questionnement sociologique", in Quéré L. (ed.), *La Théorie de l'Action: le Sujet Pratique en Eébat*, CNRS Editions, Paris, 1993.

[PUT 95] PUTNAM R.D., "Bowling alone: America's declining social capital", *Journal of Democracy*, vol. 6, no. 1, p. 65-78, 1995 (translated in BEVORT A. and LALLEMENT M., *Social Capital: Performance, Fairness and Reciprocity*, La Découverte, Paris, p. 35-50, 2006).

[RIP 06a] RIPOCHE G., SANSONNET J.-P., "Experiences in automating the analysis of linguistic interactions for the study of distributed xollectives", *Journal of Computer Supported Cooperative Work*, vol. 15, no. 2-3, pp. 149-183, 2006.

[RIP 06b] RIPOCHE G., Sur les traces de Bugzilla : vers une analyse sémantique des interactions pour l'étude des pratiques collectives distribuées, IT Thesis, LIMSI-University of South Paris.

[SAN 03] SANSONNET J.P., TURNER W.A., VALENCIA E., "Agents informationnels pour l'étude expérimentale de concepts de socio-cognition : vers une approche agent de la socio-informatique", *JFSMA 03, French Working Days on Multi-agent Systems*. Hamamet, Tunisia, 27-29 November 2003.

[SUC 94] SUCHMAN, L., "Do categories have politics? The language/action perspective reconsidered", *Journal of Computer Supported Cooperative Work*, vol. 2, p. 177-190, 1994

[TUR 06] TURNER W.A., BOWKER G., GASSER L., ZACKLAD M., "Information infrastructures for distributed collective practices", Special Issue: *Journal of Computer Supported Cooperative Work*, vol. 15, no. 2-3, pp. 93-250, 2006.

[TUR 09] TURNER W.A., MEYER J-B, DE GUCHTENEIRE P, AZIZI A. "Diaspora knowledge networks", in Kissau, K. and Hunger U., *Migration and Internet. Theoretical Approaches and Empirical Findings*, VS Verlag: Wiesbaden, 2009.

[WIN 86] WINOGRAD, T., FLORES, F., *Understanding Computers and Cognition: A New Foundation for Design.* Ablex, Norwood, NJ, 1986.

[WIT 61] WITTGENSTEIN L, *Investigations Philosophiques*, Gallimard, Paris, 1961.

Chapter 23

Limitations of Computerization of Sciences of Man and Society

23.1. Introduction

Since the publication of the report on the computerization of society by Minc and Nora [MIN 78], socio-informatics, as the accumulation of individual and collective economic and social data, has continued to develop. Databases have become huge and their access is increasingly facilitated by modern means of communication, networks (the Internet of course, but also all intranet networks) and software (search engines and database managers). The possibilities that information and communication technologies (ICT) give very easy access to extensive information and compile and exchange it, has made information availability in our society move from scarcity to abundance: previously, we had to search for the information, now we have to sort through it in most cases. The approach is not only to collect information, but also to select data in databases that are relevant and to extract the meaning from them.

To perform these operations, computer scientists and mathematicians have developed information-processing methods entirely based on computing power offered by the computer. Information and communication techniques have thus gradually changed the relationship of man to knowledge. Here it is about assessing how the combination of this accumulation of data, software necessary to analyze

---

Chapter written by Thierry FOUCART.

them and this new relation of man to knowledge changes the role of human sciences in the management and organization of society.

23.2. The scientific approach in sciences of man and society

The scientific approach in sciences of man and society is particularly to select and analyze information to understand the social and human modes of functioning. That information comes from different fields of knowledge: we find them in literature, administrative documents, history, language, traditions, political and social structures, techniques, etc. Since the mid-20th Century, investigations and surveys conducted on samples of individuals have multiplied in order to collect individual data. Administrations, banks and insurance companies, to name but a few examples, are asking for and store increasing quantities of information on all aspects of individual and collective life (marital status, family status, real estate, income, etc.). The data thus obtained are processed by quantitative methods whose results are presented in the form of percentages, graphs and numerical indices. The interpretation of these results helps us to understand the social or human fact being studied.

We commonly use these methods to analyze social facts, such as the recidivism of offenders, the impact of class size on academic achievement, conduct disorders in children and adolescents, the result of an election, violence against women and much more. A historical example of such use is the study of suicide by Durkheim [DUR 1897], which showed the role of sexual, marital or religious social structures on the increased rate of suicide in 19th Century Europe. Currently, the amount of data to be processed is such that the manual method of compiling and sorting administrative statistics followed by Durkheim to analyze them is physically impossible for one man. The mechanical processing of information has allowed progress, by performing basic processing using a punch card sorter. Technological and scientific evolution now allows analyzing socio-economic data using entirely new mathematical and computational methods. A researcher in social sciences often uses these methods to highlight the fundamental properties of the data he/she has collected. The theoretical complexity of these techniques is compensated for by the ease with which they are used, but the results they produce are becoming increasingly difficult to understand.

It is therefore about moving from the number to its meaning and vice versa. A typical example of this passage is the interpretation of a statistical relation as a causal relation, which, as we have long known, statistics can never establish. The process comes from reasoning located in a different scientific field, where it is the explanation of the observed statistical relation that gives the interpretation. The statistical approach (data gathering and statistical processing) and the interpretative approach (analysis and synthesis of results) are moreover classics in natural sciences: observation (of a star for example) calls for a scientific tool (the telescope). The resulting image approximately represents the observed object and must be correctly analyzed. This analysis depends, on one hand, on the object (the star) and the knowledge that we already have, and on the other hand on the tool (the telescope) and the interpretations of the results it gives. This duality is the source of numerous mistakes, some examples of which are given later in this chapter, either in the use of the method,or in the sense given to the results. As Alain [ALA 32] explained, "if Chaldean shepherds had our powerful telescopes, they would have learned nothing from the master science". They would not have known how to use the device and would not have understood the images obtained. Either one of these skills alone is insufficient: mastering the device does not give greater understanding of the image than examination of the latter alone.

Computerization of society and techniques, at the same time making considerable data and tools necessary to analyze them available, has put researchers in social sciences in a similar situation: they must master the technical tool – observation of data, methodologies and software – and move from the scientific field where numerical and graphical results are to the field of social sciences that give them meaning. The uncertainties of this transition are compounded by the complexity and novelty of the methods as well as by the intricacy of these approaches, which is very difficult for a single researcher to carry out.

23.3. Complexity and intricacy of mathematical and interpretative approaches

These quantitative methods are indeed not limited to computing procedures that would be useless, for the user to know or to tools whose use would be relatively simple. In astronomy, it is not only the star itself that is observed, but the light it emits, and it is by analyzing the spectrum of this light that we obtain information on the physical constitution of the star and its movement. The situation is similar in social sciences: the two approaches, mathematical and interpretative, simultaneously intervene in the understanding of a fact observed and the interpretations of results. Their intricacy becomes apparent when the quantitative methods used are based on the modeling of a studied social fact or human behavior. The modeling indeed results from *a priori* analysis of reality by the researcher. The quantitative modeling consists of representing the relations and properties coming from this analysis using mathematical formulas. We can then control *a posteriori* the consistency of the model.

Take, for example, the linear regression model often used in the sciences of man and society. We consider a data set consisting of observations of an "explained" variable Y (for example, income) and of p "explanatory" variables $X_1, X_2, ..., X_p$ (for

$p = 4$, X_1 represents the age, X_2 the political opinion, X_3 the socio-professional class, X_4 gender) on a sample of n statistical units (n voters). It is assumed that the variables X_2 (political opinion), X_3 (socio-professional class) and X_4 (gender) are digitally coded. The linear model (without going into details) consists of assuming that the variable Y (income) is equal to:

$$Y = \beta_0 + \beta_1 X_1 + \beta_2 X_2 + \beta_3 X_3 + \beta_4 X_4 + \varepsilon$$

The terms β_0, β_1, β_2, β_3 and β_4 are constants. The last term ε of the sum represents the adjustment variable for each individual. We then calculate estimates of the coefficients β_0, β_1, β_2, β_3 and β_4, using the observations made.

The assumptions necessary for the model are stronger if the designer of the model ignores the reality. It is this ignorance that imposes assumptions without which the analysis cannot be continued: we assume, for example, that the adjustment variable follows the normal distribution of mean zero, and that the relationship is linear. Other hypotheses can be envisaged, for example a non-linear relationship. These assumptions, while essential to the model, of course disturb the image of the social fact studied. This disturbance is all the more significant the more assumptions there are, and this number itself is no longer limited since the computer, by resolving all numerical difficulties, allows us to construct models of increasing complexity. We can thus multiply explanatory variables and consider 100 (or more) explanatory variables instead of four. The more complex the model is, the more *a posteriori* control is necessary. This control never shows that the model is true, but only that it is acceptable, even if we know that it is false since it is only an approximation of the reality as its author perceives it. The model is paradoxically less acceptable as it increases in complexity: it is for this reason that we favor the simplest models, following the principle of parsimony or Occam's razor[1] [EKE 00].

Social and human reality being "chaotic", according to Weber, in the same way as physical reality according to his contemporary Poincaré, the purpose of a model is to simplify it to make it intelligible, and not to represent it as accurately as possible and risk losing its meaning.

This risk is that of the empirical analysis of data which is based on the following observation: many data systematically lead to refutation of the model. This is easily explained, since modeling consists of simplifying this reality: the more the observation of the latter is accurate, the further we move from the model. The researcher, to indentify the main characteristics of a fact being studied, then proceeds in the opposite direction to the previous one: he/she accumulates

---

1 Occam's razor (14th Century) is a principle according to which, between two theories explaining the same phenomenon, we must choose the simpler one.

observations seemingly without *a priori* obtaining the best possible approximation of reality. The assumption here is that the main characteristics of data – the "factors" or "components" – must emerge from observations in an objective way because the researcher did not give them any *a priori* property. These methods (factor analysis, classification, etc.) appeared in 1965 in France led by Benzecri [BEN 73], and are still developing under the term "data mining" as computer science progresses. A simple microcomputer is now sufficient to perform such analyses on very important data.

The intricacy of the two approaches in empirical analysis is no longer that clear, but is present in spite of everything before quantitative processing, if even only in the choice of the data observed [BRE 00]. The possibility of a bias does not only concern the sample of individuals, it also exists in the choice of variables observed. By a known statistical effect, which imposes for example quotas to determine a representative sample of the population, the higher the number of observed variables is, the more the subjectivity of the researcher intervenes in his/her choice and subsequently in the factors. The latter can indeed only make what is contained in the data visible, just like Durkheim could not find explanations for the increase in suicides in the administrative statistics available to him.

This subjectivity also exists *a posteriori*, in the interpretation of results. It is easier to understand the meaning of a factor identified from the data, to "reify" it [DES 93], than to verify it if a model gives an acceptable representation of reality, especially since the results of analyses are based solely on responses as they appear. We must therefore seek the meaning of these results beyond obvious answers [BOU 01]. Durkheim did not limit himself to comparing percentages; he sought to understand them. It is qualitative analyses specific to human sciences, in particular the study of links between psychology, religion, social status, etc., that led him to define different types of suicides.

These analyses specific to the field of the data analyzed are essential in order to avoid misinterpretations of results. I asked my students the following question: *is the set of all sets infinite?* Most answered yes, the others having prudently abstained themselves. How to interpret this near unanimity, if we do not replace this question in the context of the theory of sets? Indeed, the set of all sets does not exist, according to a theorem of Russell. Their answers show their ignorance on this point. This ignorance must itself be replaced in academia where the question was raised. It is very natural as my students, in the commercial stream, have never studied the theory of sets. It would take an entirely different sense – worrying – if my students were candidates for a higher degree in mathematics. The importance of placing the interpretation of results in the fields concerned is obvious here to avoid absurd conclusions.

23.4. Difficulties in application of methods

The resulting figures have a known effect: established by computer, they present the appearance of a scientific, objective and indisputable truth. Gould [GOU 97] even speaks of "the anxiety surrounding the figures for non scientific reviewers", while Manent explains that scientific discourse "is the only publicly admissible" [MAN 01]. These two effects combine to prevent a thorough critical analysis of graphs and numerical parameters and pave the way for a more rapid and superficial interpretation that coincides with the ideas *a priori* of the researcher, with his/her immediate intuition [FOU 01]. Yet Gadamer [GAD 76] has warned against this trend, whose systematic exploitation by advertisers shows that it is widespread: "any right interpretation must be secure against the arbitrariness of meeting ideas and against the limitation that are derived from undetected habits of thoughts and direct his gaze on the same things."

Even when the quantitative methods used are not especially complex, the errors are numerous. Boudon [BOU 97, note 1. p.25] cites a US survey among doctors to whom the following question was asked: "if a test to detect a disease, whose prevalence is 1/1,000, has a false positive rate of 5%, what is the chance that a person found to have a positive result actually has the disease, assuming you know nothing about the person's symptoms?". Most doctors replied 95%, while the correct answer (2%) was given by only 18% of the doctors. To find the correct answer, we need to calculate among 100,000 people the number of positive people who are ill (100, assuming there are no false negatives[2]), positive and not ill (99,900 × 5% = 4,995). There are therefore 5,095 people detected as positive and the probability a person detected as positive is actually ill, in the absence of any other information, is the ratio of the number of positive patients to the total number of positive people: 100/5,095 = 1.96%. Poitevineau [POI 04] gives other examples of interpretation errors in the case of statistical tests applied in psychology: accepting a hypothesis for a risk of the first kind of 5% does not mean that the probability for it to be true is equal to 95%, contrary to a widespread opinion.

We can therefore be legitimately concerned about interpretation errors in the case of more elaborate methods. The linear model (see above) is typical in this respect: the assumptions on which it is based are almost never explained, let alone controlled. It is often used to identify "the proper effect" of a variable X_1 (age) on the variable Y (income). This proper effect means that the variation of the explanatory factor X_1 always acts in the same way on the explained factor Y when all other causes of variation have been eliminated. In the model considered, a proper effect of age on income is a relation meaning that an increase in age X_1 (by one year, for example) causes the same variation (then equals β_1) on income Y, the political

---

2 The presence of false negatives would change the percentage very little.

opinion X_2, socio-professional category X_3 and gender X_4 being set. This notion of proper effect, which excludes the possibility of a variation of income with age between men and women of the same socio-professional category and same political opinion, is therefore particularly rigid. It is inseparable from this model. The existence of a proper effect is moreover often accepted without the hypothesis being criticized "any other things being equal" essential to identify it but never being fully verified and challenged [CHA 01, FOU 02].

The lack of critical thinking can finally lead to internal contradictions, such as comparing the situation of a man and a woman under the assumption that "any other things being equal", while the existence of a proper effect of gender on the situation makes the equality of spouses of different sexes impossible. This existence would thus contradict the assumption on which it is based and would deprive reasoning from any logic. It is sometimes this formulation itself of the model that prevents the existence of a proper effect, which we are seeking anyway: it is the case of models that introduce an interaction among explanatory variables. Subsequently, the complexity of the model increases the need to analyze the questions raised and their answers and to ensure that it is adequate enough to respond to them.

The language here carries a negative influence on the exercise of critical thinking: the simplicity of the expression ("any other things being equal", "proper effect", etc.) conceals the complexity of the concept and mechanisms involved. Interpretation uses immediate intuition without taking this complexity into account, against the previous recommendation of Gadamer.

23.5. Quantification and loss of information

To these interpretation errors is added an undue reliance in quantitative data, in particular in the case of social and human data. We must first make the information contained in the observations compatible with mathematical and computer processing. The two most common procedures are to measure the observed facts quantitatively, to classify each observation in one of the categories on a list set *a priori* in general, and to digitally encode the qualitative variables, as we have indicated when introducing the linear model. Textual data analyses are based on the same principle: we let the people being questioned respond freely to the questions asked, and we assign the same meaning to them regardless of the person questioned, which is the same as encoding them.

To quantify a social fact or human behavior of qualitative nature, the usual procedure is to measure the quantitative effects. We assess, for example, the cost for parents to care for their children by comparing the cost generated by using an external person with an income that this allows to parents receive [LEW 01]. We measure the time spent on the children by the father and mother to evaluate how each takes care of them. The service provided is assessed in a totally anonymous way, independent of individual links between the people involved: a childminder, grandparents, father and mother play a role seen as identical to children in their care. The transformation of an individual fact to numbers causes the loss of all non-quantifiable information, the information that is of greatest interest to people.

When data are classified, the information stored is reduced to that of each group: the socio-professional categories are typical of this reduction because all members of the same category are considered identical, a doctor and a bailiff in the professions, a small farmer and a grain carrier in the farmers. Conversely, a farm worker and a skilled worker are not classified in the same category, which may be disputed because social mobility between the two professions is strong. Here we neglect the differences within the same category and ignore the similarities in different categories.

This quantification lends to observations that are in fact an illusory objectivity. It is the researcher who chooses the criteria measured, time, cost, mathematical function of the criteria, etc. For example, by choosing the cost by which to measure pleasure a person takes to pursue a leisure activity, we consider that golf practice is more fun than cycling because it is more expensive. Similarly, the loss of profits of someone who stops working to take care of a child depends on his/her previous salary: the fact that one woman is better paid than another does not obviously mean that she would better look after her baby if she left her job. Generally, measures quantifying the manifestations of a purely qualitative activity are, as numbers, ordered. This numerical order is carried on the activities themselves [FOU 04a]. We try to limit this effect by using mathematical functions taking different manifestations of the activities into account; however, measuring by a number always results – by induction – in the classification of activities evaluated on a scale of values that becomes collective, since statistical methods apply it to all.

There is moreover a collective subjectivity. Social productions are subject to various influences, as shown by the disappearance of hysteria in the classification of mental disorders [BLA 00] and that of the category of idiots following the intelligence quotient (there are now only children of average or light mental retardation). This subjectivity is also apparent in the nature of social statistics dependent on their legal definition and data collection [BER 00]. It is illegal in France to establish administrative statistics by race, religion, etc., while due to the law passed May 9, 2001 statistics by gender on the professional equality of men and women have become mandatory. In other countries like the USA, some of that information is known and identified: statistics give another perception of discrimination since it is possible to measure it. The administrative statistics of a country thus result from a characteristic choice, subject to a sort of collective political or cultural subjectivity.

We ultimately associate a cost with all aspects of individual life, time, number, category and classification. This measure, whose calculation is the same for all, can be considered as relatively objective compared to the observer, but it does not represent the tastes, activities and capabilities of each individual and does not allow us to compare two different situations. The objectivity sought is in fact only an implicit collective subjectivity, which appears in the selection of criteria, the definition of classes and categories in which each individual is placed. The very design of a model sometimes results from implicit collective choices, as is shown by Moatti [MOA 93].

This apparent objectivity considerably reduces the richness of the information processed, and may contain a kind of reciprocal of the theorem of Hume [ARN 00]: the quantitative measure of a social fact or human behavior results from subjective premises that the systematization of the measure transforms into collective norms.

23.6. Some dangers of wrongly controlled socio-informatics

The computerization of society generalizes the use of the systematic measure seen by Hayek [HAY 53] as "the source of the worst aberrations and absurdities produced by scientism in social sciences". It has other effects that are difficult to discern. We first witness the gradual disappearance of methods concerning the individual (psychoanalysis for example) in favor of collective approaches based on sorting, classification and analysis of questionnaires (cognitive). Care must be taken not to be lured by this collective approach, which is based on the assumption that individuals are identical after their classification in groups is considered homogeneous. We must not only be interested in the apparent ease of mathematical and computer processing used to analyze the data thus collected.

This approach is that taken by most statistical analyses conducted within the educational system, most often based on cohort studies [JOU 99]. By measuring the effectiveness of an educational decision (repetition, comprehensive method for learning to read, etc.) in the same way that we show the effectiveness of a fertilizer or a drug, we analyze the social fact by following the scientific approach of natural sciences. A molecule has a specific biochemical effect on a given bacteria or virus, however, while an educational method acts differently depending on each student and teacher. The case of repetition as assessed in most education sciences is particularly clear: pretend that this is a failure because it does not allow students to catch up. It leads us to consider that all repetitions are due to the same cause and that their success is always reflected by catching up with the delay, irrespective of the personality of the student (mental disorders, intelligence quotient, family environment, etc). The assumption "any other things being equal" on which the comparison of academic performance of students is based is often limited to one grade level and the same socio-professional category [TRO 05]. Ignoring the relation between parents, children, teachers and many other criteria, the results will be a long way from the reality [FOU 06].

The problem is the same in the health sector, particularly in psychiatry: the individuality of the patient, doctor and their relationship cannot be taken into account in statistical analyses. Researchers from the Institut National de la Santé et de la Recherche Médicale (INSERM) have recently evaluated the psychodynamic (or psychoanalytic) approach and the cognitive-behavioral approach[3] to compare the effectiveness of psychotherapies coming under them. Such a comparison only makes sense if we think of all things as being equal: identical diseases and patient-doctor relationships, etc. Meta-analyses that consist of compiling studies conducted on the same subject hardly bring any additional information: they are subject to similar constraints. Comparing the effectiveness of the cognitive-behavioral approach often conducted in France with the version followed especially in US does not make sense, the socioc-ultural conditions of the two countries being different when they are directly involved in therapeutic practices. Moreover, these meta-analyses only take account of published studies. However, in order to be published, an article must contain statistically significant results, and therefore any study whose statistical results do not meet this condition is ignored, even if it provides relevant information [ARM 05]. A meta-analysis is thus not at all representative of the set of studies performed. For all these reasons, concluding from a meta-analysis that the cognitive-behavioral treatment is more effective than psychodynamic treatment is unfair to science. The inverse would be equally unfair.

The objective of the sponsor of this meta-analysis (Ministry of Health) is also perhaps not only scientific [CAS 05]: we cannot ignore that analytical psychotherapies are much more expensive than cognitive-behavioral therapies and that the empirical analysis of data is, according to Le Bras [BRA 00], "more and more closely associated with government requests". It would then mean instrumenting the scientific arguments for financial purposes. This approach to promote economic projects or ideologies is much more frequent than previously thought and very rarely detected by the general public. Sometimes even some journalists adhere to it, consciously or not: the newspaper *Le Monde*, among others, systematically reported a survey showing the ordinary life of children of

---

3 The INSERM report: *Psychotherapy, Three Approaches Evaluated*, frequetnly challenged by psychoanalysts, is no longer available on the institute's website. The summary can be found at the following address: http://iis13.domicile.fr/essentiaco/Psychoth%E9rapies-%20rapport%20%E9valuation%20Inserm.htm#PresentRapport (accessed February 12, 2010).

homosexual[4] [KRE 00] parents despite critics highlighting its statistical problems and its serious psychological and medical shortcomings [FOU 04b, LAC 03]. It is not always easy to detect these diversions, which combine a numerical argument of seemingly indisputable mathematical logic with a particular form of expression: "Words are, as we said, as much the objects as the instruments of ideological and political conflicts. This is why they are, intentionally or not, used ambiguously." [SCH 94]

The same can be said about figures. This strength of belief is well captured by Fraisse, who, by considering the statistics as "the best educational tool"[5] to promote feminist theories, confuses the strength of conviction of numbers and the educational methods based on reflection and critical thinking. Furthermore, by confirming that the investigation on children of homosexual parents was conducted with "the greatest methodological rigour possible"[6], the Professor of Child Psychiatry Bouvard defends it on a point that is one of the study's main weaknesses: the language is here a roundabout means by which to prevent criticism of its obvious lack of scientificity [FOU 04b][7].

The ease with which ICT can be used thus causes the revival of some form of scientism. Hayek explained that the latter is reborn from its ashes at every noticeable progress of natural sciences [HAY 53]. It is currently the case in biology and genetics, where the influence scientism is significant, and questions the previous notion of humanity [KAH 00]. The similar questioning of society is a therefore not an easy step to perform: the apprehension of the human brain as a computer and intelligence as a software [GUI 01] naturally leads to that of the society as a computer network whose maintenance and management are the purview of sociology, modern technologies providing the necessary tools. We return to the idea of Comte and Durkheim, which, according to Dubois [DUB 94] "have admitted what Spencer has always denied: the existence in the society of an organ comparable to the brain, of a collective conscience built on the model of individual conscience."

To the cognitivist illusion reducing man to his genetic equation corresponds to the socio-cognitivist illusion reducing society to a quantifiable organism in its

---

4 This is obviously not about taking sides on this issue in society.
5 Interview with Geneviève Fraisse: http://www.diplomatie.gouv.fr/fr/france_829/monde-solidaire_19090/monde-solidaire_13697/developpement-humain_14468/numero-special-50-sup-e-sup-anniversaire-declaration-universelle-droits-homme-dudh-no34-1998_17009/droits-homme-droits-femmes-entretien-avec-genevieve-fraisse-deleguee-interministerielle-aux-droits-femmes_51623.html.
6 This quotation from Bouvard is taken from Kremer's article published in *Le Monde* [KRE 00].
7 It is obvious that this is not about giving an opinion on the adoption of children by homosexual couples.

integrality. By this thinking, a geneticist analyzing the human genome corresponds to the social technology engineer analyzing the social genome.

23.7. Socio-informatics and social technology

ICT have gradually transformed social sciences into a social technology in the sense of natural sciences. Henry [HEN 03] explained that it was already the wish of sociologists-polytechnicians between the two wards (Le Chatelier, Fayol and Coutrot). At the end of his article, Henry quotes a CEO polytechnician of a major French consulting firm: "sociology is a far too serious matter to be left to sociologists, engineers should be allowed to deal with social sciences, they are better prepared to handle the statistics than sociologists".

Specialists in human sciences cannot completely master the mathematical and computational methods they use. The danger to society, however, results primarily from the scientific illusion created by mastering these methods, common among polytechnicians and other specialists of pure and applied sciences, and from the ease with which we use the new technologies, in contradiction with the critical mind, which need they increase paradoxically.

This danger is not new. It emerged in the era of eugenics, where important statisticians, including Pearson, were convinced of the merits [LEM 85]. We saw the consequences of the forced sterilization of tens of thousands of people in the US, Canada, Germany between the two world wars and, until 1976, in Sweden. By substituting the risk of collective error to a series of individual errors, a purely quantitative social technology applied collectively presents a danger, as currently shown in France by the pitiful state of the education system [FOU 04c] and the worrying evolution of the health system. Such an application arises from an egalitarianism that we can qualify as primary and that consists of mixing up equal rights and the identity of individuals, and then social injustice and statistical inequality. Statistical inequality is inevitable, however: there is thus apparent creation of social injustices and the intervention of the non-polytechnician sociologist is all the more essential to detect that the statistics are misleading those who unquestioningly rely on them.

This does not mean however that the new technologies do not bring any benefit to the sciences of man and society: the modeling in mathematical economics [ARM 05] or political philosophy [PIC 05] have strongly contributed to economic and social development since the 19th Century. In analyses on man itself, the main purpose of modeling is to submit the relationships between variables for reflection. These analyses, such as rational choice modeling [RAS 72], improve the understanding of intellectual mechanisms and offer relevant quantitative measures whose intelligence quotient is a typical example and whose interest is clear when they are used wisely. The description of a social fact by the empirical analysis of data suggests relations among variables or confirmation of them. It refines the approach that leads from an observed property to its reification. It would be unfortunate if modern technologies were not used to deepen reflection on man and society. The main obstacle is the ease with which they produce results and that for correct interpretation the critical mind must see things that are contradictory [FOU 05]. The ideas expressed above reflect the old quarrel of universals (Occam, Abelard, 14th and 15th Centuries) and the conflict between nominalism and realism that appeared in the 19th Century [DES 93]. They have many consequences in terms of political and social life in the context of the welfare state [FOU 07].

23.8. Bibliography

[ALA 32] ALAIN, *Propos sur l'Éducation*, p. 155, PUF, Paris, 1932.

[ARM 05] ARMATTE M., "La notion de modèle dans les sciences sociales : anciennes et nouvelles significations", *Math. & Sci. Hum./Mathematics and Social Sciences*, no. 172, p. 43-65, 2005.

[ARN 00] ARNSPERGER C., VAN PARIJS P., *Ethique Economique et Sociale*, La Découverte, Paris, 2000.

[BEN 73] BENZECRI J.P., *L'Analyse des Données*, Dunod, Paris, 1973.

[BER 00] BERLIÈRE J.M., "Dix questions sur la police", *L'Histoire*, no. 240, February 2000.

[BLA 00] BLANCHON Y.C., "Éloge d'une soupçonnée, réflexions sur la disparition de l'hystérie dans les classifications des troubles mentaux", *L'Information Psychiatrique*, vol. 76, no. 6, p. 641-645, 2000.

[BOU 01] BOUDON R., "Du bon usage des sondages en politique", *Commentaire*, no. 93, 2001.

[BOU 97] BOUDON R., "L'explication cognitiviste des croyances collectives", in R. Boudon, A. BOUVIER and F. CHAZEL (eds.), *Cognition et Sciences Sociales*, PUF, Paris, p. 19-54, 1997.

[BRA 00] LE BRAS H., "Les sciences sociales entre biologie et politique", *La Recherche*, November 2000.

[BRE 00] BRESSOUX P., Modélisation et évaluation des environnements et des pratiques d'enseignement, Thesis of the authorisation to conduct researches, Pierre Mendès University France, Grenoble, 2000.

[CAS 05] CASTEL P.-H., "L'expertise INSERM sur les psychothérapies: ses dangers, réels ou supposés", *L'information Psychiatrique*, vol. 81, no. 4, pp. 357-361, 2005.

[CHA 01] CHAUVEL L., "Le retour des classes sociales?", *Review of OFCE* no. 79, p. 315-359, 2001. (Available at: http://www.ofce.sciences-po.fr/pdf/revue/9-79.pdf, accessed February 12, 2010.)

[DES 93] DESROSIÈRES A., *La Politique des Grands Nombres*, La Découverte, Paris, 1993.

[DUB 94] DUBOIS M., *La Nouvelle Sociologie des Sciences*, PUF, Paris, p. 14-15, 1994.

[DUR 1897] DURKHEIM E., *Le Suicide,* PUF, Paris, (ed. 1990), 1897.

[EKE 00] EKELAND I., *Le Meilleur des Mondes Possibles*, Le Seuil, Paris, 2000.

[FOU 01] FOUCART T., "L'interprétation statistique", *Math. & Sci. Hum./Mathematics and Social Sciences*, no. 153, p. 21-28, 2001.

[FOU 02] FOUCART T., "L'argumentation statistique dans la politique sociale", *Math. & Sci. hum./Mathematics and Social Sciences*, no. 156, p. 33-42, 2002.

[FOU 04a] FOUCART T., "La bulle providentielle ", *Sociétal*, no. 46, p. 39-44, 2004.

[FOU 04b] FOUCART T., "Statistique et idéologies scientifiques", *Idées*, no. 138, 2004.

[FOU 04c] FOUCART T., *Scènes Ordinaires de la Vie Universitaire*, Fabert, Paris, 2004.

[FOU 05] FOUCART T., "Le paradoxe de l'introduction des TICE", *Médialog*, no. 53, 2005.

[FOU 06] FOUCART T., "La statistique et les problèmes d'éducation"*, Les Cahiers Rationalistes*, no. 581, p. 18-25, 2006.

[FOU 07] FOUCART T., *Le Despotisme Administratif, ou l'Utopie et la Mort*, The Manuscript, Paris, 2007.

[GAD 76] GADAMER H.G., *Vérité et Méthode*, Seuil, Paris, p. 104, 1976.

[GOU 97] GOULD S.J., *La Mal-mesure de l'Momme*, Odile Jacob, Paris, p. 383, 1997.

[GUI 01] GUILLEBAUD J.C., "Le cerveau et l'ordinateur, une comparaison abusive", *Esprit*, no. 8, p. 32-52, 2001.

[HAY 53] HAYEK F., *Scientisme et Sciences Sociales,* Plon, Paris, p. 77, 1953.

[HEN 03] HENRY O., "De la sociologie comme technologie sociale", *Proceedings of Research in Social Sciences*, no. 153, p. 48-64., 2003.

[JOU 99] JOUTARD P., THÉLOT C., *Réussir l'École, pour une Politique Éducative*, Seuil, Paris, 1999.

[KAH 00] KAHN A., ., *Et l'Homme dans Tout ça?*, Nil Editions, Paris, 2000.

[KRE 00] KREMER P., "La vie ordinaire des enfants de parents homosexuels", *Le Monde,* 28 October 2000.

[LAC 03] LACROIX X., "Homoparentalité, les dérives d'une argumentation", *Education*, vol. 399, no. 3, pp. 201-211, 2003.

[LEM 85] LEMAINE G., MATALON B., *Hommes Supérieurs, Hommes Inférieurs*, Armand Colin, Paris, 1985.

[LEW 01] LEWIS J., "Les femmes et le workfare de Tony Blair", *Esprit*, p. 174-186, March 2001.

[MAN 01] MANENT P., *Cours Familier de Philosophie Politique*, Gallimard, Paris, p.19, 2001.

[MIN 78] MINC A., NORA S., *L'Informatisation de la Société,* Seuil, Paris, 1978.

[MOA 93] MOATTI J.-P. *et al.*, "Evaluation économique et éthique médicale", *Nature-Sciences-Sociétés*, vol. 1, no. 4, p. 300-308, 1993.

[PIC 05] PICAVET E., L'intervention du raisonnement, no. 172, p. 43-65, 2005.

[POI 04] POITEVINAU J., "L'usage des tests statistiques par les chercheurs en psychologie: aspects normatif, descriptif et prescriptif", *Math. & Sci. Hum./Mathematics and Social Sciences*, no. 167, p. 5-25, 2004.

[RAS 72] RASHED R., "La mathématisation de l'informe dans la science sociale: la conduite de l'homme bernoullien", in G. CANGUILHEM (ed), *La Mathématisation des Doctrines Informes*, Hermann, Paris, p. 73-105, 1972.

[SCH 94] SCHNAPPER D., *La Communauté des Citoyens*, Gallimard, Paris, p. 57, 1994.

[TRO 05] TRONCIN T., Repetition: Le redoublement: radiographie d'une décision à la recherche de sa légitimité, Thesis in education sciences, University of Bourgogne, Dijon, 2005.

Chapter 24

The Internet in the Process of Data Collection and Dissemination

24.1. Introduction

What is the nature of the link between the Internet and management research? The use of the Internet by French researchers in management sciences is quite recent, both in terms of information gathering and spreading of knowledge. Indeed, this practice only appeared in a national methodology manual in 1999 [THI 99], while in the US the number of works (mainly in marketing) on these questions continued to increase [SHE 99].

These studies favored surveys via email and were not very interested in the answers of companies on the basis of questionnaires. It was around the beginning of 2000 that French research was proposed on the subject [ARA 00, GAL 00, GUE 00].

It appears from these various studies that the Internet facilitates the work of quantitative data gathering [COU 00]. However, rather than seeing this media as a simple substitute to more traditional collection methods (postal, telephone, etc), it is possible to consider the Internet as a new way to investigate and return the results by introducing useful specificities.

Chapter written by Gaël GUEGUEN and Saïd YAMI.

Thus, Vaast [VAA 03] notes the emergence of a stream of thought supposing that the design and implementation of research tends to be revolutionized by information and communication technologies (ICT).

	Face-to-face	Postal	Telephone	ICT
Quantitative	++	++	+	+++
Qualitative	+++	--	+	+
Major Advantages	– Length of the questionnaire – Compliance with the sequence of questions – Possibilities of sensory measures	– Extended geographical scope – Availability of respondent	– Speed – Compliance with the sequence of questions	– Sending cost Speed – Unlimited geographical scope – Multimedia Support
Major Problems	– Cost – Availability of the investigator	– No response – Delays – Control of the identity of the respondent	– Availability	– Control of the identity of the respondent – On-solicitation – The respondent must have access to ICT

Table 24.1. *Comparison of the different methods of data collection*

By repeating the syntheses admitted in the management field [EVR 97, THI 99], we can compare the different traditional methods of data collection with those using ICT both quantitatively and qualitatively (see Table 24.1).

In this chapter, we will reflect on the question of information collection in companies in particular, by considering the improvement of the quantitative research process in management sciences; but also on the diffusion of results. Indeed, because of our experience in this field[1], we think it is possible to use the Internet in a dynamic way on five occasions:

---

1 Since 1998, we have used this medium as a real support to our research in the study of corporate strategies.

– during the construction of the questionnaire, the Internet improves the pre-test phase;

– during the administration of the questionnaire, the Internet facilitates the process of obtaining data;

– during the diffusion of the results, the Internet contributes to restitution of the answers;

– the diffusion of the results renews how new data are obtained;

– the Internet links a community of researchers through the diffusion of research.

All these inputs are identified in Figure 24.1, which represents the stages of a survey by questionnaire[2].

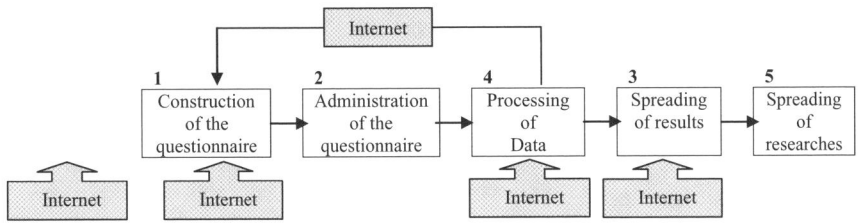

Figure 24.1. *The Internet in the various stages of investigations*

24.2. Construction of the questionnaire

During the construction phase of the questionnaire, the Internet can intervene during the pre-testing (the evaluation of the questionnaire before its final administration). The idea is to develop a measurement scale and the Internet will be used to increase its relevance. In other words, it is about mobilizing different Internet users to improve the validity of a measurement scale. We will describe, below, a method used during research in strategic management.

In our experiment, the proposed protocol considers the quality of some issues in the identification of a strategic situation faced by businesses[3]. In case tools allowing this type of situation to be measured do not exist, we have decided to create a

---

[2] Although aware of the contribution of the Internet in the processing and analysis of data [VAA 03], we will not deal directly with this question.

[3] The idea being to know whether companies operate within a particular network of other companies: an ecosystem of businesses.

specific measurement scale. Figure 24.2 shows the various stages of the protocol that we will detail:

– Creation of three hypothetical cases (scenarios) describing the situation studied exactly, a situation similar to that studied and a very different situation[4].

– Creation of a questionnaire common to the three hypothetical cases. Based on literature on the subject, 56 questions at interval scales allow us to identify the opinion of the respondent in terms of consistency with the scenario.

– Advertising on an Internet site devoted to research in management that volunteers are needed to help in the development of a measuring scale. The voluntary Internet user, then clicks on the link that takes him/her to the measure's collection page.

– The collection page of the measure consists of two parts. In the first part, a PHP[5] script will randomly select a scenario from the three possible cases to obtain answers without choosing the respondent. In the second part, the general questionnaire will be presented.

– The voluntary Internet user will then have to read a case (the scenario) and give his degree of approval with the different cases then submit his/her choice through a standard Internet questionnaire.

– Over a period of four months, a collection of the responses of voluntary Internet users is compiled. Their opinions are aggregated under the type of case to which they are submitted.

The aim of this protocol is therefore to be able to select questions that are most in tune with the situation sought and to eliminate redundant questions. We will thus use methods of scale refinement (Alpha by Cronbach, correlations, factor analysis, etc)[6] in a pre-test phase to ensure the validity and reliability of a measuring scale [EVR 97]. The Internet is therefore used to bring in a significant number of people to assist in the validation of a measuring scale without them having a link with the researcher trying to develop the measure. Some biases are thus avoided. Indeed, by bringing in people who are close, there is the risk of sharing of common knowledge, which can lead to the conditioning of certain responses.

---

4 For example, a scenario described the case of a company that maintained close relationships with other organizations.
5 PHP comes from the recursive acronym "PHP: Hypertext Preprocessor". It is an Internet programming language.
6 These are statistical techniques to ensure the coherence of different questions within a scale based on the answers obtained.

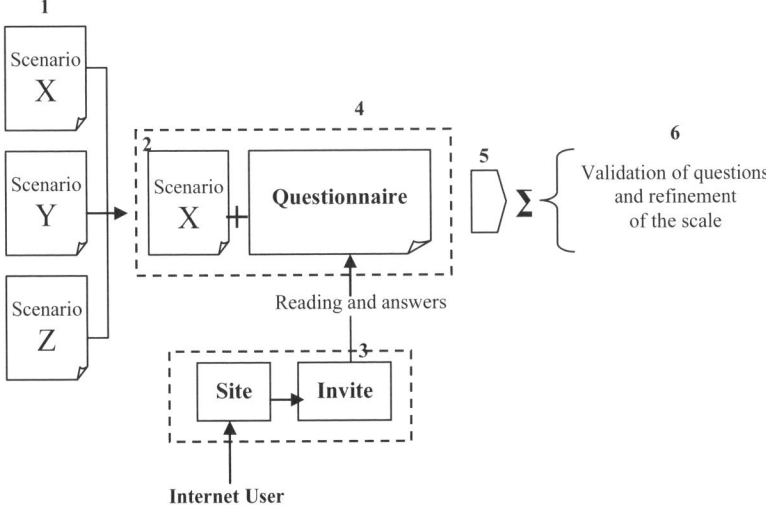

Figure 24.2. *Protocol of a pre-test of questionnaire through the Internet*

24.3. Administration of the questionnaire

Administration consists of assessing the feasibility of a quantitative survey in the opinion of respondents. The principle is to gather the answers through the Internet after initial contact using only this medium. The procedure can be achieved either by sending an email where the respondent resends the message after completing all of the fields of the questionnaire, or the archiving of the questionnaire on site from an HTML page. The idea is to replace the traditional postal questionnaire with an Internet-based questionnaire (Figure 24.3). It is this second method that we prefer in our reflection. This method of investigation (called "on site") is similar to terminal surveys and more particularly home surveys, as discussed in the literature dealing with market studies [EVR 97]. Jolibert [JOL 97] finds different advantages in these methods in terms of:

– costs and delays of survey implementation;

– monitoring of responses;

– recording response times; or

– the management of filter questions[7].

---

7 Questions to select the profile of respondents.

We should point out, however, that access to an Internet connection reduces the scope of these surveys for the individuals.

As for the relevance of the administration of questionnaires via the Internet, in terms of value of the answers, Schuldt and Totten [SCH 99] show that there are no significant differences compared to a face-to-face survey. With this analysis, we opted to use this method in our research work[8].

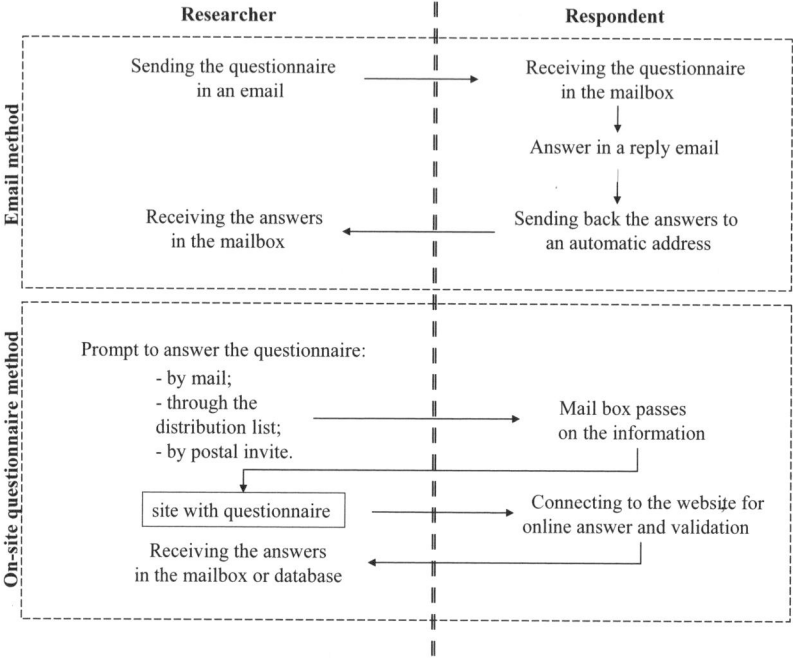

Figure 24.3. *Process of the administration of the questionnaire*

24.3.1. *Implementation of an Internet survey*

Our goal was to (digitally) contact managers of SMEs. We wanted to get two samples, on the basis of belonging to the sector of Internet activity, in order to conduct an investigation about the strategies used.

---

8 We have explored this collection method of data in particular in the Gueguen's doctoral research "Environment and strategic management of SMEs: the case of the Internet sector" [GUE 01].

24.3.1.1. *Sample composition problem*[9]

After building and launching our questionnaire online, we conducted our survey between October and December 1999, among 2,693 addresses. Of these, only 2,139 addresses were valid. We chose the questionnaire method but we randomly selected 100 business addresses to test the response rate of the email method. The general response rate obtained is 14.12%.

To identify the site visits, we divided the transmission of messages into several successive waves, boosted over a period of three months. The message calls on the company head to complete the questionnaire by indicating in the text body itself the address of the questionnaire site. When the sector of activity was clearly identified, we indicated so in the message to increase the degree of customization of the prompt.

24.3.1.2. *Methodological and epistemological implications*

The main results obtained for our survey show that the method to obtain the best response rate is that of the on-site questionnaire (see Table 24.2). This response rate is slightly higher for businesses closely linked to the Internet but not significantly. The collection method that we have used was between the postal method and traditional computer method (computer assisted personal interview). The answers were produced at the company, at the time desired by the respondent, who had no interaction with the researcher (since the questionnaire was auto-administered), but they were automatically controlled by a program facilitating statistical processing.

Epistemologically, and in the present case, the initial knowledge of surveys is very low, or absent. The studied object is only truly observed through aspects analyzed (answers to survey questions). Thus, to quote Le Moigne [LEM 95], a constructivist approach would be limited by this type of survey. This alienation from the field, while enhancing the positivist aspect of the method, allows the comparison of the researcher and respondent. Indeed, the simplicity and speed of email use allows an easier interaction. This is how we could explain 5% of total non-responses (some surveys spontaneously address the reasons for their refusal to answer). It should be noted that this was essentially either an overly academic perception of research, or a lack of confidence in the approach.

---

9 We mainly focus our discussions on the specificities of email addresses in order to consider the practical consequences for future research.

	Site	Mail	Total
Responses	24 *(20%)*	4 *(7%)*	28
Non-responses	98 *(80%)*	57 *(93%)*	155
Total	122 *(100%)*	61 *(100%)*	183

$$Khi2 = 5.40 > \chi^2(p = 0.05)$$

Table 24.2. *Number of responses according to the administration method*

24.3.2. *Reflections on the relevance of these methods and extensions*

It now seems appropriate to empirically compare the different traditional quantitative survey methods (postal or telephone) with this methodology, emphasizing the quality of answers to identify possible biases. Several works are researching this, for example Fricker *et al.* [FRI 05] are looking at the differences between the methodologies of administration by phone and Internet. Moreover, although this technique was pioneered in 1999, it is widely spread today; both at research level ([LAV 01] or [EMI 03]) and at simple commercial level. Thus, the Internet user is subject to an increase in this type of solicitation and he/she is less inclined to reply favorably. It appears that the current response rate of Internet surveys is lower than originally found.

Moreover, the latest versions of statistical software packages can easily build on-site questionnaires. Indeed, it is always possible to refine solicitation techniques – Gueguen [GUE 03] notes for example that the similarity of the first name between solicitor and the solicited favors the response rate – particularly when applying the editorial techniques of postal surveys to Internet surveys [POR 03].

We believe that this type of methodology has a bright future. Some see it as the standard data collection method of the 21st Century for researchers [SCH 94], and a method that may open new opportunities for quantitative studies. The method indeed allows the management of the questionnaire (by analyzing the response rate), a use of complementary media (sound, video, etc.), a definition of terms used, a control of unanswered questions [GUE 00, GUE 01] or use of visual elements (in favor of the response rate) [COU 04].

Beyond a reflection based mainly on the problems of collecting and processing data, some further experiments have led us to consider the use of the Internet compared to the dynamic couple (information gathering-dissemination of research results). This idea refers primarily to the form and customization of the release of results and, then, to the development of an automated specific method of analysis by comparing answers in order to recover the data and to perform analysis on a case-by-case basis. The interactivity from this data collection mode is likely to motivate and involve the respondent a bit more. We will therefore describe this procedure.

24.4. Dissemination of results

Use of the Internet by the researcher in the administration of his/her investigations can go beyond the simple collection of data. Here we will present a possible outcome from an experiment conducted on the study of the strategic behavior of companies, while questioning ourselves first on the dissemination phase.

24.4.1. Difficulty of disseminating research results

The Internet allows the dynamic broadcast of elements on the research carried out. Indeed, often the question of releasing results is raised. In most cases, the researcher produces a streamlined and simplified outline of the work conducted. This document made, it spreads among participants, accompanied in the best case by various sentences of thanks.

However, what is the scope of such broadcast? The respondent, if the form was more appealing, would be more likely to read the proposed restitution and thus could benefit from a view different from his/hers on the problems concerning him/her. The researcher, then, could have access to comments from respondents and would thus be able to shed a new light on his/her work. Why therefore not use the Internet to make this return of results more attractive?

It is to this end that we built an automated module of strategic analysis of environmental dimensions (that is, the analysis of the business context) operating entirely though the Internet and based on a diagnosis specific to the responding enterprise[10]. This technique allows a dissemination of results in relation to the respondent's situation.

24.4.2. Proposition of a solution for dissemination of results through the Internet: the development of an automated specific analysis

The process of the automated specific analysis (ASA) consists of comparing the answers of a respondent to those already available (during a preliminary work, as it

---

10 Based on answers from different enterprises, it is possible to describe a sector of activities and to thus determine which strategy seems more relevant within this sector of activities. This can be achieved from the strategic diagnosis carried out.

is a theoretical situation). The situation of the responding enterprise will thus be considered in light of the theoretical situation.

In technical terms, a program is used to retrieve data from the responding enterprise regarding, for example, its level of uncertainty and make comparisons on a case-by-case basis[11].

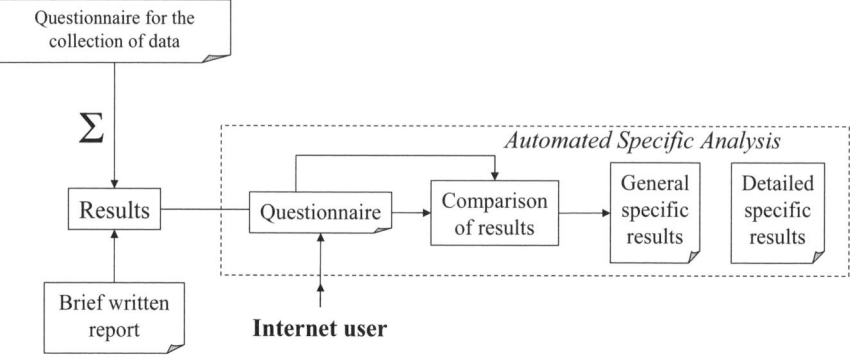

Figure 24.4. *Automated specific analysis*

The respondent is first invited to answer a series of questions drawn directly from a questionnaire used in previous research[12]. Once the various answers are obtained, the comparison process begins. Then, the person is informed of the exact nature of the measure carried out. This is to test a set of variables operated in the form of questions used classically in the literature of strategic management. Finally, the results are delivered in two stages:

– a raw analysis, summarizing the main answers and scores obtained in terms of refined measurement scales built previously. For example, "what is the degree of uncertainty that affects an enterprise?";

– a deeper analysis, allowing us to refine the understanding of the concepts used and to relativize them. For example, "what is the most relevant strategy in terms of the level of uncertainty?".

---

11 For the characteristics of the environment (considered as the set of external factors affecting the enterprise), four dimensions were measured in our example: complexity, uncertainty, dynamism and turbulence. A synthesis dimension was added: the intensity of the environment.

12 In this case, our doctoral work previously mentioned on the relation of SMEs to the environment, in strategic terms.

First of all, we situate the answers obtained against the answers previously collected. For each concept measured (uncertainty, complexity, type of strategy, etc.), a brief explanation (about 15 pages) is provided. A more detailed explanation is then provided directly from previous research. With these elements, it is possible for us to make a comparison of the respondent's situation on the basis of previous answers. Different graphs representing the position of the enterprise then allow us to summarize the discussion and therefore provide an immediate vision of the respondent's real situation. A summary of the environment/strategy relationship is also proposed with a hypothesis issued on the level of performance achieved.

Let us recall that the different analyses proposed are specific to the respondent's answers. The structuring of results, however, means that the respondent has access at to an explanation using the main theories in strategic management and to the scientific analysis the same time as the researcher. In addition, we attach an evaluation to this diagnosis questionnaire allowing the user to give his/her opinion on the proposed analysis. We thus wish to facilitate the feedback between expertise and management practice.

24.4.3. *Limitations and expectations of the process*

The first limitation is based on the ultra-positivism of the method. We reduce a complex reality (the strategic situation of the enterprise in relation to its environment) to a highly simplified model (the matching relationships among the different types of answer). However, we should remember that the main purpose of the method is to allow dynamic spreading of the results and that we voluntarily accentuate certain features of the research for convenience. We also record data using a mixed approach combining simplicity and complexity. The simplicity corresponds to the diagnosis proposed on the basis of some variables. The complexity to the refinement of proposals that the reader can find in the attached documents (added to deepen knowledge of the concepts used). A final limitation is the risks associated with a too elliptical interpretation of the proposed synthesis.

It must be noted that this method is applicable to different fields of management sciences, although we have limited the use to the field of strategic management here. Moreover, this practice can also be of interest to the dissemination of materials initially collected by qualitative methods (interviews). The researcher then needs to find proposals allowing the answers obtained through the Internet to be compared with the results obtained previously. Furthermore, this type of process is hampered by various technical constraints for the researcher related to the mastery of PHP and HTML. Here again, solutions appear gradually, such as increasingly high-performance and easy to use HTML editors or training sessions in Internet use, etc.

The simplest solution, however, is to collaborate with the IT services of the institution to which the researcher is linked.

By using the principle of ASA, we can improve the response rate to the questionnaire and thereby enhance its attractiveness. The idea is to arouse the direct interest of respondents (particularly enterprises). For example, a researcher working on strategic dimensions may propose instantly processing the respondent's answers under the principle of ASA. He/she may also plan a comparison of the answers provided by the respondent, not in relation to the synthesis of previous answers (which is impossible because we are in the collection phase) but in relation, either to similar research (we do not fully approve of this, however, as a transposition could turn out to be approximate), or an analysis and synthesis of the scientific literature on the topic. Thus the collected answers and the broad principles traditionally set out can be compared. The detected inferences would allow us to propose an analysis of the situation of the responding enterprise or of useful advice. Thus, the respondent will obtain an immediate analysis of some of his/her answers while the researcher can retrieve these same answers.

Through this method, the respective contribution is high. This operation, however, requires a special effort to simplify and restrict the object from the influence of the researcher.

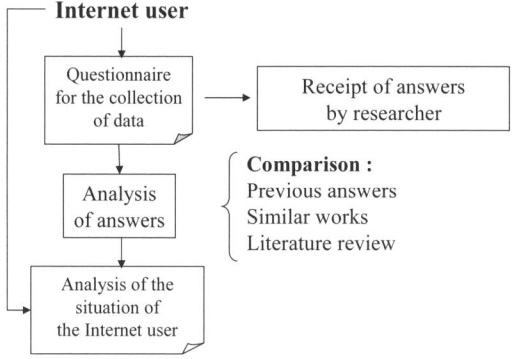

Figure 24.5. *A synchronous analysis of answers increasing the attractiveness for the respondent*

24.4.4. *Broadcast of research: "Internetized research area"*

Vaast [VAA 03] noticed that the use of email may allow collaboration among researchers and that the Internet can support discussion related to research in management and that ICT can enhance research through the specific websites of researchers or research associations. We therefore understand that the Internet can also extend research and go beyond the dissemination of its results through:

– its ability to touch, *a priori*, a large audience;

– control of costs (absence of printing or postage costs, etc.) during the broadcast of research;

– a strong responsiveness to research needs (update, upgrade);

– adaptability to the wishes of researchers (thematics, design);

– the possibility to make research, practitioners and participants of the economic life who are geographically distant, interact;

– appropriation of the media by the researcher (personal website, associative website, etc.).

These various specificities seem to lead the Internet towards the development of research communities around specific themes. Thus, the idea is to foster the development of "Internetized research areas" (IRAs), not on the basis of an individual or association, but by focusing on a specific theme of research.

For some time, we conducted strategic reflection work on the concept of a "business ecosystem", a concept developed by Moore [MOO 93, MOO 96] that corresponds to a formal and informal grouping of enterprises on the basis of the development of a shared standard. In order to foster a particular dynamic to the study of this concept, we developed an IRA with the URL: www.ecosystemedaffaires.net. On this space, solely dedicated to business ecosystems, it is possible to find:

– a presentation of the concept;

– examples from the news;

– illustrations;

– bibliographical references (with for some, direct access to the resource);

– a discussion forum, educational material (Power Point animation directly usable for a course presenting business ecosystems);

– etc.

The will that drove us was the wish to promote the theorization based on business ecosystems (in addition to published works) by using the resources that the Internet has to unite researchers, practitioners and students. We wanted access to these resources to be as free as possible and transnational. Obviously, this process raises some difficulties. We note that research published in a refereed journal is of particular interest, however, this process frees itself from such evaluation. Furthermore, it is something to be present on the Web, to be known (and visited) by Internet users.

24.5. Conclusion

Our discussion focused on quantitative approaches by trying to present techniques improving the response rate to questionnaires and the dissemination of results. But what about qualitative approaches? Can we believe that a non-directional interview can easily be replaced by interviews with Webcams, or that a discussion on an Internet forum can enable specific attitudes to be highlighted? To address these questions, Vaast [VAA 03] pondered on the difference between wealth and poverty of the media used in conducting interviews as permitted by communication theory. It is possible to think that here, again, the Internet should not be seen as a substitute or as a complement to traditional communication techniques but as a new technique in data collection.

Igalens and Benraïss [IGA 04] have also addressed the issue of qualitative studies through the Internet and seem to see many sources of interest, so much so that several precautions (ethics, synchronicity, role of the researcher, etc.) are observed upstream. Thus, the advantage of the Internet in research appears to be conclusive and it is necessary to reflect on its users and limitations. Several works moreover tend to experiment with different techniques to improve the quality of answers for this type of investigation [COU 01, OCU 04]. In this context, we can advance the idea that a "labeling" of scientific investigations through the Internet would lead to the respondent having increased confidence. It would be a level of belonging to a scientific association dedicated to Internet methodologies guaranteeing the anonymous use of data in a scientific framework. Researchers using this media could then use this label to increase the positive perception of their respondents and the legitimacy of their scientific status.

24.6. Bibliography

[ARA 00] ARAGON Y., BERTRAND S., CABANEL M., LE GRAND H., "Méthode d'enquêtes par Internet: leçons de quelques expériences", *Décision Marketing*, n°19, p. 29-37, 2000.

[COU 00] COUPER M.P., " Web survey: a review of issues and approaches", *Public Opinion Quarterly*, vol. 64, no.4, p. 464-494, 2000.

[COU 01] COUPER M.P., TRAUGOTT M.W., LAMIAS M.J., "Web survey design and administration", *Public Opinion Quarterly*, vol. 65, no. 2, p. 230-253, 2001.

[COU 04] COUPER M.P., TOURANGEAU R., KENYON K., "Picture this! Exploring visual effects in Web surveys", *Public Opinion Quarterly*, vol. 68, no. 2, p. 255-266, 2004.

[EMI 03] EMIN S., L'intention de créer une entreprise des chercheurs publics: le cas français, Doctorate Thesis in Management Sciences, Pierre Mendès University, France, Grenoble II, 2003.

[EVR 97] EVRARD Y., PRAS B., ROUX E., *Market. Etudes et Recherches en Marketing*, Nathan, Paris, 1997.

[FRI 05] FRICKER S., GALESIC M., TOURANGEAU R., YAN T., "An experimental comparison of Web and telephone surveys", *Public Opinion Quarterly*, vol. 69, no. 3, p. 370-392, 2005.

[GAL 00] GALAN J.P., VERNETTE E., "Vers une quatrième génération: les études de marché "on-line'"", *Décision Marketing*, no. 19, p.39-52, 2000.

[GUE 00] GUEGUEN G., "L'administration des enquêtes par Internet", 9th *International AIMS Conference*, Montpellier, May 24-26, 2000.

[GUE 01] GUEGUEN G., Environnement et Management stratégique des PME: le cas du secteur Internet, PhD Thesis, Montpellier I University, 2001.

[GUE 02] GUEGUEN N., JACOB C., "Solicitation by e-mail and solicitor's status: A field study of social influence on the Web", *Cyberpsychology and Behavior*, vol. 5, no. 4, p. 377-383, 2002.

[GUE 03] GUEGUEN N., "Help on the Web: the effect of the same first name between the sender and the receptor in a request made by e-mail", *The Psychological Record*, vol. 53, p. 459-466, 2003.

[IGA 04] IGALENS J., BENRAÏSS L., "Les e-recherches qualitatives", *Traversée des Frontières Entre Méthodes de Recherche Qualitatives et Quantitatives*, Proceedings of the 1st international conference ISEOR and division "research methods" of the Academy of Management (USA), Lyon, p. 1397-1414, March 18-20, 2004.

[JOL 97] JOLIBERT A., "Etudes de marché", in Simon Y. and Joffre P. (eds.), *Encyclopédie de Gestion*, 2nd edition, Economica, Paris, p. 1257-1290, 1997.

[LAV 01] LAVASTRE O., Coûts de mobilité du client vis-à-vis du fournisseur, Doctorate Thesis in Management Sciences, Montpellier I University, 2001.

[LEM 95] Le MOIGNE J.L., *Les Épistémologies Constructivistes*, "Que sais-je?" Collection, PUF, Paris, 1995.

[MOO 93] MOORE J.F., "Predators and prey: a new ecology of competition", *Harvard Business Review*, vol. 71, no. 3, p. 75-86, May-June 1993.

[MOO 96] MOORE J.F., *The Death of Competition – Leadership and Strategy in the Age of Business Ecosystems*, John Wiley and Sons, Chichester, 1996.

[POR 03] PORTER S.R., WHITCOMB M.E., "The impact of contact type on Web survey response rates", *Public Opinion Quarterly*, vol. 67, no. 4, p. 579-588, 2003.

[SCH 94] SCHULDT B.A., TOTTEN J.W., "Electronic mail vs. mail survey response rates", *Marketing Research*, vol. 6, no. 1, p. 36-39, 1994.

[SCH 99] SCHAAPER J., *La Qualité de l'Information dans une Enquête par Sondage Menée sur Internet*, WP. Center for Research in Management, no. 171, IAE of Poitiers, 1999.

[SHE 99] SHEEHAN K.B., Mc MILLAN S.J., "Response variation in e-mail surveys: an exploration", *Journal of Advertising Research*, vol. 39, no. 4, pp. 45-54, 1999.

[THI 99] THIETART R.A. *et al.*, *Méthodes de Recherche en Management*, Dunod, Paris, 1999.

[VAA 03] VAAST E., "Recherche en gestion avec TIC et recherche sur la gestion avec TIC", *French Review of Management*, vol. 29, no. 146, p. 43-58, 2003.

Conclusion

Take things for what they are: concretions that your mind produces from sensations, by comparing forms of languages most of the time. These things do not reify them more than needed, otherwise they will reify you in return, depriving you of your subjective freedom.

Tchouang-Tseu, in J.-F. Billeter, *Leçons sur Tchouang-Tseu*, Allia, Paris, p. 108, 2002.

Above all, do not conclude this book. Any form of closure, synthesis or pathway (Chapter 4) would be a return to the classic book, "root book" or its modern version "rootlet book" that Deleuze and Guattari denounced in *Mille Plateaux* [DELE 80][1] as being binary and of strong main unity, "the book as a picture of the world, anyway insipid". "The manifold, *one has to do it*, with sobriety", points out that, for those interested in defending the rhizome book, one of the main features "is to be of multiple entries" [BRO 03][2]. As an invitation to the diversity of readings and to a thought on the movement, *Digital Cognitive Technologies* provides, if only through its index, an idea of the wealth of possible routes. Indeed, the authors quoted above have put computer science in a classical book [DELE 80][3] and recognized in the detour of a sentence that they have failed in their undertaking[4].

Conclusion written by Bernard REBER and Claire BROSSAUD.
1 p. 11-13 and p. 20.
2 Another example of a book with multiple entries that has also inspired this interdisciplinary book, is Claire Brossaud's [BRO 03] PhD thesis.
3 p. 12.
4 *Ibid.*, p. 33: "For the manifold, there must be a method which effectively does it; no typographical trick, no lexical skill, combination or creation of words, no syntactical boldness can replace it. [...] We failed to do so for our purpose". (This and other quotes are a translation from French).

Some of the new technical characteristics presented in this book, such as the various forms of networking (Part III) and the distributed systems of knowledge development (Part VI), just to take these two examples, could question their restrictive classification of computer science. Deleuze and Guattari might even be able to sustain their initial project of the escape of "a hidden unit", at the risk of increasing the number of authors[5], since each could contribute to the project with different forms of writing and different media.

Hence, computer science has changed, especially with the introduction of another unique characteristic of the novelty of these cognitive and social technologies: hypertext. As stressed by several contributors, this concept is more than a metaphor of complexity, since it is also a reading and writing tool that enables the integration of some of its conceptual potentialities (Chapter 10). Indeed, the disorientation caused by ICT can be compared to the concept of noise in the paradigm of complexity. However, it can generate reorganization on a higher level, or rely on tags to support reading (Part II). Similarly, the notion of a system is where the reader, the device and the machine interact to produce knowledge (Part I). It is mainly, however, the opportunity offered to the reader to take another route via hyperlinks, which resembles a property of chaotic systems.

Deleuze and Guattari claimed that their plateaus[6], at the limit of writing could be read independently of one another. Yet they claimed that the conclusion of *Capitalism and Schizophrenia 2* must be read in the end[7]. This is not the case with our conclusion, which can be read at any time, as well as the various chapters making up this book, that can be approached according to the interests of the readers. We suggest nevertheless five thematic "navigations" or "tracks" related to different relationships established between information and communication technologies (ICT), ICT sciences (ICTS) and human and social sciences (HSS). We shall therefore present them non-hierarchically below.

---

5 Deleuze and Guattari refer to only one success, "a truly nomadic book", *Absolument Nécessaire*, by Joëlle de La Casinière [DELA 73]. However, this author, who was juxtaposing photos, calligraphies, typos, graffitis and drawings in her writings, invented an "electronic scriptorium" in 1995. See http://www.cipmarseille.com/auteur_fiche.php?id=1777 (accessed on February 12, 2010).

6 "What happens on the contrary for a book made of plateaus, communicating with one another, through micro-fissures, like a brain? We call "plateau" any multiplicity connectable with others by superficial underground branches, so as to form and extend a rhizome" [DELE 80], p. 33.

7 *Ibid.*, p. 8.

Conclusion 391

1. Of course the different contributions to this book deliver a precise presentation of ICT, in particular the Internet[8] and its measurement. They also analyze their *uses* and *impacts* in various spheres, ranging from politics (Part VII) to the very intimate and daily management of time, through the reorganization of enterprises (Chapters 2 and 3; Chapters 16 and 22), or through the development, status and dissemination of the learned document in public areas, yielding a new organization of knowledge. From Part IV, *Digital Cognitive Technologies* also studies and documents questions related to new cognitive technologies that cover several interacting social spheres (politics, everyday life, civil society, open source community, economy, HSS, natural sciences, etc.). With ICT, we are not only dealing with ICT but new types of networks of knowledge and distributed cognition.

2. Far from only giving an exclusively ancillary role to study the uses of to *HSS* technologies once established, or to help build their social acceptability, this treatise shows, on the contrary the place that these sciences can take *upstream of computer work*. Cases of cooperation (Part VI) or socio-informatics (Part VIII) illustrate this second position in an illuminating manner. We can even sometimes wonder whether technology or the human mind is the more stable. In any case, multidisciplinary editorial reflections precede the generation of ICT, including political dimensions, as if to better "equip" civil society (Chapters 17 and 20), helping dialog (Chapter 21), but also for more traditional political acts such as distance voting (Chapter 19). In these different contexts, HSS are not content with just adapting to the techniques imposed (Chapter 22).

3. Answers to certain questions, such as the measurement of the Internet (Chapter 7), the qualification of cyberspace (Chapters 5 and 6) or the possibility of an automatic reasoning (Chapter 13) require *methodological and theoretical choices* that are made within the ICTS disciplines to be explained. These concerns are as much issues of architectures and ontologies as the way to select, within HSS, elements from which the computer scientists can be inspired (Chapter 18). As for HSS, the wide disciplinary choice presented here yields two types of results. On the one hand, this diversity of disciplines illuminates the concept chosen for each part of the treatise with nuances characteristic of relevance frameworks and of the methods of each of them. On the other hand, several articles use different methodological (Chapters 9, 11and 24) and possible theoretical choices within each discipline (Chapters 15 and 23), up to exploring the requirement of translations or inventions of

---

8 And many other techniques or concepts more or less associated in the index, in particular semantic or socio-semantic Web, wikis, digitized documents for action (DFA), discussion threads, integrated development environments (IDEs), open archives, digitized learned document, online journals, Web 2.0, numerous types of software (for sociology, history, anthropology or management), blogs, crawlers, groupware (of production or consultation), platforms and numerous protocols (XML, for example).

interlanguages among themselves (Chapter 14). This is largely the case with sociology[9], in its more qualitative or quantitative versions, with the limitations that this distinction involves. The earliest discussion of methodological and epistemological pluralism finds a new force with the potential of ICT available to researchers in social sciences, assuming that they seize them. Here again, the practices are sometimes more rigid than the technologies. Furthermore, the intradisciplinary choices made in ICTS or in HSS do not exclude disciplinary hybridizations, such as anthropology and neurosciences (Chapter 12) or history, knowledge engineering and semiotics (Chapter 1).

4. In several research results presented here, the *encounter* between *ICT and HSS* appears to revolutionize a field. This is the case with anthropology in the Oceanian environment (Chapter 12). Here hypertext offers an original access possibility for people without writing, going as far as creating techniques dreamed up by the ethnologist and pre-historian Leroi-Gourhan, who envisaged the release of the hand enslaved by linear writing and mold of thought [LER 64, LER 65]. He called for greater mobility in the systems aimed at fixing thought in material supports, as a significant trait of human and social evolution [LER 64][10]. A similar observation is also made with history (Chapter 1), which is drastically changed by the digital document and ICTS, from the observation, critical analysis and formulation of hypotheses through to the publication of results. The whole methodological chain of historiography is affected. The forms of the traditional document seem to be ruined by the digital document, the latter having specific forms that require the renewal of methods for the exploitation of historic sources. Similarly, the properties of traceability of ICT have an influence upon the forms of reading and writing of history.

While some authors place themselves in the continuity of a history of writing or literary techniques of text processing and exegesis, for example, to insist on the fact that research operations are already old, they furthermore show what is made possible with ICT. This is the case with sharing interpretations or triangulating methods, making a hermeneutic pluralism effective (Chapters 11, 14 and 15). Several of them give special care to making some choices on the specific side of HSS explicit, and others that fall within software (functionalities, algorithms).

5. A final route on which we will linger is that of the *relationships between tools and concepts*, which are found as the two polarities in the title of this book. First, readers familiar with HSS may be surprised to see the *concepts* they are used to taken back by researchers in *ICTS*. The latter speak about the computer science of introspection, ontologies, intercession (Chapters 13, 16 and 18) or networks

---

9 For a reflective approach unique to this discipline, see also [DEM 06].
10 p. 41.

(Chapter 8). We can measure the variation due to these borrowings, particularly through comparison with their uses in the field of HSS (Chapter 3). We will also understand the risks of such borrowings for the evolution of our societies (Chapters 2 and23). Secondly, and in the opposite way, to express it in the words of one of the pioneers of research in computer science for HSS, Gardin: "far from sparing us any research on the representation of knowledge in our particular fields, [the new technologies] impose it strongly instead and force us to give accurate, practical, operating answers, to questions that we have been accustomed to rather address from the less sharp angle of philosophy." [GAR 94][11]

In order to understand this lack of keenness, could we not say that thinking tools, "objects invested by the mind" according to Husserl, have a debt to the *origin of our schemes of thoughts* and *concepts*? Kant himself acknowledged this in *What does Moving in the Thought Mean?*: "However high we place our concepts and to whichever degree we thus leave aside the sensitivity, *concrete representations* are always attached to them." [KAN 91][12]

However, among these concrete representations, technical objects, ranging from the simplest to the most complex, are often at work. Two thinkers on technology, Dewey and Simondon[13], one a pragmatist and the other a phenomenologist, were able to take advantage of this debt-concerning technique, moved by a philosophy driven by the question of temporality, and thus the future of techniques[14]. We could say that from the first surveyors Husserl mentions in *The Origin of Geometry* [HUS 62] to contemporary questions of digital archiving, not only techniques participate in the temporal conscience of man and extend his memory, but the technical objects, concrete arrangements of structures and functions appear in a process of invention (scheme, materialization, realization) according to the lines remarkably described by Simondon [SIM 89a][15].

Dewey is known for having taken the importance of technologies seriously in his broad theory of investigation, not content with the great division between natural sciences and HSS [DEW 67]. He wrote: "the mind and specialized knowledge of the past are embodied in instruments, utensils, systems and technologies." [DEW 03]

More radically, he developed the concept of *instrumentalism*, for want of anything better, he admitted, whose origin, in his own words would be in its turmoil with the "intellectual scandal (which seems) involved in dualism [...] between

---

11 p. 30.
12 p. 55.
13 Authors furthermore often quoted in *Mille Plateaxs* [DELE 80].
14 For the relations between time and technique, see [REB 99].
15 See in particular the first two parts of this book.

something called 'science' on one side and something called 'moral' on the other". Dewey is still dwelled on, however, by the feeling that an effective method of (logical) inquiry that would apply without brutal division between these two fields at the same time would be the solution to our theoretical needs and the resource for "our greatest practical demands" [DEW 83]. A bit like Bacon, the real founder of modern thought [DEW 03b][16], according to him, this logic of discovery is not the one interested in finding an adequacy between perceptions and reality, but in a search for intellectual tools to resolve the deadlocks of a problematic experiment. In support of his argument, he mentioned the problem of distance communication and the inventions of the telegraph and telephone: "The ideal is achieved by its own use as a tool or a method of examination, experimentation, selection and combination of natural operations."[17]

Challenges and their solutions discussed in this treatise go well beyond that of distance communication with, for example, the validation of research through the intervention of a new support (Chapter 12), hybridization of interpretations (Part V) or the sharing and evolution of knowledge (Part VI).

For his part, Simondon, in search of "ethics coming out of techniques" [SIM 89a][18], became interested in concepts of information, potential and form, to find the starting point of an axiomatization of humanities and social sciences: "There is a certain relation between a study of the technical object and the problem presented here, namely: *Form, Information and Potential.*" [SIM 89b][19]

His starting problem is that of ontogenesis and, more particularly, that of individuation. However, generally, such a research perspective grants an ontological privilege to the constituted individual, with the risk of not "operating a real ontogenesis, of not putting the individual back in the system of reality in which individuation occurs"[20]. He begins with the review of the form, the most ancient concept that has been defined by philosophers interested in the study of human problems. If the designs of forms have evolved, their ancient versions remain outstanding. Simondon highlights the way "psychologists and sociologists of Antiquity have addressed the processes of interaction and influence"[21], by using this concept of form. He convincingly shows how philosophers take their models from technical schemes. He begins his review with the Platonic archetypical form, which

---

16 p. 55-70.
17 [DEW 03b], p. 112.
18 p. 177.
19 p. 33.
20 *Ibid.*, p. 10.
21 *Ibid.*, p. 37.

considers the individual based on him/herself in a substantialist way, resistant to what is not him/herself.

For Plato, an advocate of a vertical model of interaction, the technique evoked in his conception is that of the hallmark or wedge with which coins are stamped. He could explain, for example, the relations of idea to sensitivity, such as the wedge to the coins. Simondon covers all the philosophical traditions and that of HSS with their representations of the three concepts of form, information and potential in the same way. He thus goes on until the theories of information and cybernetics of Wiener. From a technical point of view, Simondon refers to the scheme of the network: "[…] the reticular structures of the integrated techniques are not only means available for an action and abstractly transferable anywhere, usable at any time; we change tools and instruments, we can build or repair our own tool, but we do not change network, we do not construct a network ourselves: we can only connect to the network, adapt to it, participate to it; the network dominates and encloses the action of the individual being, even dominates each technical set." [SIM 89a][22]

Reading this treatise sometimes allows us to forsee the relevance of such insights in the way in which technical models inspire concepts for HSS and *vice versa* for ICTS.

Certainly, this book, the fruit of a collective work, is limited in time. Numerous questions remain open. Research on cooperation in online communication networks, for example, are still in a largely exploratory phase (Chapter 17) and their development is still driven by a better conceptual integration between social sciences and cognitive sciences. In terms of social sciences, the progress of research is dependent, for example, on demanding exchanges between the analysis of social networks (Chapter 9) and knowledge economy. The crossing between the analysis of social networks and analysis of epistemic networks will also be a significant issue in this context [ROT 05]. For all the questions addressed here, many methodological and conceptual problems must be clarified. They will not continue to develop without fruitful links between ICTS and HSS.

Bibliography

[BRO 03] BROSSAUD C., *Identification d'un "Champ" autour d'une ville: Le Vaudreuil Ville Nouvelle (Val de Reuil) et son "maginaire Bâtisseur"*, L'Harmattan, Paris, 2003.

[DELA 73] DE LA CASINIERE J., *Absolument Nécessaire*, Editions de Minuit, Paris, 1973.

---

22 p. 220-221. For development, see [REB 07].

[DELE 80] DELEUZE G., GUATTARI F., *Capitalisme et Schizophrénie 2. Mille Plateaux*, Editions de Minuit, Paris, 1980.

[DEM 06] DEMAZIERE D., BROSSAUD C., TRABAL P., VAN METER K., *Les Logiciels d'Analyse Textuelle en Actions – Usages, Résultats, Productions dans une Perspective Sociologique Comparative*, PUR, Rennes, 2006.

[DEW 67] DEWEY J., *Logique. La théorie de l'Enquête (1937)*, G. DELEDALLE (ed.), PUF, Paris, 1967.

[DEW 83] DEWEY J., "From absolutism to experimentalism", *Latter Works (1925-1953)*, Southern Illinois University Press, Carbondale, p. 156-157, 1983.

[DEW 03a] DEWEY J., *Le Public et ses Problèmes (1927)*, J. ZASK (ed.), Publications of the University of Pau Farrago/Léo Scheer, Pau, p. 199, 2003.

[DEW 03b] DEWEY J., *Reconstruction en Philosophie (1920)*, Publications of the University of Pau Farrago/Léo Scheer, Pau, 2003.

[GAR 94] GARDIN J.-C., "Informatique et progrès dans les sciences de l'homme", *Informatique et Statistique dans les Sciences Humaines Review*, vol. 30, no. 1-4, pp. 11-35, 1994.

[HUS 62] HUSSERL E., *L'Origine de la Géométrie (1954)*, introduction and translation by J. DERRIDA, PUF, Paris, 1962.

[KAN 91] KANT E., "Que signifie s'orienter dans la pensée" (1786), in *Vers la paix Perpétuelle. Que Signifie s'Orienter dans la Pensée? Qu'est-ce que les Lumières? and Other Texts*, introduction and translation by F. PROUST and J.-F. POIRIER, Flammarion, Paris, 1991.

[LER 64] LEROI-GOURHAN A., *Le Geste et la Parole I. Technique et Langage*, Albin Michel, Paris, 1964.

[LER 65] LEROI-GOURHAN A., *Le Geste et la Parole II. La Mémoire et les Rythmes*, Albin Michel, Paris, 1965.

[REB 99] REBER B., La nouveauté éthique des "nouvelles" technologies. Les techniques confrontées à l'exigence apocalyptique, Doctorate Thesis, Center for Political Research, Raymond Aron, EHESS, Paris, 1999.

[REB 07] REBER B., "Quand la nouveauté technique oblige à penser autrement: Jonas, Dewey, Simondon", in P.-A. CHARDEL and P.-E. SCHMIT (eds.), *Phénoménologie et Techniques*, Le Cercle Herméneutique/Vrin, Paris, pp. 107-126, 2008.

[ROT 05] ROTH C., BOURGUINE P., "Epistemic Communities, description and categorization", *Mathematical Population Studies*, vol. 12, no. 2, p. 107-130, 2005.

[SIM 89a] SIMONDON G., *Du Mode d'Existence des Objets Techniques (1958)*, Aubier, Paris, 1989.

[SIM 89b] SIMONDON G., *L'Individuation Psychique et Collective à la Lumière des Notions de Forme*, Information, Potentiel et Métastabilité, Aubier, Paris, 1989.

Postscript

Computer Science and Humanities

I am the author of 101 books, 421 articles (that is over 80,000 pages) and, despite my 93 years and my fatigue, I have accepted the invitation to write the postscript of this treatise. Indeed, after considering the originality of this editorial project, I feel this invitation is like a consolation from heaven for me, who is close to eternity. I am sure I am not mistaken and express my admiration for the editors of this book, Claire Brossaud and Bernard Reber. I see indeed, from reading the chapters, that there are younger researchers, skilled and of obvious creative power. They give me joy through their concern in understanding how computer science, a recent discipline, and the developments of ICT allow the methodologies and theoretical concepts of other older disciplines, up to philosophy, to deepen. It therefore seems appropriate today to reflect on humanities and social sciences themselves, often misunderstood, but also deal with ICT. By doing this, it is as if I was able to participate in this ambitious adventure.

Computer science has allowed me to perform accurate elementary analyses, from hermeneutical microanalyses to logical macrosyntheses. When Wiener's book on cybernetics appeared in the 1940s and I had just finished my thesis on the Thomistic terminology of interiority, I began to think of automatic linguistic analysis and to contemplate using the computer for a linguistic microanalysis of natural texts on a large scale. I explored the major European libraries and made more than 80 trips to the United States, including several visits to the Library of Congress. In 1946 I visited Thomas Watson, the founder of IBM, in New York, who responded to my request: "My technicians will not be able to realize the machine you want". It is true that I had to catalog the 9 million Latin words of the *Opera Omnia* of Aquinas, then 2 million words of the other medieval philosophers who have a connection with his writings (that is more than 11 million words from 18 languages, eight alphabets and very different literary genres), and that I had chosen to test the method I was developing. I then replied to him: "Is it logical to say that this research is impossible if we did not try? I am simply asking you the necessary conditions to be able to try". He looked me in the eyes and told me that would help me until the end of the project on the sole condition that we do not change "IBM into International Busa Machines...". I therefore used this machine for 30 years and was surrounded by up to 64 collaborators between 1962 and 1967.

For this challenge, I was able, for every word of all of the selected medieval texts, to attach its position and the number of characters, the specification of six topologies of speech, the lemma to which it belongs and several morphological and homographic characteristics. I then added four classifications to the electronic text. The first focuses on the signs alone, through segmentation of the 11 million Latin words into sequences of characters, identical and repeated in combination with other different sequences. The second classifies and makes an inventory of the endings and inflexions of the Latin language in general. The third groups together the 20,000 lemmas of the *Index Thomisticus* into thematic families. As for the fourth, the most remarkable one from the conceptual point of view, it allowed me to establish types of "semanticity", that is, the different relations between signifier and signified, of which all depend on the type of signified. With this typology, I wanted to affirm and document the heterogeneity of words, which, I believe, exists in all languages, although expressed differently. I am well aware that linguistic statistics are still very far from the forecast goals, precisely because processing the words of a text as if they were homogenous and undifferentiated, like numbers within the same calculation, certainly helps philology and databases, but remains very limited. Several articles on this treatise abound in this direction.

With my collaborators, throughout this research life, I have been interested in:

– the problems of computational linguistics;

– automatic translation (in particular on ancient languages into modern languages);

– concordancers[1];

– statistics;

– the history of computer science;

– computerized legal theory and technique[2]; and

---

1 See Chapter 11.

2 In Italy, the *Istituto di Teoria e Tecniche dell'Informazione Giuridica del CNR* (www.ittig.cnr.it), which was founded in 1970.

– finally, computer-assisted philosophy works, discussed in particular since 1974 at the international conferences organized every three years to address the following topics: *ordo, res, spiritus, imaginatio-phantasia, idea, ratio, sensus, experientia, machina*[3].

I also became interested in sociology, as its research expanded, and I am glad to find that, in this treatise, we are compelled to deal with ontological issues.

Let us stick with philosophy to finish. It is true that philosophy and the history of philosophy are distinct as the two cotyledons of the same seed. It is not difficult to recognize that the computer provides valuable services to the history of philosophy, particularly concerning access to the words of others and to its further examination, everything of which we do not have direct experience and which is the result of communication by the repetition of knowledge. Yet, in philosophy, there is also a first type of understanding – intuition or discovery by direct experiences – on the method of invention and originality. Both routes are reversible. If technology is an expression of being human, could we therefore see in the sublime of human manufacture something even more interesting: to think of ICT as an aid to an imaginative science. Numerous analyses and experiments offered in this interdisciplinary work document this and allow us to think about this[4].

Professor Roberto BUSA

---

3 See in particular *Companion to Digital Humanities*, edited by S. Schreibman, R. Siemens and J. Unsworth, and for which I have written a preface (Blackwell, London, 2004).

4 "Coincidence" or serendipity (see Chapter 10 for this term) meant that the translation of this postscript was due to a professor of Vietnamese origin, Professor Tran Van Doan, polytechnician, doctor in physics and then philosophy, lecturing at the National University of Taiwan(http://homepage.ntu.edu.tw/~philo/EuroPhilos/Tran.htm).

List of Authors

Henry BAKIS
University of Montpellier III
France

Luc BONNEVILLE
Communication Department
University of Ottawa
Canada

Dominique BOULLIER
UMS CNRS Lutin 2809 and
LAS, EA 2241
University of Rennes 2
France

Grégory BOURGUIN
Coastal Information Science
Laboratory (LIL)
University of the Littoral Opal Coast
Calais
France

Claire BROSSAUD
University of Lyon
France

Stéphan BRUNESSAUX
Research Centre for Information
Processing
EADS Defense and Security
Val-de-Reuil
France

Roberto BUSA
Faculty of Philosophy
Georgian University
Rome
and
Catholic Sacred Heart University
Milan
Italy

Alberto CAMBROSIO
Department of Social Studies of
Medicine
McGill University
Montreal
Canada

Dominique CARDON
Usage Sociology Laboratory
France Telecom R&D
France

Jean CLÉMENT
"Hypermedias" Department
University of Paris 8
France

Bernard CONEIN
University of Nice Sophia-Antipolis
and
ICT Uses Laboratory
Sophia Antipolos
Nice
France

Pascal COTTEREAU
AGUIDEL
Paris
France

François DAOUST
ATO
and developer of the SATO software
France

Jules DUCHASTEL
Sociology Department
University of Quebec
Montreal
Canada

Thierry FOUCART
University of Poitiers
France

Anne GOLDENBERG
University Sofia Antipolis
Nice
France
and
University of Quebec
Montreal
Canada

Sylvie GROSJEAN
Communication Department
University of Ottawa
Canada

Gaël GUEGUEN
Toulouse Business School
University of Toulouse
France

Andrea IACOVELLA
ENSIIE
Évry
France

Nicolas LARRIEU
CNS Department
LEOPART Laboratory
ENAC
Toulouse
France

Philippe LAUBLET
University of Paris 4
France

Christophe LEJEUNE
Research Center in information and
Communication Sciences
Université Libre de Bruxelles
Belgium

Arnaud LEWANDOWSKI
Coastal Information Science
Laboratory (LIL)
University of the Littoral Opal Coast
(ULCO)
Calais
France

List of Authors

Pierre MARANDA
Professor Emeritus
Anthropology Department
Laval University
Quebec
Canada

Alain MILON
Paris West University
France

Maxime MORGE
Computer Science Department
University of Pisa
Italy

Andrei MOGOUTOV
AGUIDEL
Paris
France

Philippe OWEZARSKI
LAAS-CNRS
and
University of Toulouse
France

Stefan POPOWYCZ
Reports Department
Canadian Institute of Health
Information
Canada

Christophe PRIEUR
University Paris Diderot
France

Serge PROULX
University of Quebec at Montreal
(UQAM)
Canada

Bernard REBER
Research Centre "Meaning, ethics
and society" (CERSES-CNRS)
Paris Descartes University
France

Richard ROGERS
Media Studies Department
University of Amsterdam
The Netherlands

Eddie SOULIER
Charles Delaunay Institute
University of Technology of Troyes
France

William TURNER
Computer Science Laboratory for
Mechanics and Engineering Science
CNRS-LIMSI
French National Research Council
France

Karl M. VAN METER
Bulletin de méthodologie
sociologique (BMS)
Paris
France

Tania VICHNEVSKAIA
AGUIDEL
Paris
France

Philippe VIDAL
University of Le Havre
France

Saïd YAMI
University of Montpellier I
and
EUROMED Management
Marseille
France

Manuel ZACKLAD
DICEN laboratory and
multidisciplinary laboratory
Tech-CICO
University of Technology of Troyes
France

Index

A

ACFCI, 329
active measurements, 108-109
Activity Theory, 306-307
AFE, 313
algorithms, 68, 120, 127-133
annotations, 11, 13, 172, 179, 206, 211-213, 217
attraction basins, 190-192
authentication, 306-307
authentification, 306-307, 316
automated specific analysis, 382
automatic hypertextualization, 160-161

B

ballot box, 305, 308
biomedical
　classification, 145
　research, 144
　sciences, 137
biometric, 307
blind signature, 305, 311
blogs, 121-129, 131-132
Boyd, 122-123, 129
bridges, 130-131

C

category, 47, 51, 182, 187, 190, 233-239, 241
centrality, 129, 130-131
chaos, 162-163
circles, 123, 127
clusters, 123, 128-132
co-authorship, 146
cognitive, 156-159, 165
　democracy, 318-320, 326-327
collaborative
　networks, 138, 148
　spaces, 318
communication, 5, 7, 71-75, 79, 80, 181
　tools, 121
communities, 119-120, 129, 130-131
complex networks, 128, 131
complexity, 161-166, 232-233, 237
components, 289-291, 294- 295
computer science, 125
concordances, 171, 174-177, 182, 197, 238
content, 122-125, 130
　analysis, 42, 180-182
controversies, 320-321, 325-327
conventions, 320-322
co-occurrence, 171, 177, 181

coordination, 254, 258, 264
counting, 302, 304
co-word analysis, 141
Cultural Hypermedia Encyclopedia of Oceania, 187, 200
cyberspace, 89-91, 98
CyberVote, 301-304, 312

D

data protection, 306
database, 158-160
decision-making, 331-335
density, 126-127
dialogic democracy, 332
diffusion, 279
dissemination of results, 381
distributed cognition, 317
document, 261-263
documentarization, 251, 263

E

ECHO, 187-190, 195
edges, 124-125
election, 302-309, 312-314
electronic voting, 301-309
email, 121
Enron, 121
e-Poll, 301, 311-312
EU-StudentVote, 301
experience, 288-291, 294-295

F

fakesters, 123
Flickr, 121-122, 126
formalisms, 7-11, 48
fragment, 164-165
free software, 183
freedoms, 310, 314
friends, 122-123

Friendster, 122-123, 129

G

game theory, 332
generative process, 68
geocyberspace, 72, 79, 82
geography, 74, 78-79, 195
graph, 120, 125-128, 177-179, 188, 191, 194-196

H

hazardous navigation, 61
hermeneutic loop, 133
historical, 3-15
  semantics, 8
historiography, 3-10, 13-16
history of hypertext, 158
history, 4-9, 13-15, 39-40, 173, 175, 182, 193
homomorphic encryption, 305, 311-314
hyperlinks, 91-92, 96, 98
hypermaps, 59, 67
hypermedia, 59, 62, 68, 155-156
hypermodern Man, 29, 32
hyperspace, 89-90, 94, 98
hypertext, 10-11, 155-166, 187-189, 196-197
hypertextual device, 158

I

ICT, 24-32
information
  and speed economy, 29
  technology, 29
integrity, 305, 313
intellectual history, 164
interaction, 271, 279
inter-citations, 142, 146-147

Internet, 71, 75-81, 103-133, 187-189, 195, 198-199, 315-316, 373-378, 381, 388
  surveys, 380
  voting, 304-313
internetized research area, 384
interpretation, 6-10, 15, 172, 178-179, 183, 192, 199, 232-237, 240-242
interpretation, 51
investigations, 375, 381, 386
Issuecrawler, 89-92, 94-99

K, L

knowledge modeling, 11-12, 16
links, 120-131
literary history, 159
long range distance (LRD), 114

M

map, 59-68
mapping, 89, 94-96
measure, 177
metadata, 206, 210-213
metarhetoric, 164
method, 183, 236-240
methodology, 9, 43, 45, 54, 227, 237
methods, 3-16, 42, 51, 53
metrology, 103-105, 108-116
migraine, 144, 146, 150
Milgram, 125, 127
mix-net, 311-312
model, 291
modeling, 291
multi-agent system, 331
multi-criteria analysis methods, 334
multimethod analysis, 224-225
MySpace, 121-122

N

narrative
  explanation of social processes, 37-38
  interaction, 51
narrativity, 37, 40, 43
neighborhood, 128-130
Netlocator, 90-94, 98
network, 89-99, 119-133
  analysis, 139
neurosciences, 187, 190-193
non-discrimination, 305

O, P

ontology, 9-10, 13, 37-38, 42-47, 51, 54, 206, 209-218
ontology web language (OWL), 48
passive measurements, 109
patterns, 128, 131
politicization
  of cognitive technologies, 318
  of practices, 320
  of the technique, 319
public key, 311-312
PubMed, 140-146, 149

Q

qualitative sociology, 183
quantification, 363
questionnaire, 373-386

R

reading, 6, 10-13, 16, 73, 77, 180-181, 234
real time, 72, 74
regular equivalence, 132
relational, 119-124, 128-132

relationships, 119-124, 130
ReseauLu, 137, 139, 140-143, 149-150
retrieval, 120
rhizome, 163
route, 61-64

S

schools of thought, 220, 225-227
scientometric, 140-141, 144, 150
scientometric-type, 150
secrecy, 304-305, 308
security, 301-302, 305-307, 310, 313-314
semantic
  content, 139
  network, 142
semantic Web (SW), 206-218
semiospheres, 191-193, 197-198
serendipity, 165
small worlds, 125, 127
social
  informatics, 345
  media, 121
  narrative, 45, 51
  networks, 119-120, 124-128, 132
  sciences, 121, 129, 219-224
  technology, 368
society, 23, 25, 29, 31-32
socio-informatics, 343
sociology, 178, 182
socio-semantic categories, 237, 239, 241
socio-technical spaces, 318, 320, 323
space, 6, 10, 13, 71-76, 79-82, 178-179, 197, 237, 239, 241
speed, 23-32, 78
sphere, 89, 94-95, 98
structural equivalence, 131
survey, 375-380, 386-388
systemic, 162

T

temporality, 24, 27-28, 37-38, 44
territory, 173, 238
text processing, 170-172
text-mining, 143
theory, 44, 48, 52
Theyrule.net, 91
ThinkMap Visual Thesaurus, 91
traffic
  characteristics, 108
  flow, 108, 111-114
transaction, 256-261
transactional flow, 250-256, 261
triangulation, 15

U, V, W

urgency, 25-31
uses, 119, 129
validation, 82, 231
vertices, 124-128
vote, 302-308, 311-314
voter, 302-313
Web
  of Science, 140-143, 145-146
  Stalker, 91
Web 2.0, 121, 128, 132
wikis, 317-325
writing, 3, 6, 10-11, 14, 16, 171, 187-189